Predictive Modeling of Drug Sensitivity

Predictive Modeling of Drug Sensitivity

Ranadip Pal

Texas Tech University, Lubbock, TX, United States

AMSTERDAM • BOSTON • HEIDELBERG • LONDON
NEW YORK • OXFORD • PARIS • SAN DIEGO
SAN FRANCISCO • SINGAPORE • SYDNEY • TOKYO

Academic Press is an imprint of Elsevier

ELSEVIER

Academic Press is an imprint of Elsevier
125 London Wall, London EC2Y 5AS, United Kingdom
525 B Street, Suite 1800, San Diego, CA 92101-4495, United States
50 Hampshire Street, 5th Floor, Cambridge, MA 02139, United States
The Boulevard, Langford Lane, Kidlington, Oxford OX5 1GB, United Kingdom

Library of Congress Cataloging-in-Publication Data
A catalog record for this book is available from the Library of Congress

British Library Cataloguing-in-Publication Data
A catalogue record for this book is available from the British Library

ISBN 978-0-12-805274-7

For information on all Academic Press publications
visit our website at https://www.elsevier.com/

Working together
to grow libraries in
developing countries

www.elsevier.com • www.bookaid.org

Publisher: Joe Hayton
Acquisition Editor: Tim Pitts
Editorial Project Manager: Charlotte Kent
Production Project Manager: Julie-Ann Stansfield
Cover Designer: Mark Rogers

Typeset by SPi Global, India

Contents

Preface

In recent years, the study of the predictive modeling of drug sensitivities has received a boost due to the ever-growing interest in precision medicine and the availability of large-scale pharmacogenomics datasets. However, predictive modeling in this area is confronted with significant challenges, due to the high dimensionality of feature sets and the presence of limited samples combined with the complexities of genetic diseases that includes nonlinearities and variability among individual patients. Researchers have tried to tackle these issues by refining model inference for commonly used predictive models and designing new models specific to drug sensitivity prediction, along with incorporating additional data sources, such as high-throughput data from DNA, epigenomic, transcriptomic, proteomic, and metabolomic levels. It has been observed that the accuracy and precision of a drug sensitivity predictive model is often exceedingly dependent on the approaches considered for data processing, feature selection, parameter inference, and model error estimation rather than the specific type of model used. Thus, a proper understanding of the underlying biology alongside the issues related to genomic data measurement techniques, feature extraction, accuracy estimation, and model selection is highly relevant for designing drug sensitivity predictive models that can have an actual impact on personalized medicine.

This book is an attempt to cover the basic principles underlying the predictive modeling of drug efficacy for genetic diseases, especially cancer. A significant portion of the book is based on collaborative research carried out in my laboratory for last 6 years, including a top performance in an NCI-DREAM drug sensitivity prediction challenge, and the design of a novel framework to integrate functional and genomic information in predictive modeling. The book is structured as follows: the introductory chapter discusses the current personalized medicine landscape, along with future trends. A review of molecular biology and pharmacology concepts, along with data characterization methodologies, are discussed in Chapter 2. Chapter 3 discusses techniques and issues related to extracting relevant information from available data sources. Model validation techniques are considered in Chapter 4, and are followed by a description of tumor growth models in Chapter 5. The overview of predictive modeling techniques based on genomic characterizations is presented in Chapter 6, and is followed by the specific model types of Random Forests and Multivariate Random Forests in Chapters 7 and 8, respectively. Modeling based on integrated functional and genomic characterizations are considered in Chapters 9 and 10. Both the model-based and model-free designs of combination therapy are considered in Chapter 11. Chapter 12 provides a compendium of available online resources that are relevant to drug sensitivity prediction. The book concludes with a chapter on the challenges that need to be addressed before the full potential of genome-based personalized cancer therapy is achieved.

I hope that the book provides an integrated resource to help understand the current state-of-the-art in drug sensitivity predictive modeling. This book will likely benefit students and researchers who are interested in applying mathematical and computational tools to analyze genomic and functional data for personalized therapeutics.

I would like to acknowledge the various individuals and organizations whose support and collaborations were instrumental in the development of the multiple ideas presented in this book. First of all, I would like to thank my collaborator of many years Dr. Charles Keller who has provided excellent biological insights on multiple aspects of personalized medicine, and is the person who motivated me to work in this area. I would also like to thank members of my research group, especially Dr. Noah Berlow, Dr. Saad Haider, Mr. Qian Wan, Mr. Raziur Rahman, Dr. Mehmet Umut Caglar, Mr. Kevin Matlock, and Mr. Carlos De-Niz who have diligently worked in this research topic and have been co-authors on multiple publications referred to in this book. Finally, I would like to acknowledge the National Science Foundation and the National Cancer Institute for their support of this research.

Ranadip Pal
September, 2016

Introduction

1

Personalized medicine refers to therapy tailored to an individual patient rather than a *one-size-fits-all* approach designed for an average patient. The idea of personalized medicine has been in existence for more than 2400 years with the notable example of 5th century BC Greek physician *Hippocrates* who treated patients based on their *humor* imbalances [1]. The balancing of *humors* or bodily fluids as the basis of medical practice remained as the guiding principle in medicine for about two millennia [1]. The personalized treatments were based on the convictions of the era, that is, humors which later on was discovered to be incorrect based on advanced understanding of human anatomy and physiology. The personalization therapy options for the 20th century were primarily based on the results of different tests on bodily fluids, along with advanced visualization approaches such as X-rays or magnetic resonance imaging scans.

The advent of genomic characterization of individual patients in the last decade of the 20th century opened up numerous possibilities for personalized therapy.

Predictive Modeling of Drug Sensitivity. http://dx.doi.org/10.1016/B978-0-12-805274-7.00001-4

1

The genomic characterizations deliver considerably more detailed information on a patient with a genetic disease as compared to phenotypic observations or nonmolecular tests. According to the US Food and Drug Administration (FDA), *Personalized Medicine* (or *Precision Medicine*) in the current era entails "disease prevention and treatment that takes into account differences in people's genes, environments and lifestyles." The definition of personalized medicine provided by the National Cancer Institute (NCI) involves "A form of medicine that uses information about a person's genes, proteins, and environment to prevent, diagnose, and treat disease." The *Precision Medicine* initiative launched by President Obama in 2015 considers the use of patients' unique characteristics, including genome sequence, microbiome composition, health history, lifestyle, and diet to tailor treatment and prevention strategies by healthcare providers. The US National Academy of Sciences has defined personalized medicine as "the use of genomic, epigenomic, exposure and other data to define individual patterns of disease, potentially leading to better individual treatment."

An underlying theme of *personalized medicine* currently is the incorporation of an individual's genomic information with other characteristics in therapeutic decisions. The current broader challenges in this area involve [1]:

(C1) Detecting meaningful differences in the genetic characterizations of individual patients and deciphering how that affects responses to different therapies.
(C2) Incorporation of this knowledge and methodologies in clinical practice.
(C3) Incorporation of this refined patient-specific approach into healthcare and regulatory systems that are built around older medical paradigms.

This book primarily considers advances in the broader challenge area *C1* with characterizations of tumor cultures or cell lines often being used in place of patient tumor characterizations. We consider approaches for predicting tumor sensitivity to a drug or drug combination based on genomic characterizations and available drug information. The book is structured as follows. Chapter 2 provides a brief overview of molecular biology followed by various genomic and functional characterization approaches along with a review of pharmacology concepts. Chapter 3 discusses various feature selection and extraction techniques for unearthing relevant information from the available genomic and functional characterizations. Chapter 4 examines various methodologies for validating a designed drug sensitivity prediction model. Descriptions of tumor growth models are considered in Chapter 5. The overview of drug sensitivity predictive modeling techniques based on genomic characterizations are discussed in Chapter 6, and details on the specific approach of random forest modeling is presented in Chapter 7. Chapter 8 considers the extension of the random forest framework to multivariate random forests (MRF) that incorporates the correlations among various drug responses in modeling. Chapter 9 considers an alternative strategy for drug sensitivity prediction, termed target inhibition maps, based on functional and genetic characterizations. Chapter 10 considers the problem of the designing dynamic networks based on functional (drug perturbation) data. The problem of design of combination therapy based on predictive models is considered in Chapter 11. Chapter 12 presents a compilation of various online resources relevant to drug sensitivity modeling. The final chapter of the book concludes with a discussion on the various challenges that need to be addressed before achieving the full potential of genome-based personalized cancer therapy.

1.1 CANCER STATISTICS

The application of the predictive modeling techniques discussed in this book are primarily for the genetic disease of cancer and thus it is highly relevant to discuss the extent of the disease and where we currently stand in terms of tackling the disorder. In a simplified sense, cancer consists of a group of diseases affecting various organs in the body that involves abnormal cell growth and is caused by genetic alterations. In more biological terms, cancer can be explained by the *hallmarks of cancer* comprised of biological capabilities acquired during the multistep development of human tumors [2]. The hallmark capabilities include (a) sustaining proliferative signaling, (b) evading growth suppressors, (c) resisting cell death, (d) enabling replicative immortality, (e) inducing angiogenesis, (f) activating invasion and metastasis with potential additional hallmarks of (g) reprogramming of energy metabolism, and (h) evading immune destruction [3]. According to cancer.gov, an estimated 1.68 million people in the United States will develop some form of cancer in the year 2016 and this number is predicted to increase in coming years. The mortality rate from cancer has decreased relative to earlier decades, but it is nevertheless high with a rate of 171.2 per 100,000 men and women per year. There are more than a hundred types of cancer and the most common forms of cancer in 2016 in terms of organ sites are projected to be *breast cancer, lung and bronchus cancer, prostate cancer, colon and rectum cancer, bladder cancer, melanoma of the skin, non-Hodgkin lymphoma, thyroid cancer, kidney and renal pelvis cancer, leukemia, endometrial cancer,* and *pancreatic cancer.* Cancers can vary widely in their aggressiveness and expected clinical outcomes. Some cancers such as glioblastoma [4] are uniformly fatal, while some such as prostrate [5] often do not kill the patient before they die of some other cause. It is estimated that around 39.6% of men and women in the United States will be diagnosed with cancer at some point during their lifetime. Expenditures on cancer care in the United States reached around $125 billion in 2010 and are expected to reach $156 billion in 2020. Cancer is also a leading cause of death worldwide, with 8.2 million deaths attributed to this disease in 2012. The number of new cancer cases worldwide is expected to rise to 22 million within the next two decades.

The ubiquitous nature of cancer intensifies our efforts to understand the causes, prevention, diagnosis, and effective treatment for this disease. Research efforts in the last few decades have been able to provide an improved understanding of the origins of cancer and development of new approaches for cancer therapy, but we have still to realize the full potential of personalized therapy in the clinic. Multiple challenges still exist both in the research and implementation domains that need to be addressed to significantly lower the cancer mortality rate and provide a good quality of life to cancer survivors.

1.2 PROMISE OF TARGETED THERAPIES

Treatment of cancer has been approached in various ways and the most commonly applied techniques are chemotherapy and radiation therapy. A persistent issue with chemotherapy and radiation therapy is the undesirable side effects that can significantly reduce the quality of life for the survivor. Furthermore, even with the success of new radiation therapy and chemotherapy methods in extending the life span of cancer patients, the death rate attributed to solid tumor cancers is staggeringly high at over 450,000 people in the United States alone [6]. This staggering number indicates a significant cohort of cancer patients essentially failing first- and second-line cancer therapies. Many experts suggest that *personalized medicine* is the answer for patients who have failed first- and second-line therapies. The moniker of personalized medicine is largely being defined by the rapidly emerging market of subtler therapies being led by new diagnostic biomarkers that lead to actionable therapies. Typically, biomarker specified drugs are used as a complementary treatment with radiation therapy and chemotherapy. Although, the general idea of personalized therapies as adjuvants is occasionally challenged. An example of this new view of cancer diagnostics/treatment methodology is *Gleevec*. While the biomarker for *chronic myelogeneous leukemia* (CML), the so-called *Philadelphia chromosome* and its resulting aberrant tyrosine kinase activity indicates the need to block a singular pathway is seen as an exceptional case, nonetheless *Gleevec* treatment has entirely replaced more standard treatments such as bone morrow transplants and aggressive chemotherapy. This uniqueness of the CML biomarker/treatment combination is due to the fact that unlike most other cancers, which are caused by a multitude of complex interacting genetic and environmental factors and therefore have many targets, CML is caused by a single aberrant protein related to a consistent chromosomal translocation [7]. The *Gleevec* success does point to the potential success of biomarker-based treatment. Some other cancer targeted therapies that have been FDA approved for patients with specific genetic characteristics include (a) *Trastuzumab*: Approved for breast cancer that is HER2+; (b) *Crizotinib*: Approved for *nonsmall cell lung carcinoma* patients with a chromosomal rearrangement causing a fusion of EML4 and ALK genes; (c) *Vemurafenib*: Approved for late-stage *melanoma* patients with V600E BRAF mutation. Fig. 1.1 shows the effect of Vemurafenib therapy in the form of fluorodeoxy-glucose (FDG) uptake on a melanoma patient with BRAF mutation. FDG is often used to study tumor metabolism and reduction in its uptake is usually followed by tumor regression. (d) *Dabrafenib*: Approved for *melanoma* patients with V600E BRAF mutation; (e) *Trametinib*: Approved for patients with V600E mutated metastatic melanoma.

Note that the previous list includes few examples, is in no way exhaustive, and there are numerous other targeted therapies that have been approved or in the pipeline for approval [8]. Some of the targeted therapies have been approved based on cancer stage and previous treatment response rather than specific genetic mutations in the patient, such as the use of kinase inhibitor *Sorafenib* for treatment of locally recurrent or metastatic, progressive differentiated thyroid carcinoma (DTC) resistant

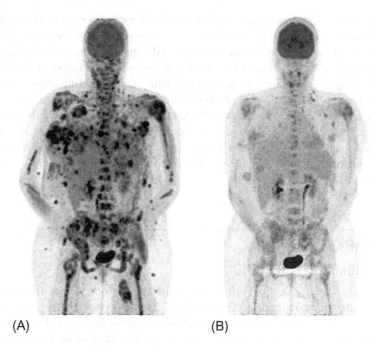

(A)   (B)

FIG. 1.1

PET scan images of a BRAF-mutant melanoma patient showing FDG uptake before and after therapy with Vemurafenib. (A) Before therapy. (B) After therapy.

(Reprinted with permission from Macmillan Publishers Ltd: Nature 467(7315):596–599. doi:10.1038/nature09454, copyright 2010.)

to radioactive iodine treatment [9]. Certain targeted therapies have been approved as part of a combination therapy, for instance *Lapatinib* has been approved for breast cancer patients in combination with *Capecitabine*.

The use of targeted therapies has also been considered for other genetic disorders besides cancer. For instance, the drug *Kalydeco* (also known as *ivacaftor*) is used to treat *cystic fibrosis* patients who are shown to have a specific genetic mutation known as G551D mutation [10]. The G551D mutation alters CFTR protein activity by replacing the amino acid *glycine* (G) in position 551 with *aspartic acid* (D). The targeted drug helps to restore the protein activity by binding to the transport channels and allowing proper flow of salt and water on the surface of the lungs.

The previous examples provide cases of approved targeted therapies that have achieved decent measures of success in treating specific disease conditions. However, significantly more can be achieved by understanding the drivers for each patient's tumor and catering therapies to meet their individual needs. Since each patient might require targeting a distinct set of proteins, the personalized approach will require either (a) the generation of a distinct drug for each patient or (b) selecting a combination of drugs from an arsenal of numerous available therapies. From a pharmaceutical development perspective, option (b) appears to be more feasible than option (a) in the current market. Current estimates suggest that the cost of developing a new prescription medicine and gaining regulatory and marketing approval stands at $2.6 billion [11,12]. In the foreseeable future, regulatory and pharmaceutical cost structure changes might usher us into the development of personalized drugs in the clinic, but the best option currently is to utilize combinations of existing drugs. In the next section, we discuss the current market trends in the area of biomarker testing, pharmaceutical developments of targeted therapies, and upcoming changes in healthcare industry.

1.3 MARKET TRENDS

1.3.1 BIOMARKER TESTING

Recent years have seen the growth of companies that provide information on some genetic traits of an individual. For instance, *23andMe* started by providing an inexpensive test for mutations (single nucleotide polymorphisms [SNP]) in specific genes for around $100 that could potentially assess the risk of an individual to some genetic diseases. Another instance is the use of four SNPs to access the need for biopsies in prostrate cancer. Compared to preventive or risk assessment options, there are other currently available tests that are geared toward treatment opportunities for cancer patients. A regularly used test for breast cancer is the evaluation of *HER2* expression which can predict success of the molecularly targeted therapy of *trastuzumab* for *HER2* positive metastatic breast cancer patients.

A sample of some commercially available tests for guiding therapeutic decisions for cancer patients is shown in Table 1.1.

The majority of these tests have been on the market for less than 5 years and they are yet to achieve high success rates in selecting efficacious treatment options for their customers. To understand the reasons behind the apparent lack of success, we need to explore the genetic traits considered by these tests. Some of these tests are based on single genetic profiles, such as a set of DNA mutations which are often not sufficient to predict drug efficacy due to the presence of latent variables and the interplay of gene expression, epigenetic factors, protein expression, and metabolites in triggering tumor proliferation. The forms of cancer that are caused by mutations on one or a few genes are typically easier to treat, but the vast majority of tumor proliferations are caused and affected by complex interactions between genome, transcriptome, proteome, and metabolome.

Some of these tests can suggest highly efficacious therapeutic options for a segment of the patient population, but will come back without any recommendations for the rest of the patients. Furthermore, crowd-based initiatives to predict tumor sensitivity to anti-cancer drugs have shown limitations in predictive accuracy [13]

Table 1.1 Examples of Companies Using Personalized Tests for Guiding Anticancer Therapies

No.	Company or Laboratory	Test Type
1	Foundation One (Foundation Medicine)	Genomic profile
2	N-of-One (Appistry)	Genomic profile
3	Perthera	Genome, protein, and phosphoprotein analysis
4	Univ. of Michigan/IntGen/Paradigm	Exome sequencing and RNA expression
5	Illumina	Whole genome sequencing
6	Caris Target Now	Multiple genomic tests
7	OHSU Knights DX Lab	Mutation screen
8	Consultative Proteomics	Protein expression
9	Weisenthal Group	Cytometric profiling
10	Rational Therapeutics	Drug screen profiling

which suggests that more advanced and innovative approaches to modeling drug sensitivity incorporating various data sources are required for effective personalized medicine.

1.3.2 PHARMACEUTICAL SOLUTIONS

A 2012 market survey showed over 530 companies with over 1300 treatments for solid tumors in various stages of development [14]. The NCI list of currently approved individual or combination drugs for cancer contained more than 300 entries [8] at the end of 2015. The push for personalized or biomarker specified treatment of solid tumors, combined with the number of new therapies that will come on line in the next 10 years will offer patients and their physicians a multitute of choices in fighting their disease. With a myriad of patients and cancer therapies; cancer therapy is quickly turning into a *Big Data* problem for care providers and patients.

1.3.3 VALUE-DRIVEN OUTCOMES

The Healthcare Industry is moving toward more value-driven outcomes (VDO) where the cost of treatment relative to outcomes will be assessed for assigning therapies [15]. Assigning a combination of anti-cancer drugs with high costs for individual drugs will have to be justified in a systematic manner before patients or insurance companies are expected to pay for the treatment. Thus it is desired that modeling approaches will specify the optimal effective stacking of therapies to help the oncology community discover the most cost effective path to achieve optimal quality adjusted life years (QALY). Furthermore, leveraging of computational process modeling and simulation in drug development can potentially assist in bringing drugs to the market faster and cheaper [16]. The ability to assure the cost-effective delivery of QALY will become increasingly important as more treatments are available and healthcare turns increasingly to VDO.

1.4 ROADBLOCKS TO SUCCESS

Irrespective of numerous available therapies and genetic testing options, there remain significant challenges to be addressed before achieving the full potential of personalized therapeutics.

1.4.1 LINKING PATIENT-SPECIFIC TRAITS TO EFFICACIOUS THERAPY

As discussed earlier, improvements to and implementation of the latest drug sensitivity predictive modeling techniques is desirable to increase therapy efficacy while minimizing side-effects for cancer patients. Testing technologies currently available in the market lack the effective integration of multiple heterogeneous data sources to increase prediction accuracy, and are often restricted to individual therapy recommendations that are prone to drug resistance. This topic is of specific relevance to the contents of this book, as predictive modeling of drug sensitivity requires the mapping of a patient's genetic makeup to therapeutic response.

1.4.2 HIGH COSTS OF TARGETED THERAPIES

The costs of targeted therapies can often be prohibitive for patients and insurance companies alike. At the lower end of the spectrum, drugs like *Lapatinib* targeting *HER2* is priced upward of $4000 per year [17]. The costs of newly approved targeted drugs have been usually higher, such as *Crizotinib* costing around $120,000 per month [18] and *Kalydeco* costing around $294,000 per year [19]. A back of the envelope calculation will show that drugs costing around $100,000 per year will put a cost burden of $168 billion a year (not including costs related to testing, oncologists, hospital visits, and associated expenditures) to only treat the 1.68 million new people diagnosed with cancer every year. The calculations are done with an assumption of costs remaining same irrespective of the number of patients, which will not likely be the case when targeted therapies become the standard of care and the advantages of *economies of scale* is incorporated into the pricing. Furthermore, as discussed earlier, integration of computational modeling and simulation from the start of the drug development process can potentially increase efficiency of the drug development pipeline and decrease costs of newly developed drugs. Current estimates suggest that only 1 in 15 oncology drugs entering clinical development in phase 1 achieves FDA approval [20]. Potential ideas to reduce the costs of approved drugs include use of adaptive clinical trial designs, flexibility with alternative surrogate endpoints, improved approaches for assessing patient benefit-to-risk, and progress in basic science, such as enhanced predictive animal modeling, earlier toxicology evaluation, biomarker identification, and new targeted delivery technologies [20].

1.4.3 RESISTANCE TO THERAPIES

Cancers treated with individual drugs have frequently developed resistance to the drugs with the consequence of tumors reappearing after few months. Tumor resistance is not limited to chemotherapy and targeted drugs, but is also observed for radiotherapy treatments. Resistance to therapies can occur through a variety of mechanisms such as poor absorption and rapid metabolism of the patient, limiting concentration of the drug in the tumor (intrinsic resistance), or acquired resistance where cancer cells may evolve specific genetic and epigenetic alternations that allows them to evade therapy [21]. Some of these evolutionary changes, such as loss of a cell surface receptor or transporter and overexpression or alteration in the drug target, induce resistance against a small number of related drugs [21]. For instance, genetic evolution of initially $EGFR^{T790M}$-negative drug-tolerant cells causes resistance to EGFR inhibitors for nonsmall cell lung cancers [22]. Further examples include (a) mutations in NRAS can provide resistance to *Vemurafenib*; (b) receptor tyrosine kinase activation and overexpression of *CRAF* or *COTMAP3K8* can produce resistance to RAF inhibitors, and (c) *Vemurafenib* resistance due to loss of PTEN or RB1 [23]. On the other hand, multi-drug resistance consists of simultaneous resistance to multiple structurally and functionally unrelated drugs [22].

1.4.4 PERSONALIZED COMBINATION THERAPY CLINICAL TRIALS

Many diseases including cancer are caused by a complex interplay of molecular and environmental factors which requires targeting of multiple mechanisms to achieve treatment efficacy [24]. Thus combination therapies are considered to be critical to developing curative strategies and avoiding drug resistance [25]. Certain combination therapies approved for cancer include *trastuzumab* (targeted drug affecting HER2) in combination with *paclitaxel* (chemotherapeutic drug) for breast cancer, and *cetuximab* (targeted EGFR inhibitor) in combination with *irinotecan* (chemotherapeutic drug) for metastatic colorectal cancer. Studies have reported that around 25% of recent oncology trials contain drug combinations [24]. A large number of these combination trials include a chemotherapy drug with a targeted drug and is applicable for a significant portion of the patients, such as the traits required for the combination therapy are observed in many patients, for instance phase II trials usually consider groups of more than 100 patients. Furthermore, the combination therapy design is often based on clinical and empirical experience [23]. We require more unbiased approaches for combination therapy design along with trials consisting of a combination of targeted drugs that can target multiple pathways' causing proliferation [23]. Regulatory authorities have to design approaches for the single patient-based drug combination therapy trials, as the complexities of cancer can often result in unique optimal drug combinations for each patient, especially for rare cancers. Measuring the success of single patient trials along with assessing the toxicity of drug combinations previously not tested are some of the challenges that need to be addressed. Additionally, *basket clinical trials* that focus on genetic mutations of the tumor, regardless of type, as compared to conventional clinical trials focusing on specific cancer types can be a step in the right direction for precision medicine [26].

1.5 OVERVIEW OF RESEARCH DIRECTIONS

Following our previous discussions on the promise of targeted therapies, current market trends, and potential roadblocks to success, we revisit the research directions that will be discussed in the subsequent chapters. This book considers various approaches that have been applied to the problem of predictive modeling of drug sensitivity based on a patient's genetic characterization and exposure information. The review of molecular biology concepts as it relates to genetic diseases along with approaches to characterize the genetic and functional behavior of tumors is considered in Chapter 2.

Significant amounts of data in terms of measurement of numerous gene and protein expressions and mutational information are typically collected in drug sensitivity studies, but the use of the entire set of input features for designing a predictive model can result in model overfitting and subsequent erroneous predictions. Thus selecting a smaller set of features relevant to the drug sensitivity problem at hand is extremely important for lowering the generalization error (error over unforeseen data) of the model. Chapter 3 considers various feature selection and extraction approaches to extricate relevant information from the available datasets. For any designed predictive model, it is important that we are able to assess the accuracy of the model for new samples. Various techniques can be applied to make the required evaluations, which are discussed in Chapter 4. Furthermore, the experimental models used for drug sensitivity model validations are also reviewed in this chapter. Dynamic models in the form of differential equations to capture the growth of a tumor without incorporation of genetic characteristics is included in Chapter 5.

The overview of various regression models to predict sensitivity of new samples based on genetic characterizations is considered in Chapter 6. We broadly categorize the approaches into groups of linear regression techniques, nonlinear regression methods, kernel approaches, ensemble methods, and dynamical models. Application of these techniques to NCI-DREAM drug sensitivity challenges is also discussed. Chapter 7 discusses ensemble-based approaches for predictive modeling. We consider commonly used random forest regression approach which is a collection of regression trees. The effect of various parameters of the forest on the predictive performance is examined along with applications to drug sensitivity studies. The chapter also considers probabilistic regression trees and their ensembles. Chapter 8 discusses the extension of random forest-based approaches to incorporate multivariate responses. Traditional random forest-based approaches infer a predictive model for each drug individually but correlation between different drug sensitivities suggest that multiple response prediction incorporating the covariance of the different drug responses can possibly improve the prediction accuracy. The chapter discusses the extension of the RF framework to MRF that incorporates the correlation between different drug sensitivities using covariance matrices and copulas.

The predictive modeling approaches based on genetic characterization alone are often restricted in their accuracy, as the genetic characterizations observed under normal growth conditions can only provide a single snapshot of the biological system. Perturbing the patient tumor culture using various known perturbations in the form of targeted drugs and observing the responses can be utilized as additional information to gain further insights on the specific patient tumor circuit. Based on this idea, Chapter 9 presents a computational modeling framework to infer target inhibition maps utilizing responses of a tumor culture to a range of targeted drugs. The target inhibition profiles of the drugs and the functional responses are utilized to infer potential tumor proliferation circuits that can assist in the generation of synergistic drug combinations. Application of this framework to synthetic pathways and experimental canine tumor cultures are also included in this chapter.

Chapter 10 considers the extension of the target inhibition map approach to infer dynamic biological networks from drug perturbation datasets. The inverse problem of possible dynamic models that can generate the static Target Inhibition Map model is considered. From a deterministic viewpoint, we analyze the inference of Boolean networks that can generate the observed binarized sensitivities under different target inhibition scenarios. From a stochastic perspective, we investigate the generation of Markov Chain models that satisfy the observed target inhibition sensitivities.

Chapter 11 considers the problem of the design of combination therapeutics under various constraints. The prohibitively high number of potential combination drugs rules out exhaustive testing of all drug combinations. We first consider a model-based combination therapy design using the target inhibition map framework where set cover and hill climbing-based techniques are used to arrive at the desired drug combinations. We also consider the scenario of design of efficacious drug combination when limited information is available to infer a model. The chapter presents a stochastic search approach to arrive at an efficacious drug combination with limited iterations.

Research in drug sensitivity modeling frequently requires access to multiple forms of information that are normally accessible from online sources. Chapter 12 provides a compendium of the online resources relevant to drug sensitivity predictive modeling. Information on online databases for pathways, genetic, and functional characterizations; various predictive modeling tools, drug synergy estimation software, and regulatory resources are presented in this chapter.

The book concludes with a chapter on the challenges facing the predictive modeling for personalized medicine research community. We touch upon the issues of tumor heterogeneity, data inconsistencies, prediction accuracy limitations, collaborative constraints, toxicity of combination therapeutics, and ethical considerations of the research.

The various topics covered in the book are shown as a flowchart in Fig. 1.2.

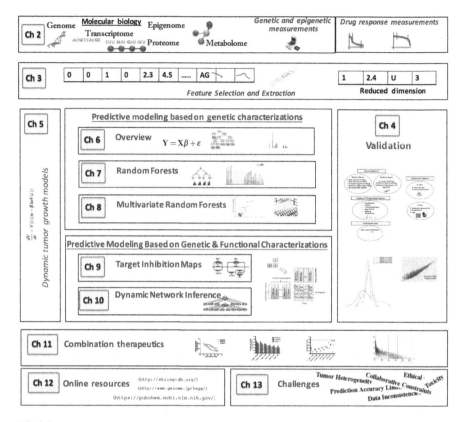

FIG. 1.2

Pictorial representation of the topics covered in this book.

REFERENCES

[1] F.R. Steele, Personalized medicine: something old, something new, Future Med. 6 (1) (2009) 1–5.

[2] D. Hanahan, R.A. Weinberg, The hallmarks of cancer, Cell 100 (1) (2000) 57–70.

[3] D. Hanahan, R.A. Weinberg, Hallmarks of cancer: the next generation, Cell 144 (5) (2011) 646–674.

[4] B. Tutt, Glioblastoma cure remains elusive despite treatment advances, 2011, http://www2.mdanderson.org/depts/oncolog/articles/11/3-mar/3-11-1.html.

[5] ACS, Survival rates for prostate cancer, 2016, http://www.cancer.org/cancer/prostate cancer/detailedguide/prostate-cancer-survival-ratesn.

[6] Centers for Disease Control and Prevention, National Center for Health Statistics, Health Data Interactive, 2014, www.cdc.gov/nchs/hdi.htm.

[7] L. Pray, Gleevec: the breakthrough in cancer treatment, Nat. Educ. 1 (1) (2008) 37.

[8] NCI, A to Z list of cancer drugs, http://www.cancer.gov/about-cancer/treatment/drugs.

[9] NCI, FDA Approval for Sorafenib Tosylate, 2013, http://www.cancer.gov/about-cancer/treatment/drugs/fda-sorafenib-tosylate.

[10] FDA, Paving the way for personalized medicine: FDA's role in a new era of medical product development, Food and Drug Administration Report, 2014.

[11] CSDD, Cost to develop and win marketing approval for a new drug is $2.6 billion, 2014, http://csdd.tufts.edu/news/complete_story/pr_tufts_csdd_2014_cost_study.

[12] A.E. Caroll, $2.6 billion to develop a drug? new estimate makes questionable assumptions, 2014, http://www.nytimes.com/2014/11/19/upshot/calculating-the-real-costs-of-developing-a-new-drug.html?_r=0.

[13] J.C. Costello, et al., A community effort to assess and improve drug sensitivity prediction algorithms, Nat. Biotechnol. (2014), http://dx.doi.org/10.1038/nbt.2877.

[14] P.R. Newswire, Therapies for solid tumors: pipelines, markets, and business considerations, 2012, http://www.thestreet.com/story/11753653/2/therapies-for-solid-tumors-pipelines-markets-and-business-considerations.html.

[15] K. Kawamoto, C.J. Martin, K. Williams, M.C. Tu, C.G. Park, C. Hunter, C.J. Staes, B.E. Bray, V.G. Deshmukh, R.A. Holbrook, S.J. Morris, M.B. Fedderson, A. Sletta, J. Turnbull, S.J. Mulvihill, G.L. Crabtree, D.E. Entwistle, Q.L. McKenna, M.B. Strong, R.C. Pendleton, V.S. Lee, Value Driven Outcomes (VDO): a pragmatic, modular, and extensible software framework for understanding and improving health care costs and outcomes, J. Am. Med. Inform. Assoc. 22 (1) (2015) 223–235.

[16] A. Shanley, Can better modeling reduce pharmaceutical development and manufacturing costs?, 2016, http://www.pharmtech.com/can-better-modeling-reduce-pharmaceutical-development-and-manufacturing-costs.

[17] C.M. Barnett, A.B. Michaud, Lapatinib: the tip of the iceberg for targeting HER2-positive breast cancer, 2007.

[18] D. Mitchell, FDA approves lung cancer drug crizotinib, costs nearly $10K monthly, http://www.emaxhealth.com/1275/fda-approves-lung-cancer-drug-crizotinib-costs-nearly-10k-monthly.

[19] A. Pollack, F.D.A. approves new cystic fibrosis drug, New York Times, 2012.

[20] M. Hay, D.W. Thomas, J.L. Craighead, C. Economides, J. Rosenthal, Clinical development success rates for investigational drugs, Nat. Biotechnol. 32 (1) (2014) 40–51.

[21] J. Foo, F. Michor, Evolution of acquired resistance to anti-cancer therapy, J. Theor. Biol. 355 (2014) 10–20.

[22] A.N. Hata, M.J. Niederst, H.L. Archibald, M. Gomez-Caraballo, F.M. Siddiqui, H.E. Mulvey, Y.E. Maruvka, F. Ji, H.C. Bhang, V. Krishnamurthy Radhakrishna, G. Siravegna, H. Hu, S. Raoof, E. Lockerman, A. Kalsy, D. Lee, C.L. Keating, D.A. Ruddy, L.J. Damon, A.S. Crystal, C. Costa, Z. Piotrowska, A. Bardelli, A.J. Iafrate, R.I. Sadreyev, F. Stegmeier, G. Getz, L.V. Sequist, A.C. Faber, J.A. Engelman, Tumor cells can follow distinct evolutionary paths to become resistant to epidermal growth factor receptor inhibition, Nat. Med. 22 (3) (2016) 262–269.

[23] B. Al-Lazikani, U. Banerji, P. Workman, Combinatorial drug therapy for cancer in the post-genomic era, Nat. Biotechnol. 30 (7) (2012) 679–692.

[24] M. Wu, M. Sirota, A.J. Butte, B. Chen, Characteristics of drug combination therapy in oncology by analyzing clinical trial data on ClinicalTrials.gov, Pac. Symp. Biocomput. (2015) 68–79.

[25] ASoC Oncology, Shaping the Future of Oncology: Envisioning Cancer Care in 2030, 2014, http://www.asco.org/about-asco/asco-vision.

[26] J. Stallard, Clinical trial shows promise of basket studiesİ for cancer drugs, 2015, https://www.mskcc.org/blog/clinical-trial-shows-promise-basket-studies-drugs.

Data characterization

2

CHAPTER OUTLINE

2.1 INTRODUCTION

A critical preliminary step in drug sensitivity modeling entails collecting observations on various entities of a biological system. This chapter provides an overview of the commonly measured information desired for predictive modeling of drug sensitivity. From a systems point of view, we discuss the ways to measure the

Predictive Modeling of Drug Sensitivity. http://dx.doi.org/10.1016/B978-0-12-805274-7.00002-6

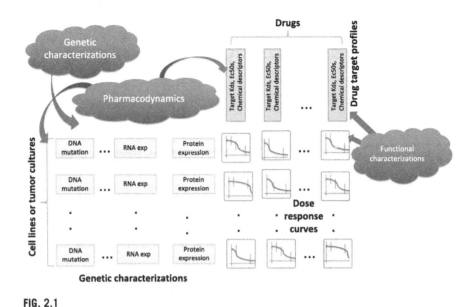

FIG. 2.1

Pictorial representation of the data types discussed in this chapter.

genetic information of a tumor that reflects the internal state of the system, along with methodologies to observe the tumor (system) response when different drugs (perturbations) are applied to the system. The chapter is organized as follows. Firstly, we provide a brief review of molecular biology required to understand the various genomic measurements. Secondly, we describe the characterizations of the genomic, transcriptomic, and proteomic levels. Finally, we describe the pharmacokinetics (PK) and pharmacodynamics (PD) observations followed by cell viability measurements. The different data types to be discussed in this chapter is pictorially represented in Fig. 2.1.

2.2 REVIEW OF MOLECULAR BIOLOGY

An important characteristic of living organisms is the ability of their cells to store, retrieve, replicate, and translate the genetic instructions essential for the development, functioning, and reproduction of the organism. The genetic instructions are carried through **deoxyribonucleic acid (DNA)** molecules [1]. A DNA molecule consists of two polynucleotide chains (commonly known as DNA strands or DNA chains) held together by hydrogen bonds and each chain is formed of four types of nucleotide subunits. A nucleotide is composed of a monosaccharide sugar (deoxyribose), a phosphate group and one of the four following bases:

(i) Adenine (**A**)
(ii) Cytosine (**C**)

(iii) Guanine (**G**)

(iv) Thymine (**T**)

Thus DNA can be considered analogous to a language with four letters in its alphabet.

The two strands of the DNA have a double helix structure and energetically favorable *complementary base pairing* where **A** pairs with **T** and **G** pairs with **C**. The base pairing between **A** and **T** is held together by two hydrogen bonds and the base pairing between **G** and **C** is held together by three hydrogen bonds. Nucleotides **A** and **G** belong to the class of nitrogen-containing bases termed *purine* that have a nine-member ring, whereas nucleotides **T** and **C** belong to the class of nitrogen-containing bases termed *pyrimidine* that have a six-member ring. The base pairings are of similar width and thus hold the sugar-phosphate backbones at an equal distance apart along the DNA molecule. As a result of the complementary base pairing, one strand of DNA is exactly complementary to the DNA of its partner strand. This characteristic plays a central role during the DNA replication process.

Different organisms have dissimilarities in the nucleotide sequences forming their DNA, resulting in carrying of different biological messages. The similarities and differences between DNA sequences can be measured in various ways. For instance, if we have two DNA sequences and try to check the differences in the nucleotides of genes that these two DNA share, we will arrive at a similarity score different from when we consider the diverse insertions, deletions, and substitutions required to transform one DNA to another DNA.

The next question is how are the biological messages encoded in the DNA? Genes are stretches of the DNA sequence that contain information to create proteins. The common flow of genetic information is in the form of DNA being converted to **ribonucleic acid (RNA)** through a process called *transcription*, and the RNA being converted (through a process called *translation*) to proteins that carry out numerous cellular functions. This common flow of information is termed *Central Dogma of Molecular Biology*. Note that other directions of information flow are also observed such as (a) DNA to DNA in the form of *DNA replication*, (b) RNA to DNA in the form of *reverse transcription*, and (c) RNA to RNA in the form of *RNA replication* [2].

Compared to DNA, RNA molecules are single stranded and the nucleotides in RNA are ribonucleotides containing the sugar ribose, rather than deoxyribose as in DNA. Furthermore, RNA contains the bases (i) Adenine (**A**), (ii) Cytosine (**C**), (iii) Guanine (**G**), and (iv) Uracil (**U**) as compared to **A**, **C**, **G**, **T** for DNA. Similar to DNA, RNA also supports base pairing where **A** pairs with **U** and **G** pairs with **C**. The single-stranded RNA can fold up to a variety of structures and the RNAs involved in the process of translation are (a) messenger RNA (mRNA) that is used to direct the synthesis of proteins, (b) ribosomal RNAs (rRNAs) form the core of ribosomes, the machinery used to translate mRNAs to protein molecules, (c) transfer RNAs (tRNAs) that form the adapters that select the amino acids and hold them in place on a ribosome for their incorporation into proteins. There also exists other forms of RNA such as **small interfering RNAs (siRNA)** that can down-regulate gene expression.

The entire DNA does not encode for proteins and there are significant portions of DNA that do not encode any protein. Coding regions in the DNA are known as *exons* interrupted by noncoding regions known as *introns*. Genes contain regulatory sequences to denote the start (promoter sequence) and end positions of transcription. The complete set of information in an organisms' DNA is known as the genome. Human genome is around 3 billion base pairs long and is estimated to carry more than 20,000 protein-encoding genes. During each cell division, a cell has to copy its entire DNA to pass it to both its daughter cells. The process of DNA replication involves (i) separating out of the two strands, (ii) each strand being used as a template for synthesizing its complementary strand, and (iii) pairing of the synthesized strand and the template strand to form a new double helix. The primary enzyme involved in this process is known as *DNA polymerase* that catalyzes the addition of nucleotides to the 3′ end of a growing DNA strand by the formation of a phosphodiester bond between this end and the 5′ phosphate group of the incoming nucleotide. The 5′ and 3′ end refers to the nomenclature used to number the carbon atoms in the sugar ring of the nucleotide. The replication is *semi-conservative* in the sense that one of the strands in each double helix is *conserved* from the previous round of replication. DNA replication begins at replication origins where the separation process of the two DNA strands is initiated. Human genome has around 10,000 replication origins. DNA polymerase has built-in *proofreading* or error correcting activity by checking the base pairing of the last created nucleotide pair. The proofreading of DNA polymerase ensures a low error rate of 10^{-7}, such as one error made for every 10^7 nucleotides copied. Additional DNA repair mechanisms further reduces the error rate of DNA replication to 10^{-9}.

The use of RNA intermediate in the production of proteins from DNA enables faster production of large amounts of a particular protein, as multiple identical RNA copies can be made from the same gene and each RNA can be used to produce multiple identical protein molecules. The production of proteins can be controlled by regulating the efficiency of transcription and translation of different genes. The process of transcription with an error rate of around 1 in every 10^4 nucleotides is comparatively less accurate than DNA replication (1 in 10^7).

Transcription takes place in the cytoplasm for *prokaryotes* (single-cell organisms without nucleus) and in the nucleus for *eukaryotes* (organisms whose cells contain nucleus). Before the RNA produced in the nucleus of prokaryotic cells, termed *primary transcript*, is released to the cytoplasm for protein synthesis, the following processing steps are carried out (a) Addition of *methyl guanine (methyl G) cap* at the 5′ end of the transcript for mRNA, (b) *polyadenylation*: 3′ ends are trimmed at specific nucleotide sequences, and a series of adenine (**A**) nucleotides are added at the cut end. The capping and addition of poly-A tail increases the stability of the mRNA transcript and provides a measure for the protein synthesis machinery to confirm that the message is complete based on the presence of both ends.

In eukaryotic cells, the entire RNA transcript does not code for generation of proteins. The noncoding portions of the RNA transcript (*introns*) are removed by *RNA splicing* during the generation of the final RNA product with the help of splicing enzymes, termed *small nuclear ribonucleoprotein particles* (**snRNPs**). *Alternative splicing* refers to the production of different mRNAs based on various ways of splicing a primary transcript consisting of multiple exons (coding regions).

TRANSLATION

Translation requires the decoding of nucleotides in an mRNA molecule to produce the appropriate protein molecules. Living organisms use a map termed **Genetic Code** for converting the nucleotide-based language of mRNA to the amino acid-based language of proteins. Proteins are formed of sequences of 20 amino acids. Since the RNA code is four letters long, a minimum of triplet code is required to map RNA nucleotides to protein amino acids. In nature, sequences of nucleotides in the mRNA molecule are read in groups of three where each group specifies either the start of a protein, the end of a protein, or an amino acid. Each such nucleotide triplet is termed as **codon**. The genetic code is shown in Fig. 2.2, where **UAA, UAG, UGA** are stop codons denoting the end of translation and **AUG** specifies the amino acid *methionine* that signifies the start of a protein. The remaining 60 triplets code for some amino acid. The genetic code is **degenerate** in the sense that 20 different amino acids are coded by 61 codons which ensures multiple codons coding for the same amino acid. All known life on earth uses the same universal genetic code to map nucleotides to amino acids. Since each triplet of nucleotide codes for an amino acid, an mRNA sequence can be converted to amino acids in one of the three **reading frames** based on where the decoding process begins.

MUTATION

Despite the existence of elaborate DNA repair mechanisms, a permanent change in the DNA, known as *a mutation*, can sometimes occur. Mutations can occur as a result of replication errors or DNA damage due to external agents, such as sunlight, chemicals, etc. Some common forms of DNA damage include (a) **depurination** where a purine base (**A** or **G**) is removed, (b) **deamination** where an amino group is lost, such as Cytosine changing to Uracil, or (c) formation of a covalent bond between adjacent thymine bases termed as **Thymine dimer**.

Some mutations can be beneficial to an organism by providing new capabilities, and the frequency of these variants increases based on *natural selection*. The large variety of living species that we see on earth are results of genetic mutations that improved an organisms' chances of survival in a specific environment. Some examples of beneficial mutations observed in recent times include: (a) increase in bone density due to a mutation in LDL-receptor-related protein 5 (LRP5) [3]; (b) mutations in a hemoglobin gene can provide resistance to malaria [4]; (c) *CCR5*Δ32 mutation can increase resistance to HIV infection [5]; or (d) extra copies of tumor suppressor gene *p*53 in elephant genome reduces its chances to develop cancer [6].

	U		C		A		G		
U	UUU	Phe (F)	UCU	Ser (S)	UAU	Tyr (Y)	UGU	Cys (C)	U
	UUC		UCC		UAC		UGC		C
	UUA	Leu (L)	UCA		UAA	Stop	UGA	Stop	A
	UUG		UCG		UAG	Stop	UGG	Trp (W)	G
C	CUU		CCU	Pro (P)	CAU	His (H)	CGU	Arg (R)	U
	CUC		CCC		CAC		CGC		C
	CUA		CCA		CAA	Gln (Q)	CGA		A
	CUG		CCG		CAG		CGG		G
A	AUU	Ile (I)	ACU	Thr (T)	AAU	Asn (N)	AGU	Ser (S)	U
	AUC		ACC		AAC		AGC		C
	AUA		ACA		AAA	Lys (K)	AGA	Arg (R)	A
	AUG	Met (M)	ACG		AAG		AGG		G
G	GUU	Val (V)	GCU	Ala (A)	GAU	Asp (D)	GGU	Gly (G)	U
	GUC		GCC		GAC		GGC		C
	GUA		GCA		GAA	Glu (E)	GGA		A
	GUG		GCG		GAG		GGG		G

FIG. 2.2

Genetic code.

However, mutations can also be catastrophic for the organism by enabling harmful cellular characteristics such as unrestricted growth. Cancer is a disease that can result from the accumulation of DNA mutations.

For describing mutations, some terminologies need to be mentioned

- An *allele* is a variant of a gene.
- A *wild-type* allele is the kind normally present in the population.
- A *mutant* allele is one that differs from the wild type.
- A genomic alteration that does not manifest in a change at the macroscopic level will be termed as *genotypic* change, while a characteristic that manifests itself at the observational level is called a *phenotypic* change.
- When a phenotypic change requires mutation of both copies of the gene, the mutation is known as a *recessive mutation*. Whereas, mutations that show phenotypic manifestation based on mutation of one copy of the gene are termed *dominant mutation*.
- When the two alleles of a particular gene in a genome are of the same type, the genome is said to be *homozygous* for that gene, whereas a genome is termed *heterozygous* if the two alleles are of different types.

Mutations in DNA sequences can be of various forms. The common forms of mutations types are described next.

Nonsynonymous substitutions are nucleotide mutations that changes the amino acid sequence of a protein. They are of two types:

- **Missense mutations**: Missense mutations consist of single nucleotide mutations that changes the codon to code for a different amino acid. An example of *missense mutation* is shown in Fig. 2.3 where a change from **A** to **C** in the third 3-bp coding for amino acid changes the amino acid from *Asn* (AAU) to *Thr* (ACU).
- **Nonsense mutations**: Nonsense mutations are similar to missense mutations but they change the codon to a premature stop codon which results in truncation of the protein.

An example of *nonsense mutation* is shown in Fig. 2.4 where a change from **U** to **A** in the third 3-bp coding for amino acid changes the amino acid from *Tyr* (UAU) to the stop codon UAA.

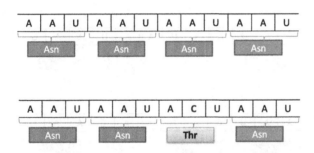

FIG. 2.3

An example of *missense mutation.*

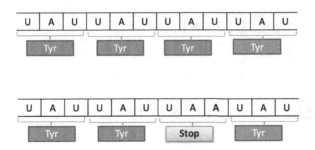

FIG. 2.4

An example of *nonsense mutation.*

Synonymous substitutions does not alter the amino acid sequence. This is feasible due to the degeneracy in the generation of amino acids where multiple 3-bp codons can represent a single codon. Note that we have 20 amino acids and a 4-letter alphabet of length 3 that can produce $4^3 = 64$ possible distinct combinations. Synonymous substitutions are not necessarily harmless as some codons coding for the same amino acid can have different translational efficiency [7]. An example of *synonymous mutation* is shown in Fig. 2.5 where a change from **U** to **A** in the third 3-bp coding for amino acid does not alter the amino acid as both **AAU** and **AAC** code for amino acid *Asn*.

A **frameshift mutation** is caused by insertion, deletion, or duplication of a number of nucleotides in a DNA sequence that changes a gene's reading frame. An example of *frameshift mutation* is shown in Fig. 2.6 where a shift of one nucleotide the amino acids from *Asn, Glu, Gly, Asn* to *Thr, Lys, Val, His*.

Insertion mutation refers to addition of a sequence of nucleotides to a DNA. An example of *insertion mutation* is shown in Fig. 2.7 where an insertion of nucleotide **C** in the third 3-bp code for amino acid changes the last two amino acids to *Ser* (**UCA**) and *Leu* (**UUA**).

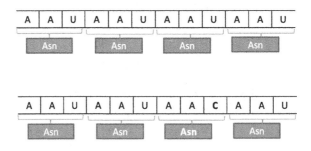

FIG. 2.5

An example of *synonymous mutation*.

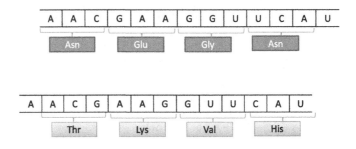

FIG. 2.6

An example of *frameshift mutation*.

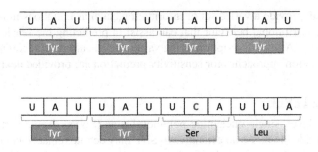

FIG. 2.7

An example of *insertion mutation*.

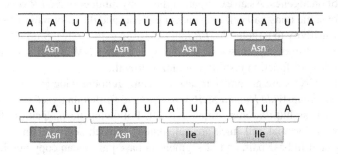

FIG. 2.8

An example of *deletion mutation*.

Deletion mutation refers to removal of a piece of DNA.

An example of *deletion mutation* is shown in Fig. 2.8 where a deletion of nucleotide **A** in the third 3-bp code for amino acid changes the last two amino acids to *Ile* (**AUA**) and *Ile* (**AUA**).

Duplication mutation involves abnormal copying of a piece of DNA multiple number of times.

Repeat expansion mutation: Nucleotide repeats consists of short DNA sequences that are repeated a few number of times in a row. The repeats can be n-bp sequences such as $n = 3$, which is termed a trinucleotide repeat and $n = 4$ is termed a tetranucleotide repeat.

2.3 GENOMIC CHARACTERIZATIONS

Genomic characterization data can be derived from different levels of the cell such as genome, transcriptome, proteome, or metabolome with each level playing a role in the translation of information coded in DNA to functional activities in the cell.

The levels can provide a varied set of information, such as mutations in the genome or altered transcriptional behavior that can assist in predicting the sensitivity of the tumor to a drug. A brief description of the various levels along with commonly used data quantification approaches for sensitivity prediction are provided next.

2.3.1 DNA LEVEL

At the DNA level, single nucleotide polymorphisms (SNPs) denote variations in a single nucleotide block. As an example, an SNP may denote replacement of cytosine (C) by thymine (T) in a stretch of DNA. Copy number variations (CNV) signify deletions or duplications of stretches of DNA and can denote number of copies of a gene. SNPs and CNVs can play an important role in analyzing response to drugs as over-amplification of some genes, as captured by CNV measurements can represent activation of oncogenes. As an example, higher copy number of *EGFR* gene has been observed for some nonsmall cell lung cancer patients and the information can be used to predict sensitivity to targeted *EGFR* inhibitor drug *gefitinib* [8]. Similarly, deletion or mutations in tumor suppressor genes, as captured by CNV or SNP measurements may lead to loss of function resulting in tumor growth.

SNPs and CNVs are generally measured using genome wide arrays such as the *Affymetrix Genome-Wide Human SNP Array 6.0* and further analyzed using software platforms such as http://aroma-project.org [9]. Following normalization, the CNV as compared to normal cell lines are calculated for each segment (in the form of start of segment in base pair, end of segment in base pair, mean copy number of the segment). For our purposes, it is preferable to have the genomic information as a numerical feature vector for each cell line and characterization type. Since the start and stop segments can vary between different cell lines, it is advisable to convert the CNV information of start and stop parts of a segments to gene-level variations which makes it seamless to map different cell lines. One potential program to achieve the conversion is using **R** Package *CNTools*. *The Cancer Genome Atlas* (TCGA) project mutation data are stored in *.maf* format. For details on the *Mutation Annotation Format* (maf), readers are referred to https://wiki.nci.nih.gov/display/TCGA/Mutation+Annotation+Format+%28MAF%29+Specification. Another common approach for measuring variations in DNA level is through *Exome Sequencing* (such as using *Agilent SureSelect*) that measures the variations in the coding regions of the genome. Similar to whole genome sequencing SNPs, the mismatch at different parts of the genes observed through exome sequencing has to be converted to a common set of features for all tumor cell lines.

The large amounts of mutation data spread across multiple chromosomes are sometimes viewed or presented as circular plots for representing the data in compact form using software tool *CIRCOS* (http://circos.ca/).

2.3.2 EPIGENETIC LEVEL

At the epigenetic level, DNA methylation is process of methyl groups attaching to DNA that can alter the transcription process. It is usually measured using arrays such as *Illumina Human Methylation Bead Array* where the proportion of methylated groups are generated based on the intensities of methylated and unmethylated probes.

2.3.3 TRANSCRIPTOMIC LEVEL

Current transcriptomic profiling techniques include DNA microarray, cDNA amplified fragment length polymorphism (cDNA-AFLP), expressed sequence tag (EST) sequencing, serial analysis of gene expression (SAGE), massive parallel signature sequencing (MPSS), RNA-Seq, etc.

Among the above-mentioned technologies, DNA microarray [10] is the most widely used one. But its application is dependent on the availability of complete genome sequence or knowledge of significant amount of transcript sequence. This technique has evolved from Southern blotting [11] and has been widely accepted as an inexpensive analog technique for high-throughput transcriptomic profiling. cDNA-AFLP [12] is a highly sensitive method which allows the detection of low-abundance mRNAs. Recent examples of cDNA-AFLP-based transcriptomic studies are documented in [13,14]. EST[1] sequencing is another approach for transcriptomic profiling which has been used in a large number of transcriptomic studies (e.g., [15,16]). SAGE [17] is a RNA-sequencing-based transcriptomic profiling method that can be used to analyze large number of transcripts quantitatively and simultaneously (e.g., [18,19]). MPSS [20] is another sequence-based approach for profiling transcriptomic data which is somewhat similar to SAGE but with a substantial difference in sequencing approach and with different approach to biochemical manipulation (e.g., [21,22]).

The most recent technology for transcriptomic profiling is RNA-Seq [23] which is considered as a revolutionary tool for this purpose. Eukaryotic transcriptomic profiles are primarily analyzed with this technique and it has been already applied for transcriptomic analysis of several organisms including *Saccharomyces cerevisiae*, *Schizosaccharomyces pombe*, *Arabidopsis thaliana*, mouse, and human cells [24–29].

RNA-Seq technology shows clear advantages over existing profiling technologies in terms of amount of sequence coverage, revealing new transcriptomic insights, accuracy of defining transcription level, etc. However, existing microarray technology still remains reliable to many researchers for various reasons (explained in an article by Blow [30]). Overall comparison of existing technologies and most recent RNA-Seq technology can be found in recent reviews by Roy et al. [31] and Schirmer et al. [32]. Use of different transcriptomic technologies and their success in Amyotrophic Lateral Sclerosis study was discussed in recent review [33]. Also, the omics-era technologies for systems-level understanding of Streptomyces has been discussed in a recent review [34]. Genome-wide copy number analysis [35] is another area where extensive use of different transcriptomic technologies is exercised.

---

[1]http://www.ncbi.nlm.nih.gov/About/primer/est.html.

2.3.4 **PROTEOME LEVEL**

Current *state-of-the-art* proteomic technologies include: two-dimensional difference gel electrophoresis (2D DIGE), matrix-assisted laser desorption/ionization (MALDI) imaging mass spectrometry, electron transfer dissociation (ETD) mass spectrometry, and reverse-phase protein array.

Two-dimensional DIGE is a form of gel-electrophoresis that can label three different samples of proteins with fluorescent dyes. This method overcomes the limitations due to inter-gel variation in traditional 2D gel electrophoresis technique (2D-GE) [36] of proteomic profiling. Despite the limitation in 2D-GE method, it is still a mature proteomic profiling technique backed by three decades of research. Examples of proteomic study using 2D-GE can be found in [37,38]; whereas [39,40] provide examples of using 2D-DIGE technique in proteomic study. A detailed comparison between these two techniques can be found in the article by Marouga et al. [41]. MALDI imaging mass spectrometry [42] is a unique technique for identification of biomarkers in different diseases. Studies of proteomics profiling using this technique include [43,44]. Mass spectrometry-based quantitative proteomic analysis is another form of proteomic profiling which is followed by 2D-GE. Here, intensity of protein stain is measured to find the existence and amount of protein present in a sample. Liquid chromatography mass spectrometry (LC-MS) (example studies [45,46]), liquid chromatography-tandem mass spectrometry (LC-MS/MS) (example studies [47,48]), and in-gel tryptic digestion followed by liquid chromatography-tandem mass spectrometry (geLC-MS/MS) (example studies [49,50]) are different versions of mass spectrometry techniques used in proteomic profiling. ETD mass spectrometry [51] is another form of proteomic study which is a method of fragmenting ions in a mass spectrometer. Molina et al. [52] and Swaney et al. [53] are examples of proteomic studies that use ETD. *Reverse-phase protein array* [54] is a protein microarray technology that has use in quantitative analysis of protein expressions in various kinds of cells, including cancer cells, body fluids, and tissues (example studies [55,56]). RPPA data were provided for the NCI-DREAM drug sensitivity prediction challenge [57]. RPPA data can provide measurements for native proteins as well as phosphorylated isoforms.

Use of several technologies stated above on prognosis and outcome of the treatment of breast tumor was discussed in a recent review paper [58].

2.3.5 **METABOLOME LEVEL**

Metabolites in a cell before or after drug application can be measured using techniques such as LC-MS [59]. Because metabolism of cancerous cells is different from normal cells, the metabolic profiles of cancerous cells can provide unique insights on the potential sensitivity of drugs.

Examples of quantification techniques at various genetic levels are represented as a flowchart in Fig. 2.9.

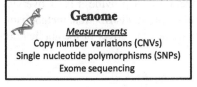

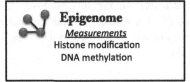

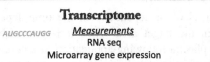

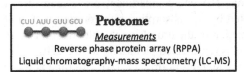

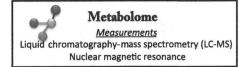

FIG. 2.9

Various levels of genomic information.

2.3.6 MISSING VALUE ESTIMATION

The genomic characterizations can have missing values due to experimental uncertainties that are provided in the form of variables, such as level of background noise for RNA expression, cross-reactivity for proteins, or average base quality for exome sequencing. Features with missing values either need to be removed from the analysis, or the missing values have to be estimated for continuing with the subsequent analysis steps. One of the basic techniques for missing value estimation is to approximate it based on the weighted average of closely related cell lines. The weights are inversely proportional to the Euclidean distance of the cell line containing the missing feature to cell lines with the feature present [60].

2.4 PHARMACOLOGY

Because the ultimate goal of predictive modeling of drug sensitivity is the design of effective therapeutic regimes for a patient, it is essential to understand how a drug works on a patient and how the effects of a drug on a patient are modeled and characterized. For this purpose, this section will provide a brief overview of PK and PD. PK studies the time-dependent absorption, distribution, metabolism, and excretion of a drug applied to the body. PD studies the effect of the applied drug to the body.

2.4.1 PHARMACOKINETICS

The effect of a drug is often dependent on the time-dependent concentration of the drug achieved in the target area. Thus it is worthwhile to model the drug concentration in the plasma following application of a certain amount of drug. The drug concentration is often modeled using a one-compartment model with extravascular administration. The one-compartment model considers a rate of absorption in the blood denoted by K_a (absorption rate constant with unit as inverse of time) followed by elimination of the drug from the blood with rate constant K_e. The one-compartment model is represented pictorially in Fig. 2.10.

The change in the amount of drug ($Y_a(t)$) with time at the site of administration due to absorption is given by [61]:

$$\frac{dY_a(t)}{dt} = -K_a Y_a(t) \tag{2.1}$$

with solution $Y_a(t) = Y_a(0)e^{-K_a t}$. The rate of change of amount of drug in the blood is given by

$$\frac{dY(t)}{dt} = K_a Y_a(t) - K_e Y(t) \tag{2.2}$$

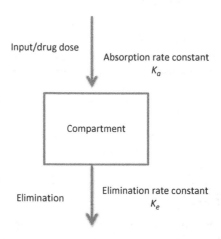

FIG. 2.10

One-compartment model for extravascular administration.

where $Y(t)$ denotes the amount (mass) of drug in the body at time t. Since $Y(0) = 0$, the solution of Eq. (2.2) is given by

$$Y(t) = \frac{K_a Y_a(0)}{K_a - K_e}(e^{-K_e t} - e^{-K_a t}) \tag{2.3}$$

To derive the concentration of drug in the blood ($C_p(t)$) at time t, we divide the amount of drug by volume V.

$$C_p(t) = \frac{K_a Y_a(0)}{V(K_a - K_e)}(e^{-K_e t} - e^{-K_a t}) \tag{2.4}$$

Fig. 2.11 shows an example time-dependent blood concentration curve with parameters $K_a = 10$ per h, $K_e = 4$ per h, $V = 10$ mL, $Y_a(0) = 200$ ng.

C_{max} denotes the peak plasma concentration achieved at time t_{max}. At t_{max}, $K_a Y_a(t_{max}) = K_e Y(t_{max})$ and solving for t_{max} gives

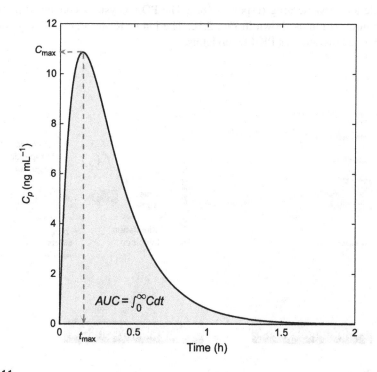

FIG. 2.11

Example drug concentration curve.

$$t_{\max} = \frac{\ln\left(\frac{K_a}{K_e}\right)}{K_a - K_e} \tag{2.5}$$

An objective of effective dosing is to achieve plasma drug concentrations between minimum toxic concentration (MTC) and minimum effective concentration (MEC).

PK also consider multicompartment models, such as a two-compartment model with a central compartment (where distribution and elimination occurs) and a peripheral compartment (representing tissues and other organs) [62–64].

The parameters of the one-compartment model are often estimated based on experimental data collected from blood plasma of patients following drug delivery. The estimation can be patient-based or population-based. The nonlinear mixed-effects approach is a commonly used technique for population studies where the population characteristics are described by the fixed-effects parameters and the individual variability is captured by the random-effects parameters [65–67].

2.4.2 PHARMACODYNAMICS

PD studies the time-dependent effect of a drug on the body. PD modeling and analysis based on experimental data allows us to quantify the system-drug interactions for desirable and adverse drug responses [68]. The PD analysis is conducted based on the PK information as shown in Fig. 2.12. The joint PK and PD modeling is often termed in the literature as PK/PD modeling.

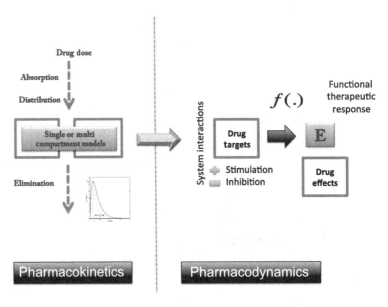

FIG. 2.12

Components of a PK/PD model.

2.4.2.1 Modeling techniques

Simple direct effect models: The model considers the relationship between the response of the system and the plasma drug concentration and is given by Eq. (2.6) [68,69]:

$$E(C) = E_0 + \frac{E_{max} \times C^{\gamma}}{EC_{50}^{\gamma} + C^{\gamma}} \tag{2.6}$$

where C denotes the plasma drug concentration, E_0 denotes the baseline effect with no drug, E_{max} denoting the maximal effect, EC_{50} denotes the drug concentration producing $E(EC_{50}) = 0.5E_{max}$, and γ denotes the Hill coefficient. The model is also know as sigmoid E_{max} model.

Further simplifications are sometimes applied when there is a concern about the limited number of experimentally tested concentrations to properly fit a model, such as in Eq. (2.6). When $\gamma = 1$, the model is known as E_{max} model. When the tested concentrations are significantly less than EC_{50} for the E_{max} model, $EC_{50} + C$ can be approximated by EC_{50} and Eq. (2.6) can be approximated by the following *linear model*:

$$E(C) = E_0 + S \times C \tag{2.7}$$

where S denotes the effect of one unit of drug concentration or the slope of the relationship. When $E(C)$ is expected to be between 20% and 80% of maximal effect, Eq. (2.6) for $\gamma = 1$ can be approximated by the following *Log-linear* model:

$$E(C) = E_0 + S_l \times \log(C) \tag{2.8}$$

where S_l denotes the change in response per one unit of log concentration.

Biophase time-dependent models: The pharmacological effects can lag behind the plasma drug concentrations and the previously presented models cannot capture those delays. Thus a compartmental model similar to PK is often applied to link the plasma drug concentration to drug effects [68,70]. The model is described by Eq. (2.9).

$$\frac{dC_e}{dt} = k_{eo}(C_p - C_e) \tag{2.9}$$

where C_e denotes the concentration at the drug site of action, C_p denotes the plasma drug concentration, and k_{eo} is a first-order rate constant.

Indirect response models

An indirect response model considers modeling of drugs whose mode of actions consists of inhibition or stimulation of a process. The response (R) of a process in the absence of drug is given by the following equation:

$$\frac{dR}{dt} = k_{in} - k_{out} \times R \tag{2.10}$$

where k_{in} and k_{out} denotes the rate constant for production and loss of the response, respectively. In the presence of the drug, four potential scenarios can arise (i) inhibition of k_{in}, (ii) inhibition of k_{out}, (iii) stimulation of k_{in}, and (iv) stimulation of k_{out}. The inhibition is modeled similar to a direct-effect model as shown in Eq. (2.6) with parameters $E_0 = 1$, $E_{max} = -I_{max}$, $\gamma = 1$ and considering IC_{50} rather than EC_{50}. IC_{50} denotes the concentration producing 50% inhibition. Thus the differential equation representing scenario (i) is as follows:

$$\frac{dR}{dt} = k_{in}\left(1 - \frac{I_{max} \times C}{IC_{50} + C}\right) - k_{out} \times R \qquad (2.11)$$

where C represents the drug plasma concentration. Similarly for scenario (ii), we have

$$\frac{dR}{dt} = k_{in} - k_{out}\left(1 - \frac{I_{max} \times C}{IC_{50} + C}\right)R \qquad (2.12)$$

Stimulation scenarios are also represented similarly by Eqs. (2.13), (2.14).

$$\frac{dR}{dt} = k_{in}\left(1 + \frac{E_{max} \times C}{EC_{50} + C}\right) - k_{out} \times R \qquad (2.13)$$

Similarly for scenario (iv), we have

$$\frac{dR}{dt} = k_{in} - k_{out}\left(1 + \frac{E_{max} \times C}{EC_{50} + C}\right)R \qquad (2.14)$$

Further descriptions on various modeling techniques for PK/PD modeling is available at [68,69,71–74].

2.4.3 SOFTWARE PACKAGES

There are multiple software packages available for analyzing PK and PD data. One such shareware example is *Pmetrics* from University of Southern California [75]. *Pmetrics* is written in **R** that allows parametric and nonparametric population model estimation and simulation. A subsequent program such as *BestDose* (http://www.lapk.org/bestdose.php) can utilize the population models to design optimal doses for individual patients. For Microsoft Excel-driven analysis, a plugin such as *PKSolver* [76] can be utilized for basic PK and PD data analysis. Whole body PK simulations can be performed by software such as *PK-Sim* [77].

PK/PD simulations in *Matlab* can be conducted using the *SimBiology* package. It provides a block diagram-based editor as well as capability to model programmatically and includes a library of standard PK/PD models.

2.4.4 DRUG TOXICITY

Drug toxicity refers to the adverse effects of a drug that can include minor side effects to lethal consequences. Toxicology is the study of the adverse side effects of a drug.

Pharmaceutical drugs are designed to create desirable therapeutic responses, but they can also cause drug-induced detrimental effects. In recent years, a number of public computer-readable databases have been created that contain information on drug side effects, such as Side Effect Resource (SIDER) [78] or information on the absorption, distribution, metabolism, excretion, and toxicity (ADMET) properties of different drugs such as ChEMBL [79].

For gathering information on toxicity of combination drugs, the Drug Combination Database (DCDB) [80] can be accessed that contains information on more than 1000 different drug combinations. In vitro combination screens have also been studied [81] that have empirically shown in specific cases that low-toxicity, high therapeutic relevance combinations do exist.

The first step in the experimental procedure to estimate the toxicity of a drug or drug combination usually consists of assessing the drug toxicity in cultured normal tissue cells for organs of key interest such as kidney, liver, and muscle cells. Examples of specific cells that might be tested are Human Primary Kidney Glomerular Endothelial Cells (HPKGEC), Human Skeletal Muscle Cells (SkMc), Osteoblasts from human bone cells (HOb), or hepatocytes from human liver (Hu4239). Following the in vitro tests, toxicity tests in animal models are conducted followed by clinical trials in humans.

Other than the experimental approaches, researchers have also tried to design modeling approaches for drug toxicity prediction. Existing computational methods for drug toxicity prediction can be broadly categorized as *expert-driven systems* that depend on the knowledge of human experts and *data-driven methods* that depend on supervised learning from experimental data. Toxicity is often predicted based on the quantitative structure-activity relationship (QSAR) models, using various chemical descriptors of the drugs [82]. A number of modeling techniques such as Partial Least Squares, SVMs, Neural Nets, Decision trees, and K-nearest neighbor have been applied for the toxicity prediction problem [83,84].

2.5 FUNCTIONAL CHARACTERIZATIONS

In this section, we discuss the primary measurement techniques relevant to cell line or tumor culture-based drug sensitivity studies. The majority of drug sensitivity predictive modeling approaches are designed based on cell line studies due to limitations on the number of samples for animal or human studies. The behaviors over in vitro cell lines are considered as a surrogate for in vivo behavior, but responses can be different in controlled environments due to lack of surrounding entities present in the living organism. Furthermore, cell lines can potentially have changes in their genomic landscape with time, and a closer alternative can be primary cell cultures that consist of the cells extracted from the tumor biopsy. As compared to cell lines, primary cell cultures are more heterogeneous and also contain a small fraction of normal stromal cells, which can be an advantage in drug sensitivity studies because tumor response is often influenced by interplay of the tumor cells and the nonmalignant cells in the tumor microenvironment.

The section is organized as follows: First, we describe the cell viability measurement approaches for cell lines or tumor cultures in Section 2.5.1 followed by characterizations of a tumor drug in Section 2.5.2.

2.5.1 CELL VIABILITY MEASUREMENTS

The drug responses are usually observed using pharmacological assays that measures metabolic activity in terms of reductase-enzyme product or energy-transfer molecule ATP levels following 72–84 h of drug delivery. The cell viability drug screenings are usually conducted using a robotic system where drugs are delivered to small wells containing a portion of the cell culture of interest. Example commercial sellers of such robotic screening systems include *GNF Systems* (http://gnfsystems.com) and *Wako Automation* (http://www.wakoautomation.com/).

Each well is expected to contain more than 100 cells. Experiments are conducted with multiple drug concentrations (around 5–10 concentrations) and the Luminescence in each well is measured at the steady state to assess cell viability. A dose-response curve for each cell line and specific drug is generated by observing the cell viability at different drug concentrations and fitting a curve through the observations as shown in Fig. 2.13.

An example model to fit observed drug responses as used in the cancer cell line encyclopedia study [85] was the following four-parameter (A_t, A_b, H, EC_{50}) sigmoidal model

$$y = A_b + \frac{A_t - A_b}{1 + \left(\frac{x}{EC_{50}}\right)^H}$$
(2.15)

where A_t and A_b denote the top and bottom asymptotes of the response, respectively, H is the hill slope, and EC_{50} denotes the concentration at which the curve response is midway between A_b and A_t.

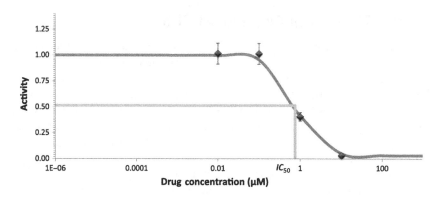

FIG. 2.13

Example dose-response curve.

Commonly used univariate features to represent dose-response curve include IC_{50} (drug concentration required to reduce cell viability to 50%) and AUC (area under the dose-response curve). The IC_{50}s are usually converted to sensitivities between 0 and 1 using a logarithmic mapping function such as $y = 1 - \frac{\log(IC_{50})}{\log(MaxDose)}$ [86,87]. The different features of a dose-response curve as used in Cancer Cell Line Encyclopedia (CCLE) database is shown graphically in Fig. 2.14.

While describing cell viability measurements, it is important to discuss the role of automation played in large-scale pharmacology study. Consider the example of CCLE database (http://www.broadinstitute.org/ccle/home) that contains Pharmacologic profiles for 24 anticancer drugs across 504 cell lines. To generate the data, eight drug concentrations were tested (0.0025, 0.0080, 0.025, 0.080, 0.25, 0.80, 2.53, and 8 μM) for each drug and each cell line in replicates of three. Thus a total of $8 \times 504 \times 24 \times 3 = 290,304$ experiments had to be conducted for generating the drug-response data. This sort of large-scale studies is thus not conducive for manual experimentation and will require more than 72 man-years, assuming an estimate of 30 min for each experiment preparation and set-up. Thus robotic screening systems are used that can accurately deposit tissue samples and drugs in separate wells in required concentrations, speeding up the process significantly. For instance, the robotic high-throughput screening (HTS) system at National Institutes of Health's Chemical Genomics Center can generate more than 2 million drug-response curves per year [88]. The HTS technology at UCLA can screen more than 100,000 compounds in a single day. However, the high-throughput facilities are still expensive to set up and thus are only available to select universities and research laboratories. Advances in microfluidics provide hope for further speed up of the process and

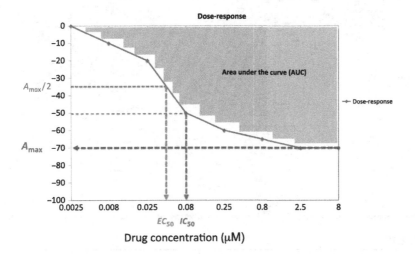

FIG. 2.14

Commonly used features of a drug-response curve based on CCLE dataset.

reduction in cost. For instance, an ultra HTS platform using drop-based microfluidics has been proposed to speed up the screening process by 1000 times and producing the results in a fraction of the cost [89].

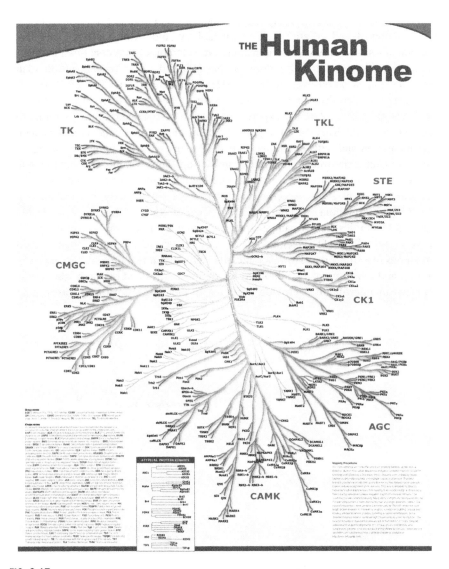

FIG. 2.15

Phylogenetic tree of human protein kinases.

(Illustration reproduced courtesy of Cell Signaling Technology, Inc. (www.cellsignal.com).)

2.5.2 DRUG CHARACTERIZATIONS

Independent of the cellular response, a drug can be characterized based on its chemical structure or the targets being inhibited or stimulated by the drug. The chemical structure is frequently captured by Chemical Descriptors of a drug that are designed from the molecular structure of the compound. The chemical descriptors can be used to predict the biological response based on regression approaches and the mapping from chemical structures to bioactivity is termed quantitative structure-activity relationships (QSARs) [90,91].

The generation of numerical descriptors of a drug for QSAR modeling can be done with the help of software packages such as *PaDEL* and *Dragon*. *Dragon 6* can generate *4885* descriptors of a chemical compound, such as molecular weight, atomic bond descriptions, description of the rings in the compound, functional groups, fragment counts, topological indices, charge descriptors, and many more. For detailed description of chemical descriptors, readers are referred to [92].

Other than the description of a drug provided by the chemical descriptors, we can also characterize the drugs in terms of the targets that it inhibits or stimulates. In cancer therapy, this is highly relevant for targeted drugs that target-specific proteins [93–97]. For anticancer targeted drugs, the targets of interest are usually the kinases and the human Kinome consists of around 518 proteins [98]. The phylogenetic tree depicting the relationships between the human protein kinases is shown in Fig. 2.15.

The level of inhibition of targets by each drug as measured by EC_{50}s or K_ds can be obtained from earlier studies in the literature [99,100] or from databases such as *pubchem* (http://pubchem.ncbi.nlm.nih.gov/), *Library of Integrated Network-based Cellular Signatures* (**LINCS**) (http://commonfund.nih.gov/lincs/). The half-maximal concentration (EC_{50}) value is directly related to the notion of inhibition of a target; in particular, the EC_{50} values correspond to the amount of a compound needed to deactivate via phosphorylation 50% of the population of the associated target. Thus, for a drug, a target with a lower EC_{50} is the one that will be heavily inhibited at low drug concentration levels.

REFERENCES

[1] B. Alberts, A. Johnson, J. Lewis, M. Raff, D. Bray, K. Hopkin, K. Roberts, P. Walter, Essential Cell Biology, second ed., Garland Science/Taylor and Francis Group, New York and London, 2003.

[2] M.M. Lai, RNA replication without RNA-dependent RNA polymerase: surprises from hepatitis delta virus, J. Virol. 79 (13) (2005) 7951–7958.

[3] L.M. Boyden, J. Mao, J. Belsky, L. Mitzner, A. Farhi, M.A. Mitnick, D. Wu, K. Insogna, R.P. Lifton, High bone density due to a mutation in LDL-receptor-related protein 5, N. Engl. J. Med. 346 (20) (2002) 1513–1521.

[4] D. Modiano, G. Luoni, B.S. Sirima, J. Simpore, F. Verra, A. Konate, E. Rastrelli, A. Olivieri, C. Calissano, G.M. Paganotti, L. D'Urbano, I. Sanou, A. Sawadogo, G. Modiano, M. Coluzzi, Haemoglobin C protects against clinical *Plasmodium falciparum* malaria, Nature 414 (6861) (2001) 305–308.

[5] A.D. Sullivan, J. Wigginton, D. Kirschner, The coreceptor mutation CCR5Delta32 influences the dynamics of HIV epidemics and is selected for by HIV, Proc. Natl Acad. Sci. USA 98 (18) (2001) 10214–10219.

[6] E. Callaway, How elephants avoid cancer, Nature 1038 (2015) 18534.

[7] H. Gingold, Y. Pilpel, Determinants of translation efficiency and accuracy, Mol. Syst. Biol. 7 (2011) 481.

[8] F. Cappuzzo, F.R. Hirsch, E. Rossi, S. Bartolini, G.L. Ceresoli, L. Bemis, J. Haney, S. Witta, K. Danenberg, I. Domenichini, V. Ludovini, E. Magrini, V. Gregorc, C. Doglioni, A. Sidoni, M. Tonato, W.A. Franklin, L. Crino, P.A. Bunn, M. Varella-Garcia, Epidermal growth factor receptor gene and protein and gefitinib sensitivity in non-small-cell lung cancer, J. Natl Cancer Inst. 97 (9) (2005) 643–655.

[9] H. Bengtsson, K. Simpson, J. Bullard, K. Hansen, aroma.affymetrix: a generic framework in R for analyzing small to very large Affymetrix data sets in bounded memory, Tech. Rep. 745, Department of Statistics, University of California, Berkeley, 2008.

[10] M.J. Heller, DNA microarray technology: devices, systems, and applications, Annu. Rev. Biomed. Eng. 4 (1) (2002) 129–153.

[11] E.M. Southern, Blotting at 25, Trends Biochem. Sci. 25 (12) (2000) 585–588.

[12] P. Vos, R. Hogers, M. Bleeker, M. Reijans, T. van de Lee, M. Hornes, A. Friters, J. Pot, J. Paleman, M. Kuiper, M. Zabeau, AFLP: a new technique for DNA fingerprinting, Nucl. Acid Res. 23 (21) (1995) 4407–4414.

[13] P. Sojikul, P. Kongsawadworakul, U. Viboonjun, J. Thaiprasit, B. Intawong, J. Narangajavana, M.R.J. Svasti, AFLP-based transcript profiling for cassava genome-wide expression analysis in the onset of storage root formation, Physiol. Plant. 140 (2) (2010) 189–298.

[14] M. Claverie, M. Souquet, J. Jean, N. Forestier-Chiron, V. Lepitre, M. Pr, J. Jacobs, D. Llewellyn, J.-M. Lacape, cDNA-AFLP-based genetical genomics in cotton fibers, TAG Theor. Appl. Genet. 124 (2012) 665–683.

[15] G. Xiaomeng, C. Weihua, S. Shuhui, W. Weiwei, H. Songnian, Y. Jun, Transcriptomic profiling of mature embryo from an elite super-hybrid rice LYP9 and its parental lines, BMC Plant Biol. 8 (2008) 114.

[16] M.D. Adams, J.M. Kelley, J.D. Gocayne, M. Dubnick, M.H. Polymeropoulos, H. Xiao, C.M. Merril, A. Wu, B. Olde, R.F. Moreno, et al., Complementary DNA sequencing: expressed sequence tags and human genome project, Science 252 (5013) (1991) 1651–1656.

[17] V.E. Velculescu, L. Zhang, B. Vogelstein, K.W. Kinzler, Serial analysis of gene expression, Science 270 (5235) (1995) 484–487.

[18] C.D. Hough, C.A. Sherman-Baust, E.S. Pizer, F.J. Montz, D.D. Im, N.B. Rosenshein, K.R. Cho, G.J. Riggins, P.J. Morin, Large-scale serial analysis of gene expression reveals genes differentially expressed in ovarian cancer, Cancer Res. 60 (22) (2000) 6281–6287.

[19] S.P. Gygi, Y. Rochon, B.R. Franza, R. Aebersold, Correlation between protein and mRNA abundance in yeast, Mol. Cell. Biol. 19 (1999) 1720–1730.

[20] S. Brenner, M. Johnson, J. Bridgham, G. Golda, D.H. Lloyd, D. Johnson, S. Luo, S. McCurdy, M. Foy, M. Ewan, R. Roth, D. George, S. Eletr, G. Albrecht, E. Vermaas, S.R. Williams, K. Moon, T. Burcham, M. Pallas, R.B. DuBridge, J. Kirchner, K. Fearon, J. Mao, K. Corcoran, Gene expression analysis by massively parallel signature sequencing (MPSS) on microbead arrays, Nat. Biotechnol. 18 (6) (2000) 630–634.

[21] D. Erdner, D. Anderson, Global transcriptional profiling of the toxic dinoflagellate *Alexandrium fundyense* using massively parallel signature sequencing, BMC Genomics 7 (1) (2006) 88.

[22] R. Natrajan, A. Mackay, M.B. Lambros, B. Weigelt, P.M. Wilkerson, E. Manie, A. Grigoriadis, R. A'Hern, P. van der Groep, I. Kozarewa, T. Popova, O. Mariani, S. Turajlic, S.J. Furney, R. Marais, D.-N. Rodruigues, A.C. Flora, P. Wai, V. Pawar, S. McDade, J. Carroll, D. Stoppa-Lyonnet, A.R. Green, I.O. Ellis, C. Swanton, P. van Diest, O. Delattre, C.J. Lord, W.D. Foulkes, A. Vincent-Salomon, A. Ashworth, M.H. Stern, J.S. Reis-Filho, A whole-genome massively parallel sequencing analysis of BRCA1 mutant oestrogen receptor-negative and -positive breast cancers, J. Pathol. 227 (1) (2012) 29–41.

[23] Z. Wang, M. Gerstein, M. Snyder, RNA-Seq: a revolutionary tool for transcriptomics, Nat Rev. Genet. 10 (1) (2009) 57–63.

[24] U. Nagalakshmi, Z. Wang, K. Waern, C. Shou, D. Raha, M. Gerstein, M. Snyder, The transcriptional landscape of the yeast genome defined by RNA sequencing, Science 320 (5881) (2008) 1344–1349.

[25] B.T. Wilhelm, S. Marguerat, S. Watt, F. Schubert, V. Wood, I. Goodhead, C.J. Penkett, J. Rogers, J. Bhler, Dynamic repertoire of a eukaryotic transcriptome surveyed at single-nucleotide resolution, Nature 453 (7199) (2008) 1239–1243.

[26] A. Mortazavi, B.A. Williams, K. McCue, L. Schaeffer, B. Wold, Mapping and quantifying mammalian transcriptomes by RNA-seq, Nat. Methods 5 (7) (2008) 621–628.

[27] R. Lister, R.C. O'Malley, J. Tonti-Filippini, B.D. Gregory, C.C. Berry, A.H. Millar, J.R. Ecker, Highly integrated single-base resolution maps of the epigenome in arabidopsis, Cell 133 (3) (2008) 523–536.

[28] N. Cloonan, A.R.R. Forrest, G. Kolle, B.B.A. Gardiner, G.J. Faulkner, M.K. Brown, D.F. Taylor, A.L. Steptoe, S. Wani, G. Bethel, A.J. Robertson, A.C. Perkins, S.J. Bruce, C.C. Lee, S.S. Ranade, H.E. Peckham, J.M. Manning, K.J. McKernan, S.M. Grimmond, Stem cell transcriptome profiling via massive-scale mRNA sequencing, Nat. Methods 5 (7) (2008) 613–619.

[29] J.C. Marioni, C.E. Mason, S.M. Mane, M. Stephens, Y. Gilad, RNA-seq: an assessment of technical reproducibility and comparison with gene expression arrays, Genome Res. 18 (9) (2008) 1509–1517.

[30] N. Blow, Transcriptomics: the digital generation, Nature 458 (7235) (2009) 239–242.

[31] N.C. Roy, E. Altermann, Z.A. Park, W.C. McNabb, A comparison of analog and next-generation transcriptomic tools for mammalian studies, Brief. Funct. Genomics 10 (3) (2011) 135–150.

[32] K. Schirmer, B.B. Fischer, D.J. Madureira, S. Pillai, Transcriptomics in ecotoxicology, Anal. Bioanal. Chem. 397 (3) (2010) 917–923.

[33] A. Henriques, J.L.G. De Aguilar, Can transcriptomics cut the gordian knot of amyotrophic lateral sclerosis?, Curr. Genomics 12 (7) (2011) 506–515.

[34] Z. Zhou, J. Gu, Y.-L. Du, Y.-Q. Li, Y. Wang, The -omics Era- toward a systems-level understanding of streptomyces, Curr. Genomics 12 (6) (2011) 404–416.

[35] S.M. Rothenberg, J. Settleman, Discovering tumor suppressor genes through genome-wide copy number analysis, Curr. Genomics 11 (5) (2010) 297–310.

[36] T. Rabilloud, C. Lelong, Two-dimensional gel electrophoresis in proteomics: a tutorial, J. Proteomics 74 (10) (2011) 1829–1841.

[37] E. Piruzian, S. Bruskin, A. Ishkin, R. Abdeev, S. Moshkovskii, S. Melnik, Y. Nikolsky, T. Nikolskaya, Integrated network analysis of transriptomic and proteomic data in psoriasis, BMC Syst. Biol. 4 (2010) 41.

[38] D. Greenbaum, R. Jansen, M. Gerstein, Analysis of mRNA expression and protein abundance data: an approach for the comparison of the enrichment of features in the cellular population of proteins and transcrips, Bioinformatics 18 (4) (2002) 585–596.

[39] E. Com, E. Boitier, J.-P. Marchandeau, A. Brandenburg, S. Schroeder, D. Hoffmann, A. Mally, J.-C. Gautier, Integrated transcriptomic and proteomic evaluation of gentamicin nephrotoxicity in rats, Toxicol. Appl. Pharmacol. 258 (1) (2012) 124–133.

[40] J.X. Yan, A.T. Devenish, R. Wait, T. Stone, S. Lewis, S. Fowler, Fluorescence two-dimensional difference gel electrophoresis and mass spectrometry based proteomic analysis of *Escherichia coli*, Proteomics 2 (12) (2002) 1682–1698.

[41] R. Marouga, S. David, E. Hawkins, The development of the DIGE system: 2D fluorescence difference gel analysis technology, Anal. Bioanal. Chem. 382 (2005) 669–678.

[42] J. Franck, K. Arafah, M. Elayed, D. Bonnel, D. Vergara, A. Jacquet, D. Vinatier, M. Wisztorski, R. Day, I. Fournier, M. Salzet, MALDI imaging mass spectrometry, Mol. Cell. Proteomics 8 (9) (2009) 2023–2033.

[43] M.R. Groseclose, P.P. Massion, P. Chaurand, R.M. Caprioli, High-throughput proteomic analysis of formalin-fixed paraffin-embedded tissue microarrays using MALDI imaging mass spectrometry, Proteomics 8 (18) (2008) 3715–3724.

[44] M. Elsner, S. Rauser, S. Maier, C. Schne, B. Balluff, S. Meding, G. Jung, M. Nipp, H. Sarioglu, G. Maccarone, M. Aichler, A. Feuchtinger, R. Langer, U. Jtting, M. Feith, B. Kster, M. Ueffing, H. Zitzelsberger, H. Hfler, A. Walch, MALDI imaging mass spectrometry reveals COX7A2, TAGLN2 and S100-A10 as novel prognostic markers in Barrett's adenocarcinoma, J. Proteomics 75 (15) (2012) 4693–4704.

[45] G. Yue, Q. Luo, J. Zhang, S.-L. Wu, B.L. Karger, Ultratrace LC/MS proteomic analysis using 10-m-i.d. porous layer open tubular poly(styrenedivinylbenzene) capillary columns, Anal. Chem. 79 (3) (2007) 938–946.

[46] D.A. Elias, M.E. Monroe, M.J. Marshall, M.F. Romine, A.S. Belieav, J.K. Fredrickson, G.A. Anderson, R.D. Smith, M.S. Lipton, Global detection and characterization of hypothetical proteins in *Shewanella oneidensis* MR-1 using LC-MS based proteomics, Proteomics 5 (12) (2005) 3120–3130.

[47] L. Nie, G. Wu, F.J. Brockman, W. Zhang, Integrated analysis of transcriptomic and proteomic data of *Desulfovibrio vulgaris*: zero-inflated Poisson regression models to predict abundance of undetected proteins, Bioinformatics 22 (13) (2006) 1641–1647.

[48] R.V. Polisetty, P. Gautam, R. Sharma, H.C. Harsha, S.C. Nair, M.K. Gupta, M.S. Uppin, S. Challa, A.K. Puligopu, P. Ankathi, A.K. Purohit, G.R. Chandak, A. Pandey, R. Sirdeshmukh, LC-MS/MS analysis of differentially expressed glioblastoma membrane proteome reveals altered calcium signalling and other protein groups of regulatory functions, Mol. Cell. Proteomics 11 (6) (2012) M111.013565.

[49] N. Delmotte, C.H. Ahrens, C. Knief, E. Qeli, M. Koch, H.-M. Fischer, J.A. Vorholt, H. Hennecke, G. Pessi, An integrated proteomics and transcriptomics reference data set provides new insights into the *Bradyrhizobium japonicum* bacteroid metabolism in soybean root nodules, Proteomics 10 (2010) 1391–1400.

[50] L. Sennels, M. Salek, L. Lomas, E. Boschetti, P.G. Righetti, J. Rappsilber, Proteomic analysis of human blood serum using peptide library beads, J. Proteome Res. 6 (10) (2007) 4055–4062.

[51] L.M. Mikesh, B. Ueberheide, A. Chi, J.J. Coon, J.E.P. Syka, J. Shabanowitz, D.F. Hunt, The utility of ETD mass spectrometry in proteomic analysis, Biochim. Biophys. Acta 1764 (12) (2006) 1811–1822.

[52] H. Molina, D.M. Horn, N. Tang, S. Mathivanan, A. Pandey, Global proteomic profiling of phosphopeptides using electron transfer dissociation tandem mass spectrometry, Proc. Natl Acad. Sci. USA 104 (7) (2007) 2199–2204.

[53] D.L. Swaney, C.D. Wenger, J.A. Thomson, J.J. Coon, Human embryonic stem cell phosphoproteome revealed by electron transfer dissociation tandem mass spectrometry, Proc. Natl Acad. Sci. USA 106 (4) (2009) 995–1000.

[54] B. Spurrier, S. Ramalingam, S. Nishizuka, Reverse-phase protein lysate microarrays for cell signaling analysis, Nat. Protocols 3 (11) (2008) 1796–1808.

[55] S. Nishizuka, L. Charboneau, L. Young, S. Major, W.C. Reinhold, M. Waltham, H. Kouros-Mehr, K.J. Bussey, J.K. Lee, V. Espina, P.J. Munson, E. Petricoin, L.A. Liotta, J.N. Weinstein, Proteomic profiling of the NCI-60 cancer cell lines using new high-density reverse-phase lysate microarrays, Proc. Natl Acad. Sci. USA 100 (24) (2003) 14229–14234.

[56] S.M. Cowherd, V.A. Espina, E.F. Petricoin III, L.A. Liotta, Proteomic analysis of human breast cancer tissue with laser-capture microdissection and reverse-phase protein microarrays, Clin. Breast Cancer 5 (5) (2004) 385–392.

[57] J.C. Costello, et al., A community effort to assess and improve drug sensitivity prediction algorithms, Nat. Biotechnol. 32 (12) (2014) 1202–1212.

[58] Y. Baskin, T. Yigitbasi, Clinical proteomics of breast cancer, Curr. Genomics 11 (7) (2010) 528–536.

[59] F. Li, F.J. Gonzalez, X. Ma, LCMS-based metabolomics in profiling of drug metabolism and bioactivation, Acta Pharm. Sin. B 2 (2) (2012) 118–125.

[60] O.G. Troyanskaya, M.N. Cantor, G. Sherlock, P.O. Brown, T. Hastie, R. Tibshirani, D. Botstein, R.B. Altman, Missing value estimation methods for DNA microarrays, Bioinformatics 17 (6) (2001) 520–525.

[61] M.J. Ratain, W.K. Plunkett, Principles of Pharmacokinetics, Holland-Frei Cancer Medicine, sixth ed., BC Decker, Hamilton, 2003.

[62] S. Dhillon, A. Kostrzewski, Clinical Pharmacokinetics, Pharmaceutical Press, London, 2006.

[63] G. Levy, M. Gibaldi, W.J. Jusko, Multicompartment pharmacokinetic models and pharmacologic effects, J. Pharm. Sci. 58 (4) (1969) 422–424.

[64] H. Derendorf, B. Meibohm, Modeling of pharmacokinetic/pharmacodynamic (PK/PD) relationships: concepts and perspectives, Pharm. Res. 16 (2) (1999) 176–185.

[65] L.B. Sheiner, B. Rosenberg, V.V. Marathe, Estimation of population characteristics of pharmacokinetic parameters from routine clinical data, J. Pharmacokinet. Pharmacodyn. 5 (5) (1977) 445–479.

[66] M.E. Spilker, P. Vicini, An evaluation of extended vs weighted least squares for parameter estimation in physiological modeling, J. Biomed. Inform. 34 (5) (2001) 348–364.

[67] P.L. Bonate, Recommended reading in population pharmacokinetic pharmacodynamics, AAPS J. 7 (2) (2005) E363–E373.

[68] M.A. Felmlee, M.E. Morris, D.E. Mager, Mechanism-based pharmacodynamic modeling, Methods Mol. Biol. 929 (2012) 583–600.

[69] G. Dheeraj, P. Gomathi, Pharmacokinetic/Pharmacodynamic (PK/PD) modeling: an investigational tool for drug development, Int. J. Pharm. Pharm. Sci. 4 (2012) 30–37.

[70] W. Colburn, Simultaneous pharmacokinetic and pharmacodynamic modeling, J. Pharmacokinet. Biopharm. 9 (3) (1981) 367–388.

[71] B. Meibohm, H. Derendorf, Basic concepts of pharmacokinetic/pharmacodynamic (PK/PD) modelling, Int. J. Clin. Pharmacol. Ther. 35 (10) (1997) 401–413.

[72] C. Csajka, D. Verotta, Pharmacokinetic-pharmacodynamic modelling: history and perspectives, J. Pharmacokinet. Pharmacodyn. 33 (3) (2006) 227–279.

[73] M. Danhof, E.C. de Lange, O.E.D. Pasqua, B.A. Ploeger, R.A. Voskuyl, Mechanism-based pharmacokinetic-pharmacodynamic (PK-PD) modeling in translational drug research, Trends Pharmacol. Sci. 29 (4) (2008) 186–191.

[74] D.E. Mager, W.J. Jusko, Development of translational pharmacokinetic-pharmacodynamic models, Clin. Pharmacol. Ther. 83 (6) (2008) 909–912.

[75] M. Neely, M. van Guilder, W. Yamada, A. Schumitzky, R. Jelliffe, Accurate detection of outliers and subpopulations with Pmetrics, a non-parametric and parametric pharmacometric modeling and simulation package for R, Ther. Drug Monit. 34 (4) (2012) 467–476.

[76] Y. Zhang, M. Huo, J. Zhou, S. Xie, PKSolver: an add-in program for pharmacokinetic and pharmacodynamic data analysis in Microsoft Excel, Computer Methods Programs Biomed. 99 (3) (2010) 306–314.

[77] S. Willmann, J. Lippert, M. Sevestre, J. Solodenko, F. Fois, W. Schmitt, PK-Sim: a physiologically based pharmacokinetic whole-body model, BIOSILICO 1 (4) (2003) 121–124.

[78] M. Kuhn, M. Campillos, I. Letunic, L.J. Jensen, P. Bork, A side effect resource to capture phenotypic effects of drugs, Mol. Syst. Biol. 6 (2010) 343.

[79] A. Gaulton, L.J. Bellis, A.P. Bento, J. Chambers, M. Davies, A. Hersey, Y. Light, S. McGlinchey, D. Michalovich, B. Al-Lazikani, J.P. Overington, ChEMBL: a large-scale bioactivity database for drug discovery, Nucl. Acids Res. 40 (2012) D1100–D1107.

[80] Y. Liu, Q. Wei, G. Yu, W. Gai, Y. Li, X. Chen, DCDB 2.0: a major update of the drug combination database, Database (Oxford) 2014 (2014) bau124.

[81] J. Lehar, et al., Synergistic drug combinations tend to improve therapeutically relevant selectivity, Nat. Biotech. 27 (2009) 659–666.

[82] W. Muster, A. Breidenbach, H. Fischer, S. Kirchner, L. Mller, A. Phler, Computational toxicology in drug development, Drug Discov. Today 13 (7–8) (2008) 303–310.

[83] Y.Z. Chen, C.W. Yap, H. Li, Current QSAR Techniques for Toxicology, John Wiley & Sons, Inc., Hoboken, NJ, 2006, pp. 217–238.

[84] R. Franke, A. Gruska, General Introduction to QSAR, CRC Press, Boca Raton, FL, 2003, pp. 1–40.

[85] J. Barretina, et al., The Cancer Cell Line Encyclopedia enables predictive modelling of anticancer drug sensitivity, Nature 483 (7391) (2012) 603–607.

[86] N. Berlow, S. Haider, Q. Wan, M. Geltzeiler, L.E. Davis, C. Keller, R. Pal, An integrated approach to anti-cancer drugs sensitivity prediction, IEEE/ACM Trans. Comput. Biol. Bioinform. 11 (6) (2014) 995–1008.

[87] N. Berlow, L.E. Davis, E.L. Cantor, B. Seguin, C. Keller, R. Pal, A new approach for prediction of tumor sensitivity to targeted drugs based on functional data, BMC Bioinform. 14 (2013) 239.

[88] S. Michael, D. Auld, C. Klumpp, A. Jadhav, W. Zheng, N. Thorne, C.P. Austin, J. Inglese, A. Simeonov, A robotic platform for quantitative high-throughput screening, Assay Drug Dev. Technol. 6 (5) (2008) 637–657.

[89] J.J. Agresti, E. Antipov, A.R. Abate, K. Ahn, A.C. Rowat, J.C. Baret, M. Marquez, A.M. Klibanov, A.D. Griffiths, D.A. Weitz, Ultrahigh-throughput screening in drop-based microfluidics for directed evolution, Proc. Natl Acad. Sci. USA 107 (9) (2010) 4004–4009.

[90] A. Tropsha, Best practices for QSAR model development, validation, and exploitation, Mol. Inform. 29 (6–7) (2010) 476–488.

[91] C. Nantasenamat, C. Isarankura-Na-Ayudhya, V. Prachayasittikul, Advances in computational methods to predict the biological activity of compounds, Expert Opin. Drug Discov. 5 (7) (2010) 633–654.

[92] R. Todeschini, V. Consonni, Molecular Descriptors for Chemoinformatics, Wiley-VCH Verlag GmbH & Co. KGaA, Weinheim, 2010.

[93] C. Sawyers, Targeted cancer therapy, Nature 432 (2004) 294–297.

[94] M.R. Green, Targeting targeted therapy, N. Engl. J. Med. 350 (21) (2004) 2191–2193.

[95] B.J. Druker, Molecularly targeted therapy: have the floodgates opened?, Oncologist 9 (1) (2004) 357–360.

[96] A. Hopkins, J. Mason, J. Overington, Can we rationally design promiscuous drugs?, Curr. Opin. Struct. Biol. 16 (1) (2006) 127–136.

[97] Z.A. Knight, K.M. Shokat, Features of selective kinase inhibitors, Chem. Biol. 12 (6) (2005) 621–637.

[98] G. Manning, D.B. Whyte, R. Martinez, T. Hunter, S. Sudarsanam, The protein kinase complement of the human genome, Science 298 (5600) (2002) 1912–1934.

[99] M.W. Karaman, S. Herrgard, D.K. Treiber, P. Gallant, C.E. Atteridge, B.T. Campbell, K.W. Chan, P. Ciceri, M.I. Davis, P.T. Edeen, R. Faraoni, M. Floyd, J.P. Hunt, D.J. Lockhart, Z.V. Milanov, M.J. Morrison, G. Pallares, H.K. Patel, S. Pritchard, L.M. Wodicka, P.P. Zarrinkar, A quantitative analysis of kinase inhibitor selectivity, Nat. Biotechnol. 26 (1) (2008) 127–132.

[100] P.P. Zarrinkar, R.N. Gunawardane, M.D. Cramer, M.F. Gardner, D. Brigham, B. Belli, M.W. Karaman, K.W. Pratz, G. Pallares, Q. Chao, K.G. Sprankle, H.K. Patel, M. Levis, R.C. Armstrong, J. James, S.S. Bhagwat, AC220 is a uniquely potent and selective inhibitor of FLT3 for the treatment of acute myeloid leukemia (AML), Blood 114 (14) (2009) 2984–2992.

Feature selection and extraction from heterogeneous genomic characterizations

3

CHAPTER OUTLINE

3.1 INTRODUCTION

A significant number of current drug sensitivity studies uses high-throughput technologies to collect information from various genomic levels, resulting in extremely high dimensional genomic characterization dataset. For instance, the *Cancer Cell Line Encyclopedia* Study [1] has >50,000 features representing mRNA expressions and mutational status of thousands of genes. Because only a small set of these features are important for drug sensitivity prediction and the number of samples for training is significantly less than the number of available features, direct application of machine learning tools on all these features can result in model overfitting. If we train a model based on N samples from a population distribution and test it on the actual population distribution, the prediction error of the model over the population distribution (termed generalization or validation error) can often follow a pattern as shown in Fig. 3.1 when n is small compared to the number of features and model complexity is high. The curves shown in Fig. 3.1 are drawn based on a synthetic example created with $N = 25$ random training samples with 20 features, where the response Y is only dependent on the first 10 of these 20 features. The error curves reflect mean absolute error for linear regression models fitted using i features where $1 \leq i \leq 18$. We note that the error of the linear regression model on the 25 samples used for estimating the model parameters (termed training error) keeps decreasing as we add features (marked by crosses). The error on separate 1000 hold out samples following the same dependencies on Y (termed validation or generalization error) shows an initial trend of decrease followed by an increase in error. Since our output response is only dependent on the first 10 features, we note that the validation error keeps increasing when the number of features is increased beyond 10. After a point, the addition of more features in modeling can be detrimental from the prediction accuracy perspective. Fig. 3.1 illustrates the problem of overfitting where model complexity (in this case, the number of features) is high relative to the number of training samples resulting in the estimated model to fit the training data precisely, but producing higher error when predicting unseen data (generalization error). Note that in drug sensitivity studies, the number of features are usually significantly higher (in the range of 20,000–150,000) as compared to the number of samples (in the range of 60–1000) and thus overfitting is a highly pertinent problem.

The issue of overfitting can be addressed in various ways. We first discuss the three categories of data-driven techniques for avoiding overfitting.

Firstly, few of the modeling approaches are suitable to handle large numbers of features through the use of *regularization*, such as penalizing the norms of the feature weights in *Elastic Net* [2]. Random forest regression [3] is also generally suitable to handle large number of features by increasing the number of regression trees in training. We will discuss these approaches that have built-in capacity to handle large number of features in later chapters when we discuss regression techniques. This kind of variable selection as part of the learning process is often know as *embedded feature selection*.

Secondly, the number of features to be used in model generation can be reduced by the use of feature selection. Feature selection refers to selecting a subset of features from the full set of input features based on some design criteria. Feature selection methods are broadly categorized as *filter* and *wrapper* techniques; the former do not interact with the final designed model, whereas the latter use model design in the search itself [4,5]. In the filter approach, the features are rated based on general characteristics such as statistical independence or correlation of individual features with output response. On the other hand, wrapper techniques evaluate subsets based on their predictive accuracy based on a particular model. Filters are faster, but they tend to introduce bias and sometimes miss the multivariate relationships among features. A feature may not perform well individually, but in combination with other features can generate a high accuracy model. Wrapper methods, even though slow to

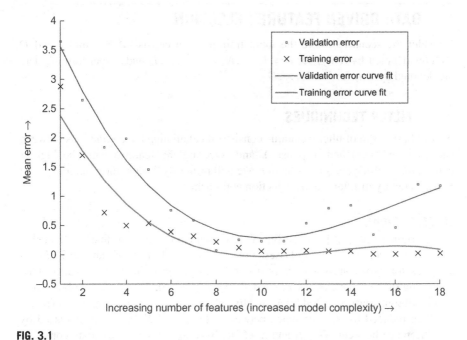

FIG. 3.1

Behavior of mean validation and training error with increase in number of features. The number of training samples is fixed.

run, tend to capture the feature combinations with higher model accuracy, but have potential to overfit the data as compared to filter approaches. In wrapper methods, the goodness of a particular feature subset S_m for feature selection is evaluated using an objective function, $J(S_m)$ which can be model accuracy measured in terms of Root Mean Square Error or correlation coefficient between predicted and experimental responses. Embedded feature selection approaches are similar to wrapper methods with lower computational burden, but are specific to a learning machine.

Finally, dimensionality reduction can also be approached based on extracting features from the input data where input data are mapped to new coordinates using different functions. One common example of feature extraction approach is Principal Component Analysis (PCA).

The chapter is organized as follows: Section 3.2 presents data-driven feature selection approaches with emphasis on Relief category of algorithms as an example of Filter Methodology in Section 3.2.1 and Sequential Feature Selection-based algorithms as an example of Wrapper Methodology in Section 3.2.2. Section 3.3 explains PCA as an example of feature extraction methodology. Section 3.4 provides specific examples of feature selection and extraction from data consisting of different genomic characterizations. The emphasis is on integrated analysis of transcriptomic and proteomic data.

3.2 DATA-DRIVEN FEATURE SELECTION

Consider the scenario where the input training data consist of N samples and D features denoted by $X(i,j)$ for $i = 1,\ldots,N$, $j = 1,\ldots,D$ and output training data are denoted by $Y(i)$ for $i = 1,\ldots,D$.

3.2.1 FILTER TECHNIQUES

A most basic form of filter technique consists of calculating the correlations between each feature $X(i,:)$ and response Y and selecting the features with the highest correlation coefficients. In this section, we will primarily discuss the commonly used *Relief* category of filter feature selection approaches.

3.2.1.1 *Relief*

Since original *Relief* [6] was designed for classification, we will first consider that entries of Y belongs to two classes. For a random selection of sample S_i, *Relief* searches for two nearest neighbors using D-dimensional Euclidean distance: One belonging to the same class as $Y(S_i)$, denoted as *Near Hit H_i* and another belonging to the different class, denoted by *Near Miss M_i*. The relevance or weight of a feature j is then reduced by the difference between $X(S_i,j)$ and $X(H_i,j)$ and increased by the difference between $X(S_i,j)$ and $X(M_i,j)$. Note that the weight update considers that the value of relevant features belonging to the same class should be similar and thus it penalizes the relevance by the difference between $X(S_i,j)$ and the nearest hit. On the other hand, the weight update increases the weight of a feature if the feature has differences from the near miss sample. The pseudo code for the *Relief* algorithm is provided in Algorithm 3.1. The dissimilarity measure considered in [6] was the Euclidean distance square, that is, $dis(X(S_i,j),X(H_i,j)) = (X(S_i,j) - X(H_i,j))^2$. Other distance measures such as Manhattan distance ($|X(S_i,j) - X(H_i,j)|$) is also frequently used.

ALGORITHM 3.1 ALGORITHMIC REPRESENTATION OF ORIGINAL RELIEF [6]

INPUT: Training Data X of dimension $N \times D$ and Y of dimension $N \times 1$
OUTPUT: Weight vector W of length D
Initialize $W[1:D] = 0$
for all $i = 1:m$ **do**
    Select random sample S_i
    Find nearest hit H_i and nearest miss M_i
    for all $a = 1:D$ **do**
        $W[a] = W[a] - \frac{dis(X(S_i,a),X(H_i,a))}{m} + \frac{dis(X(S_i,a),X(M_i,a))}{m}$
    end for
end for

Example to illustrate Relief
Let

$$X = \begin{bmatrix} 1 & 2 \\ 1 & 3 \\ 1 & 5 \\ 2 & 3 \\ 2 & 5 \\ 2 & 6 \end{bmatrix} \quad Y = \begin{bmatrix} 1 \\ 1 \\ 1 \\ 0 \\ 0 \\ 0 \end{bmatrix}$$

Consider, $m = 2$ and the first sample selected is $S_1 = 2$, that is, $X(2, :) = [1 \ 3]$ the nearest hit is $H_1 = 1$ and nearest miss is $M_1 = 4$. Thus after the first sample, $W[1] = 0 - (1-1)^2/2 + (1-2)^2/2 = 0.5$ and $W[2] = 0 - (3-2)^2/2 + (3-3)^2/2 = -0.5$. Let the second sample selected be $S_2 = 5$ and nearest hit is $H_2 = 6$ and nearest miss is $M_2 = 3$. Thus after the second sample, $W[1] = 0.5 - (2-2)^2/2 + (2-1)^2/2 = 1$ and $W[2] = -0.5 - (5-6)^2/2 + (5-5)^2/2 = -1$. After two samples, we note that the weight of the first feature is 1 and the weight of the second feature is -1, indicating that the first feature is useful in separating the two classes, which is clearly the case.

3.2.1.2 Relief-F
The original *Relief* algorithm was extended to *Relief-F* [7] that considers $k > 1$ nearest neighbors and can deal with multiclass problems. For a random selection of sample S_i, *Relief-F* searches for k nearest neighbors from the same class (denoted $H_{i,1}, \ldots, H_{i,k}$) and k nearest neighbors from each of the different classes (denoted $M_{i,1,C}, \ldots, H_{i,k,C}$ for each class C). The weight vectors are updated based on the Nearest Hits and Nearest Misses similar to original *Relief* but averaging over the hits and misses. The pseudo code for the *Relief-F* algorithm is provided in Algorithm 3.2.

ALGORITHM 3.2 ALGORITHMIC REPRESENTATION OF RELIEF-F [7]

INPUT: Training Data X of dimension $N \times D$ and Y of dimension $N \times 1$
OUTPUT: Weight vector W of length D
Initialize $W[1 : D] = 0$
for all $i = 1 : m$ **do**
    Select random sample S_i
    Find k nearest hits $H_{i,1}, \ldots, H_{i,k}$
    for all Class C not equal to $class(S_i)$ **do**
        Find k nearest misses $M_{i,1,C}, \ldots, H_{i,k,C}$ for class C
        for all $a = 1 : D$ **do**

$$W[a] = W[a] - \sum_{j=1}^{k} \frac{dis(X(S_i,a),X(H_{i,j},a))}{mk} + \sum_{C \neq class(S_i)} \frac{\frac{P(C)}{1-P(class(S_i))} \sum_{j=1}^{k} dis(X(S_i,a),X(M_{i,j,C},a))}{mk}$$

        end for
    end for
end for

3.2.1.3 R-Relief-F

For regression problems, the *Relief* category of algorithms has been extended to *RReliefF*. Note that normalized weight updates for *Relief* can be considered as a measure of the difference in (a) probability that a feature discriminates between the samples with different class values and (b) probability that a feature discriminates between the samples with same class values [7,8]. Thus the weight of a feature *a* for classification problem can be considered as the following approximation of the difference in conditional probabilities:

$$W[a] = \text{Probability (Different Value of } a|\text{Nearest Instance from Different Class)}$$
$$- \text{Probability (Different Value of } a|\text{Nearest Instance from Same Class)} \quad (3.1)$$

In regression problems, we cannot have nearest misses or nearest hits as the response is continuous and thus the problem can be modified by considering the following probabilities:

$P_{dA} = $ Probability (Different Value of a|Nearest Instances)

$P_{dC} = $ Probability (Different Prediction|Nearest Instances)

and

$P_{dC|dA} = $ Probability (Different Prediction|Different Value of A and Nearest Instances)

Using Bayes Rule, we obtain [7,8]:

$$W[a] = \frac{P_{dC|dA}P_{dA}}{P_{dC}} - \frac{(1 - P_{dC|dA})P_{dA}}{1 - P_{dC}} \quad (3.2)$$

The above idea is captured in Algorithm 3.3. The differences in output responses, features, and joint output and feature is stored in N_{dC}, N_{dA}, and N_{dCdA}, respectively. The $d(i,j)$ terms provides a relative measure of distance between sample S_i and sample $H_{i,j}$ when compared to all the k nearest neighbors. We have $\sum_{j=1}^{k} d(i,j) = 1$ and

$$d(i,j) = \frac{e^{-\left(\frac{rank(S_i, H_{i,j})}{u}\right)^2}}{\sum_{l=1}^{k} e^{-\left(\frac{rank(S_i, H_{i,l})}{u}\right)^2}} \quad (3.3)$$

where $rank(S_i, H_{i,l})$ denotes the rank of $H_{i,l}$ in a ordered list based on distance from sample S_i and u is a user-defined control parameter. A higher value of u will lower the difference between $d(i,j_1)$ and $d(i,j_2)$ for $j_1 \neq j_2$ while a lower value of u will accentuate the difference. Ranks are used rather than actual numerical differences to avoid problem-dependent normalization constraints.

ALGORITHM 3.3 ALGORITHMIC REPRESENTATION OF RRELIEF-F [7,8]

INPUT: Training Data X of dimension $N \times D$ and Y of dimension $N \times 1$
OUTPUT: Weight vector W of length D
Initialize $W[1:D] = 0$, $N_{dC} = 0$, $N_{dA} = 0$, $N_{dCdA} = 0$,
for all $i = 1 : m$ **do**
    Select random sample S_i
    Find k nearest instances $H_{i,1}, \ldots, H_{i,k}$ to S_i
    for all $j = 1 : k$ **do**
        $N_{dC} = N_{dC} + dis(Y(S_i), Y(H_{i,j})) \cdot d(i,j)$
        for all $a = 1 : D$ **do**
            $N_{dA} = N_{dA} + dis(X(a, S_i), X(a, H_{i,j})) \cdot d(i,j)$
            $N_{dCdA} = N_{dCdA} + dis(Y(S_i), Y(H_{i,j})) \cdot dis(X(a, S_i), X(a, H_{i,j})) \cdot d(i,j)$
        end for
    end for
end for
for all $a = 1 : D$ **do**
    $W[a] = \frac{N_{dCdA}[a]}{N_{dC}} - \frac{N_{dA}[a] - N_{dCdA}[a]}{m - N_{dC}}$
end for

Example to illustrate regression ReliefF

Let us consider our previously discussed example containing $D = 20$ features where the output response Y is dependent on the first 10 of the features, that is, $Y(i)$ is dependent on $X(i,j)$ with $j = 1, \ldots, 10$. Each individual entry of X was generated randomly using a uniform distribution $unif(0, 1)$. The model used to generate Y was

$$Y(i) = X(i, 1:10) * [2.05\ 2.87\ 2.47\ 2.34\ 2.85\ 2.14\ 2.39\ 2.87\ 2.20\ 2.26]^T + \epsilon \qquad (3.4)$$

where ϵ is a random variable following uniform distribution $unif(0, 8)$. The weights of the 20 individual features using R-Relief-F algorithm with $N = 50$ samples and $k = 10$ is shown in Table 3.1. We note that only 4 (features 5, 2, 7, 6) of the first 10 features are selected by R-Relief-F in the top 10 features. The primary reason behind this is the small number of samples. If we increase the training samples for R-Relief-F to $N = 150$, 8 out of top 10 features selected by the algorithm is from the 10 relevant features (results shown in Table 3.2). Similarly, if we increase the number of samples to $N = 1000$, the top 10 weighted features selected by R-Relief-F shown in Table 3.3 consists of the first 10 relevant features $[2, 7, 9, 8, 5, 3, 1, 10, 6, 4]$ and the next 10 features that do not influence Y have significantly lower weights.

While applying any feature selection approach, it is important to realize that the results can differ significantly by changing the number of training samples. Furthermore, *R-Relief-F* being a filter approach, features that become relevant only in combination with other features may not be selected by the algorithm.

Table 3.1 Feature Weights Using
R-Relief-F Algorithm With $N = 50$

Feature Number	Weight
5	0.0508
16	0.0108
15	0.0095
2	0.0091
19	0.009
13	0.0078
7	0.0075
20	0.0047
14	0.0044
6	0.0036
18	0.0035
3	−0.0018
12	−0.002
4	−0.0024
9	−0.0071
10	−0.008
11	−0.0143
1	−0.017
17	−0.017
8	−0.0231

3.2.2 WRAPPER TECHNIQUES

As discussed earlier, wrapper feature selection uses model design in the search of
the features itself [4,5]. The optimal feature set can be selected based on exhaustive
search, but that is often computationally unattainable. Thus suboptimal wrapper
feature selection techniques are often used. Commonly used techniques for wrapper
feature selection are based on sequential selection and we discuss some of the
sequential feature selection techniques in this section.

3.2.2.1 Sequential forward search

At each iteration, sequential forward search (SFS) considers the selection of one
additional feature from the remaining features that maximizes the reward R (or
minimizes the cost). The reward R is a measure of the performance of the model
designed from the selected features. A common choice for R is negative of the
prediction error (model error estimation is discussed in later chapters of this book).
The pseudo code for SFS technique is shown in Algorithm 3.4.

Table 3.2 Feature Weights Using
R-Relief-F Algorithm With $N = 150$

Feature Number	Weight
10	0.0148
2	0.0124
11	0.0122
3	0.0117
7	0.0098
4	0.0065
16	0.0054
5	0.0048
8	0.0037
9	0.0007
17	0.0005
13	0.0005
20	−0.0009
1	−0.0018
6	−0.003
14	−0.006
19	−0.0062
12	−0.0073
18	−0.0092
15	−0.0173

ALGORITHM 3.4　ALGORITHMIC REPRESENTATION OF SFS

INPUT: Training Data X of dimension $N \times D$ and Y of dimension $N \times 1$
OUTPUT: S_k set of k features
INITIALIZE $j = 0$, $S_j = \emptyset$, $S_{\text{All}} = [1, 2, \ldots, D]$
while $j < k$ **do**
　$x_+ := \arg\max_{x \in S_{\text{All}} \setminus S_j} R(S_j \cup x)$; (the most significant feature with respect to S_j)
　$S_{j+1} := S_j \cup x_+$; $j := j + 1$
end while

3.2.2.2 Sequential floating forward search

The sequential floating forward search (SFFS) is similar to SFS search but includes the option of removing a selected feature. In SFS, a feature once selected cannot be removed in later iterations, whereas SFFS checks at each iteration, if removal of any selected feature can increase the reward objective. The pseudo code for SFFS methodology is presented in Algorithm 3.5 [9].

Table 3.3 Feature Weights Using R-Relief-F Algorithm With $N = 1000$

Feature Number	Weight
2	0.008
7	0.0055
9	0.0052
8	0.0051
5	0.0037
3	0.0022
1	0.0022
10	0.002
6	0.0006
4	0.0003
17	−0.0001
19	−0.0003
18	−0.0005
14	−0.0008
12	−0.0015
15	−0.0019
16	−0.0025
13	−0.0026
20	−0.0031
11	−0.0043

ALGORITHM 3.5 ALGORITHMIC REPRESENTATION OF SFFS

INPUT: Training Data X of dimension $N \times D$ and Y of dimension $N \times 1$
OUTPUT: S_k set of k features
INITIALIZE $j = 0$, $S_j = $, $S_{All} = [1, 2, \ldots, D]$
while $j < k$ **do**
    Step 1 *Inclusion*
    $x_+ := \arg\max_{x \in S_{All} \setminus S_j} R(S_j \cup x)$; (the most significant feature with respect to S_j)
    $S_{j+1} := S_j \cup x_+; j := j + 1$
    Step 2 *Conditional Exclusion*
    $x_- := \arg\max_{x \in X_j} R(X_j \setminus x)$; (the least significant feature in X_j)
    if $R(X_j \setminus x_-) > R(X_{j-1})$ **then**
        $X_{j-1} := X_j \setminus x_-; j := j - 1$
        Go to **Step 2**
    else
        Go to **Step 1**
    end if
end while

Example to illustrate SFFS

Consider $\mathbf{X}$ to be four-dimensional with components $X_{f1}, X_{f2}, X_{f3}, X_{f4}$ and the reward function R is defined as follows:

$$R(\mathbf{X}) = 10 \times 1_{X_{f1}}(\mathbf{X}) \times 1_{X_{f2}}(\mathbf{X}) + 2 \times 1_{X_{f3}}(\mathbf{X}) + 3 \times 1_{X_{f1}}(\mathbf{X}) + 5 \times 1_{X_{f2}}(\mathbf{X}) + 6 \times 1_{X_{f4}}(\mathbf{X}) \quad (3.5)$$

where $1_{X_{fi}}(\mathbf{X})$ denotes the indicator function which is 1 when $\mathbf{X}$ contains feature X_{fi} and it is 0 otherwise. For the first *Inclusion* stage of the algorithm, we compute $R(X_{f1}) = 3, R(X_{f2}) = 5, R(X_{f3}) = 2, R(X_{f4}) = 6$ and thus X_{f4} is selected as the first feature and $\mathbf{X}_1 = \{X_{f4}\}$. For the second iteration, we consider $R(X_{f4}, X_{f1}) = 9, R(X_{f4}, X_{f2}) = 11, R(X_{f4}, X_{f3}) = 8$ and thus X_{f2} is selected as the next feature and $\mathbf{X}_2 = \{X_{f4}, X_{f2}\}$. The conditional exclusion feature at this stage yields no change and we move to the next inclusion stage where $R(X_{f4}, X_{f2}, X_{f1}) = 24, R(X_{f4}, X_{f2}, X_{f3}) = 13$ and thus $\mathbf{X}_3 = \{X_{f4}, X_{f2}, X_{f1}\}$. At this stage, we move to the conditional exclusion phase where we calculate $R(X_{f2}, X_{f1}) = 18, R(X_{f4}, X_{f1}) = 9, R(X_{f4}, X_{f2}) = 11$. Because $R(X_{f2}, X_{f1})$ is $> R(\mathbf{X}_2)$, we remove X_{f4} and $\mathbf{X}_2$ becomes $\{X_{f1}, X_{f2}\}$. Note that the significant combination of X_{f1} and X_{f2} could not have been captured by regular SFS, but the conditional exclusion principle added to the search enabled us to locate the important combination. The SFFS algorithm continues after this depending on the desired number of features or reward reaching a set threshold. The above example is represented pictorially in Figs. 3.2 and 3.3. Fig. 3.2 shows the inclusion steps until $k = 3$ with the feature sets computed being shown as gray boxes and the selected feature is marked as a darker gray box. Fig. 3.3 shows the conditional exclusion stage where one feature is removed at a time to calculate the reward. Note that both the figures contain all possible feature combinations for 1, 2, or 3 features,

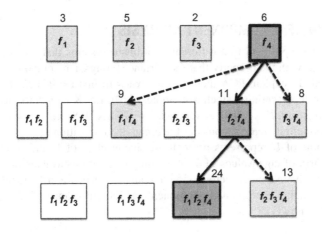

FIG. 3.2

Inclusion stages of SFFS algorithm till $k = 3$.

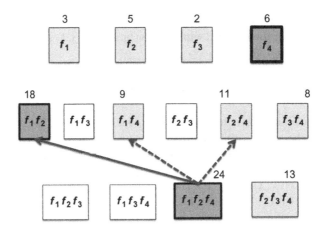

FIG. 3.3

Conditional exclusion at $\mathbf{X}_3$ stage.

but the SFFS algorithm computed a subset of these combinations. The combinations that are not computed are shown as white boxes.

3.3 DATA-DRIVEN FEATURE EXTRACTION

In this section, we consider approaches to extract features from the data by using mapping functions. We will focus on methods based on eigen decomposition techniques especially PCA as it is commonly used in multiple scenarios.

3.3.1 PRINCIPAL COMPONENT ANALYSIS

PCA maps the input data to a different coordinate systems that are orthogonal to each other and with the property that the variance along each component projection is maximized. To explain PCA methodology, we will first explain *Karhunen-Loeve* transform for representing a stochastic process [10]. Let $\mathbf{X}$ be a zero mean random vector of dimension $D \times 1$ with covariance $V = \mathbb{E}[\mathbf{X}\mathbf{X}^T]$. Since V is symmetric and positive semi-definite matrix, consider the eigen decomposition of V as $V = U\Lambda U^T$ where columns of U represents normalized eigenvectors of V and Λ represents a diagonal matrix of eigenvalues of V. Assuming distinct eigenvalues of V, we have $U^T = U^{-1}$ as distinct eigenvectors of a covariance matrix are orthogonal. Note that $Z = U^T X$ contains uncorrelated components as the covariance of Z is a diagonal matrix as shown in Eq. (3.6).

$$\mathbb{E}[\mathbf{Z}\mathbf{Z}^T] = \mathbb{E}[U^T\mathbf{X}\mathbf{X}^T U] = \mathbb{E}[U^T U \Lambda U^T U] = \Lambda \tag{3.6}$$

Thus the random vector X can be represented as a linear combination of uncorrelated vectors (columns of U) with the coefficients being uncorrelated random variables

$$\mathbf{X} = U\mathbf{Z} = \sum_{i=1}^{D} \mathbf{u_i}Z(i) \tag{3.7}$$

The above representation is known as *Karhunen-Loeve* expansion where the eigenvectors of the covariance matrix is used as the basis vectors and the coefficients of the basis vectors reflect uncorrelated random variables. Since $\mathbb{E}[\mathbf{ZZ}^T] = \Lambda$, the variances of the indices of Y (i.e., $\mathbb{E}[(Z(i))^2]$ as $\mathbb{E}[Z(i)] = 0$) are ordered according to the eigenvalues of V.

Consider the approximation of X by a fewer set of basic vectors, that is, let $\hat{X} = \sum_{i=1}^{k} \mathbf{u_i}Z(i)$ where $k < D$. The expectation of the squared error of the approximation is

$$\mathbb{E}(\mathbf{X} - \hat{\mathbf{X}})^2 = \mathbb{E}\left(\sum_{i=k+1}^{D} \mathbf{u_i}Z(i)\right)^2$$

$$= \sum_{i=k+1}^{D} \mathbf{u_i}^T\mathbf{u_i}\mathbb{E}(Z(i))^2 \quad \text{(Since } Z(i)\text{s are uncorrelated)} \tag{3.8}$$

$$= \sum_{i=k+1}^{D} \lambda_i$$

where λ_i refers to the eigenvalue for eigenvector $\mathbf{u_i}$. Thus the approximation error is minimized when we leave out the eigenvectors corresponding to the lowest eigenvalues.

In PCA, we apply the *Karhunen-Loeve* approximation technique using sample covariance matrices. Thus, if we have N samples of input $\mathbf{x}_1, \ldots, \mathbf{x}_n$ each of dimension $D \times 1$ with zero mean, the sample covariance matrix S_c is given by

$$S_c = \frac{1}{N-1}\sum_{i=1}^{N} \mathbf{x}_i\mathbf{x}_i^T \tag{3.9}$$

We select the first component to be $\mathbf{w_1}^T\mathbf{X}$ where $\mathbf{w_1}$ corresponds to the eigenvector corresponding to the largest eigenvalue of S_c. The second component is selected such that $\mathbf{w_2}^T\mathbf{w_1} = 0$ and it reflects the eigenvector corresponding to the second largest eigenvalue of S_c.

Example to illustrate PCA

Consider the 500 points shown in Fig. 3.4. Since the samples are centered at [0 0], the sample covariance is equal to

$$Sc = \frac{1}{499}\sum_{i=1}^{500} \mathbf{x}_i\mathbf{x}_i^T = \begin{pmatrix} 220.6 & 104.69 \\ 104.69 & 59.69 \end{pmatrix} \tag{3.10}$$

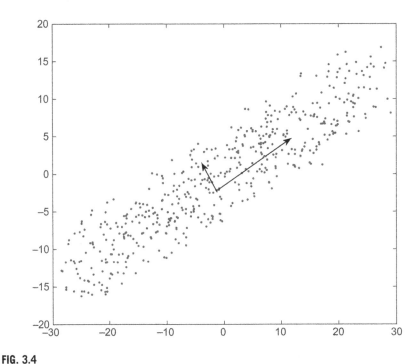

FIG. 3.4

500 samples of two-dimensional vectors represented as a scatter plot for PCA.

The eigenvectors for Sc are $\mathbf{u}_1 = [0.897\ 0.442]^T$ and $\mathbf{u}_2 = [-0.442\ 0.8970]^T$ for eigenvalues $\lambda_1 = 272.18$ and $\lambda_2 = 8.11$, respectively. The two basis vectors for the expansion are $\mathbf{u}_1$ and $\mathbf{u}_2$. The transformed variable $\mathbf{Z_1} = 0.897X_1 + 0.442X_2$ is the projection of the data in the direction of the highest variance and it can explain $\frac{100 \times 272.18}{272.18 + 8.11} = 97.11\%$ of the total variance in X.

3.4 MULTIOMICS FEATURE EXTRACTION AND SELECTION

This section provides an overview of techniques that are used for integrating multiple forms of genomic information [11]. The emphasis of this section is on integrating transcriptomic and proteomic datasets. We will not cover integration of other genomic domain information, such as copy number variation, methylation, metabolomics, but minor modifications of the techniques discussed here can potentially be applied for such integrations. For specific approaches to integrate DNA methylation, RNA expression data, or Copy Number Variation data, readers are referred to some recent works in this area [12–17].

With respect to observing expressions of transcripts and proteins, transcriptomic profiles are often measured through techniques such as microarray, RNA-Seq, etc., and proteomic profiles through techniques such as gel electrophoresis and mass spectrometry. Some of the data measurement techniques may involve destruction of the living cell and thus joint measurement of both transcripts and proteins in a single cell is not feasible by such methods. Furthermore, some approaches may provide expression data on the average behavior of a collection of cells and not the expression distribution of the cells. Thus understanding the limitations and assumptions in the data measurement techniques used for measuring the transcriptomic and proteomic profiles is essential before conducting a joint analysis of the two data sources. Developing a joint model of the two domains involves comprehending the differences in the expression of the mRNAs and proteins. Studies [18–22] have shown that there can be poor correlation between mRNA and protein expression data from same cells under similar conditions.

Furthermore, the type of extracted transcriptomic and proteomic data and the ultimate goal of analysis dictate the manner of the joint analysis of the two domains. We categorize the available techniques for integrated analysis of transcriptomic and proteomic profiles into seven groups and provide an overview of the categories along with illustration of the techniques with specific example studies [11].

3.4.1 CATEGORY 1: UNION OF TRANSCRIPTOMIC AND PROTEOMIC DATA

This type of technique can be considered as one of the most basic type of integration. The techniques in this category generally consider a union of two different datasets (such as proteomic and transcriptomic data from different samples) and create a reference dataset. The reference datasets often show new insights and reveal previously undetected phenomenon or support a new phenomenon as compared to individual datasets. A number of published studies such as [23,24] can be considered belonging to this technique. A work on *Bradyrhizobium japonicum* bacteroid metabolism in soybean root nodules by Delmotte et al. [25] is one specific example of this method which is described next.

Example: *Bradyrhizobium japonicum* is a gram negative, rod-shaped, nitrogen-fixing bacterium that communicates with its host plant and develops a symbiotic partnership with its host. The soybean plant *Glycine max* is the host considered in [25] (details of this symbiotic relation is available at: http://web.mst.edu/~djwesten/Bj.html). The complete genome sequence of *B. japonicum* was identified by Kaneko et al. [26] where 8317 potential protein-coding genes were found; 66,153 protein-coding loci have been identified in the genome sequence of *G. max*.

Delmotte et al. [25] built a database by combining the above-mentioned 8317 proteins of *B. japonicum*, 62,199 of the above-mentioned 66,153 proteins of *G. max*, and 258 contaminating proteins. Delmotte et al. [25] searched the combined database to locate the experimental protein extracts of *B. japonicum* soybean bacteroids with the protein being measured by GeLC-MS/MS technique. A probability-based protein identification algorithm [27] was employed to identify the proteins from mass spectrometry data by searching the sequence database. The use of combined database was advantageous because of the fact that soybean proteins present in the nodule extracts of *B. japonicum* and *G. max* might have symbiotic relationships; 2315 proteins in the experimental dataset were also reported to be present in the combined database.

The use of transcriptomic expression profiling alone has some limitations (such as limitations of the array, i.e., not all genes are present in the array, concealment of true expression levels due to bias of the probeset, etc.) that can equally be applicable to proteomic expression profiling. Thus the authors chose to integrate (through a simple union method) both the datasets and use the union of genes as reference dataset for bacteroid expression. In total, 3587 transcriptomic (A) and protein (B) expressions in soybean bacteroid were recorded in the union set $(A \cup B)$. The number of elements in the set $A \cap B$ was 1508; 807 proteins were identified to be expressed by only the proteomics approach $(B \setminus A)$ and 1272 genes that have been identified as expressed only in the transcriptomic study $(A \setminus B)$. Among the set $A \setminus B$, 47 were RNAs (45 tRNAs, rnpB, ssrA2) and the remaining 1225 were protein encoding genes.

A list of 15 gene functional categories were observed for the 3 different datasets: (i) the dataset X by Kaneko et al. [26] consisting of 8317 protein encoding genes, (ii) the reference datasets $A \cup B$ consisting of 3540 $(3587 - 47)$ protein encoding genes, and (iii) the dataset $X \setminus (A \cup B)$ consisting of 4777 protein encoding genes, that is, the protein encoding genes that are not detected by Delmotte et al. [25]. The number of genes present in the three datasets for each category was detected. The number of genes/proteins present in each category was divided by the total number of genes/proteins in that dataset and relative frequency of each category for the three datasets were established. In only 4 among the 15 categories, it has been observed that the relative frequency in the reference dataset $(A \cup B)$ is less than the relative frequency in the Kaneko dataset (X).

The reference dataset revealed novel insights regarding some aspects of bacterial metabolism (e.g., nitrogen metabolism, carbon metabolism, nucleic acid metabolism) and also regarding translation and posttranscriptional regulation. For example, (i) some key regulatory proteins (e.g., GlnA, GlnB, GlnK, GlnII) of N metabolism was identified in the reference dataset. (ii) The authors reported that all the enzymes related to C4 metabolism were detected for the first time as well as almost the entire set of gluconeogenesis-related enzymes was identified in the combined reference dataset. (iii) In a study of global protein expression pattern of *B. japonicum* bacteroids by Sarma and Emerich [28], nucleic acid metabolism-related proteins were reported to be lacking in the total protein expression pattern of the nodule bacteria. But, in this study, the authors have found almost all enzymes related to *de novo* nucleoside and nucleotide biosynthesis either in the gene or in the protein level or in both. (iv) The reference dataset also comprises a large number of proteins related to transcriptional and posttranscriptional regulation. Additionally, the enzymes related to protective response (to reactive oxygen species) under stress were discovered in the reference dataset.

3.4.2 CATEGORY 2: EXTRACTION OF COMMON FUNCTIONAL CONTEXT OF TRANSCRIPTOMIC AND PROTEOMIC FEATURES

The features (here *feature* refers to different genes for transcripts and proteins) for transcriptomic and proteomic data may not exactly overlap for a variety of reasons [29]. However, the features on transcriptomic and proteomic level could share the same functional context. These functional contexts may denote different biological processes or pathways in which features from both transcripts and proteins are enriched. In this category of methods, the common functional contexts are extracted through the analysis of both transcriptomic and proteomic datasets on the level of protein interaction networks. An instance of this methodology was reported in Paul et al. [30] which is discussed later. A similar kind of approach (functional analysis) was applied for integrating transcriptomic and proteomic evaluation of gentamicin nephrotoxicity in rats by Com et al. [31]. The functional analysis was conducted using Gene Ontology annotation tool (GO-Browser) of Ingenuity Pathway Analysis software. Based on the functional analysis, several gene ontology (GO) biological processes that were enriched by the features of the transcriptomic and proteomic dataset with Fisher P value ≤ 0.05 were selected. This integration by functional analysis revealed a putative model of toxicity [31] in the kidney of rats.

Example: The *common functional context* category is explained through the approach presented in [30]. Three publicly available studies on chronic kidney disease (CKD) by Schmid et al. [32], Baelde et al. [33], and Rudnicki et al. [34] were used for identifying deregulated features on the mRNA level; 697 differentially regulated genes were selected from the three studies creating the transcriptomic dataset. The proteomic dataset was extracted from the online Human Urinary Proteom database (HUPDB v2.0). Based on CKD and normal samples, 37 proteins were identified as differentially abundant. HUPDB was selected as the only source to avoid heterogeneity of datasets. Swiss-Prot annotation tool was used in this study (as HUPDB uses Swiss-Prot names as identifiers) to map the proteins to the gene symbols. Swiss-Prot [35] or UniprotKB is a protein sequence database that provides known relevant information about a particular protein. The details about Swiss-Prot entry annotation can be found in http://www.uniprot.org/faq/45.

Perco et al. [30] consider the following five different analysis procedures to elucidate the correspondence between transcriptomic and proteomic data:

1. **Direct feature overlap**: Differentially expressed features (genes for transcripts and proteins) present in both transcriptomic and proteomic lists were identified. Genes of 4 proteins (out of 37) were reported to be differentially (upregulated or downregulated) expressed in the transcriptomic dataset.
2. **Functional overlap**: PANTHER (Protein ANalysis THrough Evolutionary Relationships) classification system [36,37] classifies proteins and their genes to facilitate high-throughput analysis. The classification of proteins were done according to *family and subfamily*, *molecular function*, *biological process*, and *pathway*.

Multiple gene lists can be uploaded in the PANTHER system and jointly compared against a reference dataset to search for under- and over-represented functional categories based on *chi-squared test* or *binomial statistics* tool. Perco et al. [30] used PANTHER to identify enriched biological processes. Fully annotated sets of human genes were used as a reference dataset and a chi-squared test with a P-value < 0.05 was used to identify significantly enriched or depleted biological processes. They identified 27 biological processes that are relevant to the transcriptomic and proteomic datasets. Both the transcriptomic and proteomic feature set enriched four of the processes, whereas other processes were enriched by either transcriptomic or proteomic features.

3. **Joint pathway analysis**: The Laboratory of Immunopathogenesis and Bioinformatics (LIB) developed the DAVID (the Database for Annotation, Visualization and Integrated Discovery) tool [38] to provide functional interpretation of large set of genes derived from different genomic studies. KEGG pathway database is used as a repository for applying DAVID tool in [30]. Seven pathways were uncovered to be significantly enriched in deregulated transcripts and proteins using Fisher exact test with P-value < 0.05. Among these seven pathways, three pathways were enriched by both the transcriptomic and proteomic features and four pathways were enriched by either transcriptomic or proteomic feature.

4. **Protein dependency graph analysis**: PANTHER and KEGG does not cover all the features and thus omicsNET [39], an undirected protein interaction network database, was utilized to analyze the data. omicsNET creates edges between the nodes with edge weights referring to dependency measures between the pair of nodes. The dependency measures were determined using Gene Expression Omnibus Human Body Map, the MicroCosm database, GO data on molecular processes and functions, PANTHER, KEGG, and IntAct databases. Perco et al. [30] uncovered 65 strong dependencies in omicsNET between the features of transcriptomic and proteomic datasets. The features that were involved in the dependency graph included 21 features from the transcriptomic dataset, 21 features from the proteomic dataset, and 2 features from both the datasets.

5. **Direct edges between transcripts and proteins**: MAPPER (http://mapper.chip.org/) (Multi-genome Analysis of Positions and Patterns of Elements of Regulation) is a platform for identifying transcription factor binding sites (TFBSs) in multiple genomes [40]. Binding sites of 4 transcription factors (TFs) were identified in open reading frame (ORF) regions of the 37 proteins using MAPPER. Two of these four TFs showed up-regulation and two exhibited down-regulation in mRNA level. At least 1 of these 4 TFBSs was present in 13 proteins of the protein dataset revealing some direct edges between transcripts and proteins.

Among the five different analysis techniques mentioned earlier, *direct feature overlap* provided limited results, but this minimal overlap was increased when enriched biological processes were identified using the other analysis techniques.

Several biological processes were identified significantly enriched with both transcriptomic and proteomic features. Mapping transcriptomic and proteomic features on different KEGG pathways also revealed significant involvement of both transcriptomic and proteomic features.

The primary advantage of functional analysis of transcriptomic and proteomic data is that various pathways and processes for the genes under analysis become evident. While omicsNET can produce dependency measures between transcripts and proteins, a major shortcoming of omicsNET is the inability to create a dynamic model involving transcripts and proteins.

3.4.3 CATEGORY 3: TOPOLOGICAL NETWORK-BASED TECHNIQUES

Topological network methods (over-connection analysis, hidden node analysis, rank aggregation, and network analysis) have been used to elucidate common regulators (transcriptional factors and receptors) from two different types of datasets (transcriptomic and proteomic) by Piruzian et al. [41]. This category of approach refers to locating upstream regulators of mRNA and proteins individually and collecting the common regulators in both the networks for a combined signaling pathway. Topological and network analysis was used in finding individual TFs of mRNAs and proteins. The TFs that were not common in transcriptomic and proteomic profiles were ignored, and the common TFs were used to find the most influential receptors that could trigger maximal possible transcriptional response. Among the receptors discovered from joint analysis, some of them were never reported as *psoriasis* markers in earlier studies, while others have been reported previously. In another study [42], an integrated quantitative proteomic, transcriptomic, and network analysis approach was discussed that revealed molecular features of tumorigenesis and clinical relapse.

Example: This section provides an example of integrated analysis based on topological networks that was used to reveal the similarities and differences between transcriptomic- and proteomic-level perturbations in *psoriatic lesions* in [41]. The transcriptomic data related to psoriatic lesions contained 462 overexpressed transcripts and the proteomic data contained 10 abundantly expressed proteins. Unlike other studies, this study revealed high concordance between the proteomic and transcriptomic datasets, as 7 out of the 10 protein encoding genes were also overexpressed in the transcriptomic dataset. But the significant differences in the normalization of the two datasets limits the use of direct correlation analysis. Rather than analyzing the correlation, Piruzian et al. [41] applied topological network analysis approach to discover regulatory TFs, receptors and their ligands to reconstruct the network between them. Their approach produced biologically meaningful results and revealed unknown regulatory receptors that may be related to psoriatic lesions.

The range of methods considered in [41] are described next:

1. **Interactome overconnectivity analysis**: This approach is based on the assumption that the expression values of transcripts and proteins follow

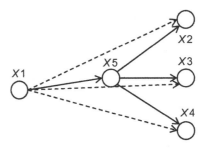

FIG. 3.5

Hidden node analysis reveals new node $X5$. The *dotted line* represents the connectivity before hidden node analysis and the *solid line* represents the connectivity after hidden node analysis [11].

hypergeometric distribution. The method for finding overconnected regulators (TFs) of a target dataset is described in the supplementary material of publication by Nikolsky et al. [43]. Overconnection analysis primarily ranks TFs by assigning a score to it. The score or significance of a TF (taken from global gene database or manually curated gene database like MetaCore http://www. genego.com/metacore.php) is a function of *hypergeometeric distribution probability mass function* [41,43].

2. **Hidden node analysis**: We demonstrate the hidden node analysis algorithm using a simple network example. For detailed description of the algorithm, readers are referred to Dezso et al. [44]. Fig. 3.5 demonstrates four genes (nodes) x_1, x_2, x_3, and x_4 that are overexpressed or abundant in a transcriptomic or proteomic dataset. Hidden node analysis reveals node x_5 that was not present in the experimental data, but is the key to regulate downstream effects of targets x_2, x_3, and x_4. The members of the hidden nodes may be derived from a global database or manually curated databases like MetaCore.

3. **Rank aggregation**: As the name suggests, Rank Aggregation can be used to combine multiple ordered lists. Rank aggregation can be formulated as an optimization problem [45] with the objective function being the weighted sum of distances of the original list from the combined list, such as $Cost(L) = \sum_{i=1}^{m} w_i \times distance(L, L_i)$, where L denotes the combined list, L_i denotes the individual lists, m denotes the number of lists, and *distance* denotes a distance metric between lists. An example of the distance function is the *Spearman footrule distance* [46], which is the absolute sum of the differences in the ranks of the unique elements of the individual list and the combined list. The optimization can be carried out using approaches such as Cross-Entropy Monte Carlo stochastic search [47] or genetic algorithms.

4. **Network analysis**: Network-based analysis is expected to select biologically connected meaningful subnetworks with relevant objects. For example, 10 overexpressed proteins were considered as relevant objects [41] and published literature was used to come up with regulatory TFs, receptors, and kinases.

For the integrated study, 10 common TFs (top ranked) of transcriptomic and proteomic data types were identified using overconnection analysis and hidden node analysis. Prior to the integrated analysis, the study identified 20 and 5 TFs for each data type based on topology and network analysis approaches, respectively. The 10 common nodes in each data type exhibit resemblance with the TFs generated from literature search. Consequently, a hidden node algorithm was used to find the most influential 44 membrane receptors that are present in the same signaling pathway with 1 of the 10 common TFs, and whose genes or corresponding ligands were 2.5-fold (or greater) overexpressed in the experimental data. Among the 44 discovered membrane receptors, 22 have been reported earlier in the literature as being relevant to psoriatic lesions, whereas the remaining represents new hypothesis on relevant receptors to be tested further.

Thus topological analysis can be applied to identify common regulators from two different datasets. The common regulatory machinery can be applied to arrive at a biologically meaningful signaling pathway that can be verified through new experiments or existing reported results in literature.

3.4.4 CATEGORY 4: MISSING VALUE ESTIMATION OF PROTEOMIC DATA BASED ON NONLINEAR OPTIMIZATION

This category of integration uses nonlinear or linear optimization to predict missing values of proteomic data. It maximizes an objective function to find out the connections between transcriptomic and proteomic networks. However, this does not result in a dynamic model with the ability to predict the abundance of next time point but rather, they are able to predict the protein expression at the same time point. A good example of nonlinear optimization is a method described in Torres-Garcia et al. [48] for a study of *Desulfovibrio vulgaris* published in 2009. The method is based on stochastic gradient boosting tree (GBT) proposed by Friedman [49]. Stochastic GBT optimization technique was also used in a study of *Shewanella oneidensis* in 2011 [50]. Artificial neural network approach was applied to find the missing values of the proteins using the relations between transcriptomic and proteomic data in a separate study published in 2011 [51].

Example: Torres-Garcia et al. [48] implemented stochastic GBT methodology to infer nonlinear relationships between mRNA and protein expression data for estimating missing protein expressions. mRNA expression data of around 3500 genes and protein expression data of around 800 genes of *D. vulgaris* were considered. After uncovering the nonlinear relationships and estimating the missing protein expression values, they validated the result using knowledge from the literature. The total procedure is shown as a flowchart in Fig. 3.6.

Gradient boosting tree: GBT is a nonlinear regression technique that generates a prediction model in the form of decision trees [49,52]. The method uses a training set $(x_1, y_1), \ldots, (x_n, y_n)$ to find an approximation $\hat{G}(x)$ to a function $G(x)$ that minimizes the expected value of a specified loss function $\phi(y, G(x))$.

$$\hat{G}(x) = \operatorname*{argmin}_{G(x)} E_{y,x}\phi(y, G(x)) \tag{3.11}$$

Several loss criteria can be used including least squares: $\phi(y, G) = (y-G)^2$, least absolute deviation: $\phi(y, G) = |y - G|$, etc. Torres-Garcia et al. used least squares criteria.

The total procedure consists of α iterations where each iteration generates a regression tree. In each tree, pseudo residuals of the dependent variable (here, proteomic expression) of the training dataset were located. The total input space was subsequently divided into β disjoint regions using least square splitting criterion [53]. These regions are the leaves of the tree. In each region, a multiplier value $\eta_{\alpha\beta}$ is calculated where α denotes the α tree and β denotes the β leaf of that tree.

Thus the approximation function $\hat{G}(x)$ primarily depends on the splitting variables and splitting points for each tree and also on the value of η in each leaf. Using the training dataset, these variables are estimated and used in future prediction.

Algorithm 3.6 demonstrates the idea behind GBT approach [11]. Here ξ ($0 < \xi < 1$) denotes the *shrinkage parameter* that controls the learning rate of the algorithm (a smaller ξ is expected to reduce prediction error while increasing computational complexity [52]).

ALGORITHM 3.6 ALGORITHM FOR IMPLEMENTATION OF GBT METHOD (MODIFIED VERSION *TREEBOOST*)

$G_0(x) = \operatorname*{argmin}_{g} \sum_{i=1}^{n} \phi(y_i, g)$,

for $a = 1 : \alpha$ **do**

(i) Compute pseudo residuals of the dependent variable of the training dataset:

$$\tilde{y}_{ia} = -\left[\frac{\partial\phi(y_i, G(x_i))}{\partial G(x_i)}\right]_{G(x)=G_{a-1}(x)} \quad \text{for } i = 1, \ldots, n$$

(ii) Divide the training data space into β different regions $R_{1a}, R_{2a}, \ldots, R_{\beta a}$ using pseudo residuals. Least square splitting criterion is used to split the region.

(iii) Compute multiplier η_{ba} for each region b ($b \in 0, 1, 2, \ldots, \beta$) by solving the following optimization:

$$\eta_{ba} = \operatorname*{argmin}_{\eta} \sum_{x_i \in R_{ba}} \phi(y_i, G_{m-1}(x_i) + \eta)$$

(iv) Update the model:

$$G_m(x) = G_{m-1}(x) + \xi \cdot \sum_{b=1}^{\beta} \eta_{ba} I(x \in R_{ba}), \quad \text{where } I(.) \text{ is the indicator function.}$$

end for
Output $F_\alpha(x)$.

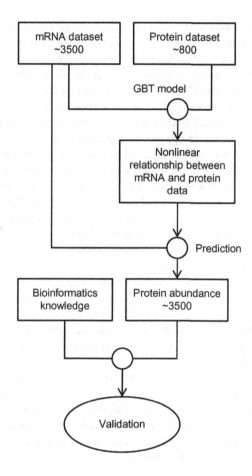

FIG. 3.6

Flowchart of the method used by Haider and Pal [11] and Torres-Garcia et al. [48].

Stochastic gradient boosting tree: A small change in the GBT method can make it stochastic. For Stochastic GBT, a random subset of training dataset is used in each iteration rather than using the total training dataset. The incorporation of randomness and the use of training data subset improves the performance of prediction as well as reduce computational complexity [52].

Validation process: The validation of the predicted missing values is important for performance analysis. Torres-Garcia et al. used existing biological knowledge to validate their results. The biological knowledge of *D. vulgaris* includes (i) functional categories of all genes (20 categories found from Comprehensive Microbial Resource [54]); (ii) sequence length, protein length, molecular weight, guanine-cytosine, and triple codon counts of all gene; (iii) a total of 609 operons of *D. vulgaris* consisting of 2–13 genes in each operon; (iv) Regulons of *D. vulgaris*; and (v) 92 metabolic pathways (KEGG pathways) for microbial genomes.

The coefficient of variation (CV) for different operons and regulons are calculated using the predicted values of proteins by calculating the ratio of standard deviation (SD) to mean expression value of each operon/regulon/pathway group. Consider an operon/regulon/pathway group consisting of n genes, then a random set consisting of n genes was created and CV (CV_{random}) calculated for that set of genes. This process was repeated for 1000 times for each operon/regulon/pathway and mean CV_{random} was then compared against the CV of the original operon. The idea is that the variability of genes in operons or regulons will be less than the variation of random set of genes because the genes in an operon or regulon are supposed to be expressed together and a relationship exists between their expressions. If the CV of an operon is found to be less than the mean CV_{random}, then it is concluded that the predicted values of that operon is somehow close to accurate. The results found by Torres-Garcia et al. show that a large portion of the operons/regulons/pathway groups indeed has less variability than the variability of randomly created group of genes. Another measure used in this study for understanding the lower dispersion of gene expressions in operons/regulons/pathways is the *percentile score* which denotes the percentage of 1000 set of random genes for each operon/regulon/pathway group that have CV values (CV_{random}) less than the CV of original operon/regulon/pathway.

Correlation between protein and mRNA expression was measured for each operon/regulon/pathway group and also for all the genes. It was found that the correlation was stronger in most of the individual operons and pathway groups than the correlation for all genes.

Note that the earlier validation process provides a measure of accessing the overall protein expression estimation accuracy but does not necessarily imply a good prediction. A poor expression prediction for all the genes in an operon might produce CV less than the mean CV_{random} if the overall prediction for other operons are also similarly poor. Cross-validation approaches can potentially mitigate this problem but it will reduce the size of the training dataset.

3.4.5 CATEGORY 5: MULTIPLE REGRESSION ANALYSIS TO PREDICT CONTRIBUTION OF SEQUENCE FEATURES IN mRNA-PROTEIN CORRELATION

Protein abundance is not only related to corresponding mRNA abundance but also depends on other biological and chemical factors (termed as *covariates*). For this reason, the idea of multiple regression analysis is used to relate characteristics of different covariates of each individual gene with the mRNA-protein correlation. The multiple regression approach can possibly provide a better explanation of protein variability than traditional univariate regression techniques. The effect of multiple sequence features on mRNA-protein correlation was discussed by Nie et al. [55] using multiple regression analysis. Poisson's linear regression model was employed by Nie et al. [56] to elucidate the relationship model of transcriptomic and proteomic networks.

Example: The effect of sequence features in different translational stages on the correlation of mRNA expression and protein abundance has been discussed by Nie et al. [55] for analyzing a study on *D. vulgaris*. Multiple regression analysis [57] was applied to predict the contribution of different sequence features on the correlation of mRNA and protein abundance.

Sequence features in translational stages: A sequence feature can be defined as an entity or data located in DNA or RNA sequence that is responsible for different biological phenomena. For example, *Shine Dalnargo* sequence is a sequence feature that is mainly a ribosomal binding site in mRNA and it helps the ribosome to start synthesis of protein. Other examples of sequence features are *start codon* (*AUG*), *stop codon* (*UAG, UAA, UGA*), *codon usage*, etc. In prokaryotes, translation can be divided into three stages: (i) initiation of translation, (ii) elongation of translation, and (iii) termination of translation. Lithwick and Margalit [58] demonstrates hierarchy of sequence features related to prokaryotic translation. *Shine Dalgarno sequences, start codon identity*, and *start codon context* are examples of *initiation* feature; *codon usage* and *amino acid usage* are examples of *elongation* feature; and *stop codon identity* and *stop codon context* are examples of *termination* feature.

Multiple regression analysis: A simple regression analysis can be expressed through the following equation:

$$Protein_i = A + B \times mRNA_i \qquad (3.12)$$

The target is to find A and B that relates the two variables. Here, $mRNA_i$ and $Protein_i$ are logarithm of the mRNA and protein expression of gene i, respectively. Nie et al. [57] reported that only 20–28% of protein variability can be captured by simple regression analysis. The following multiple regression analysis was subsequently considered:

$$Protein_i = A + (mRNA_i \times B) + \sum_{j=1}^{k}(B_j \times Covariate_{ij}) \qquad (3.13)$$

where $Protein_i$ and $mRNA_i$ are the protein abundance data and the mRNA expression level for the ith gene, respectively. $Covariate_{ij}$ refers to the jth covariate of the ith gene. B_j represents the slope for the jth covariate. Nie et al. [57] found that 52–61% variability of protein can be captured by this multiple regression analysis.

In this study of effect of sequence features, Nie et al. [55] used the sequence features in different translational stages as covariates and performed multiple regression analysis to locate the sequence features that have the highest effect on the mRNA-protein correlation. Multiple regression analysis for each type of sequence feature (i.e., sequence features related to initiation stage, elongation stage, and termination stage) was conducted separately and also for a combination of sequence features. The results showed that the sequence features are significantly responsible for the variation in mRNA-protein correlation. It was also observed that the elongation stage

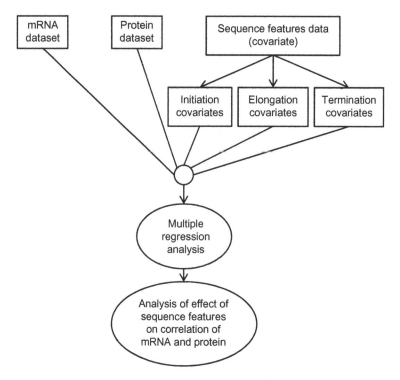

FIG. 3.7

Finding effect of sequence features on mRNA-protein correlation by multiple regression analysis [11].

features significantly affected the mRNA-protein correlation. The method of finding the effect of covariates in mRNA-protein correlation using one of the three datasets can be visualized by the flowchart shown in Fig. 3.7.

Three different datasets containing transcriptomic and proteomic measurements were used in this *sequence feature analysis* of *D. vulgaris*. The three datasets were expression levels for three different growth conditions (lactate, formate, and lactate-stationary). Partial correlation coefficient R_p^2 was applied to generate the contribution of specific sequence features in the variability. The partial correlation coefficient can be interpreted as:

$$R_p^2 = \frac{R_2^2 - R_1^2}{1 - R_1^2} \tag{3.14}$$

where R_1^2 is the Pearson correlation coefficient using only simple regression; R_2^2 is the Pearson correlation coefficient when multiple regression model is used with sequence features included. Standard F-test [59] was used to examine the significance (P-value for the F-test) of each covariate.

The R_p^2 for different sequence features varied; for *SD sequence*: 1.9–3.8%, for sum of *start codon*: 0.1–0.7%, for *start codon context*: 0.3–2.6%, for sum of *codon usage*: 5.3–15.7%, for sum of *amino acid usage*: 5.8–11.9%, for sum of *stop codon*: 1.3–2.3%, and for sum of *stop codon context*: 3.7–5.1%.

The sum of the individual R_p^2 values for all the sequence features ranged from 21.8% to 39.8% whereas R_p^2 for sequence features together in a single multiple regression ranged from 15.2% to 26.2%.

The analysis illustrated that, among multiple sequence features, *amino acid usage* and *codon usage* are the top factors that affect mRNA-protein correlation. The results were validated by conducting similar analysis where sequence features were kept the same for all the genes, but protein values were randomly assigned to the mRNA values. The resulting *P*-value in this validation stage analysis was found to be less than the original *P*-values indicating better statistical significance of the inferred model. This study provided another proof of the fact that mRNA expression alone is not sufficient to predict protein expression. Complex biological factors such as sequence features related to translational stages can have a significant role in predicting protein expressions.

3.4.6 CATEGORY 6: CLUSTERING-BASED TECHNIQUES

Clustering mRNA and protein abundance datasets individually and locating similarities (and hence correlation) between the individual clusters fail to produce promising results. This failure leads to the assumption that concatenating the proteomic and transcriptomic datasets and then clustering the concatenated dataset may not be a good idea either. Based on these observations, a new clustering method called coupled clustering was implemented by Rogers et al. [60,61]. Coupled clustering creates certain number of proteomic and transcriptomic clusters and provides the conditional probability of a gene to be in a protein cluster given that it is in an mRNA cluster. These conditional probabilities can reveal the relational complexity of mRNA and protein data. Rogers et al. used time series transcriptomic and proteomic data extracted under same experimental conditions. Note that this type of approach is also not a dynamic modeling approach that can provide temporal predictions.

Example: The mRNA or protein expression of a random set of genes is likely to show multiple different levels of expression, but genes involved in similar functions or having similar effects on cellular regulation is expected to exhibit similar expression levels. A mixture model [62] generally clusters such datasets into a predefined number of subsets in an unsupervised manner. For example, *Gaussian mixture model* is an unsupervised clustering algorithm that is able to create soft boundaries among the clusters, such as points in the space can be present in any cluster defined by a given probability. This is primarily a mixture of a certain number of Gaussian distributions with unknown parameters where each Gaussian distribution fits its corresponding cluster. Estimation maximization (EM) algorithm [63] is used to find the parameters of the Gaussian distribution and the cluster probabilities.

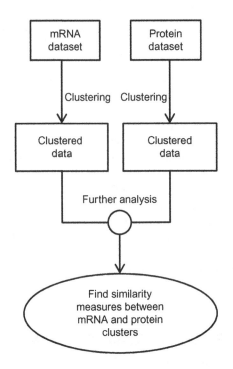

FIG. 3.8

Method for clustering individually [11].

Rogers et al. [60] proposed a coupled mixture model to investigate the correspondence between transcriptomic and proteomic expressions. The dataset consisted of transcriptomic and proteomic profiles of 542 human genes from the Human Mammary Epithelial Cell line (HMEC). Measurements were taken between 0 and 24 h after the cells were stimulated with epidermal growth factor (EGF). There were a total of six transcriptomic (mRNA) measurements (1, 4, 8, 13, 18, and 24 h) and seven proteomic (proteins) measurements (15', 1, 4, 8, 13, 18, and 24 h).

The mRNA and protein datasets were clustered individually using *Gaussian mixture model* and the similarity between the two sets of clusters were determined using *Rand index* [64]. *Rand index* ranges from 0 to 1 where 1 denotes that the two cluster sets are exactly the same. It was observed in the study that the two cluster sets showed very little similarity. The large dissimilarity suggested that if the two datasets were clustered after concatenating them into a single dataset, the number of clusters could have been as large as $20 \times 15 = 300$, which was impractical as the total number of genes was 542. Furthermore, comparing GO enrichment analysis showed that the individual clustering produced different biologically meaningful clusters that were lost when clusters were created after concatenation. The failure of individual and concatenated clustering lead to the implementation of coupled mixture models described in [60]. The ideas of clustering individually and clustering after concatenation are shown in Figs. 3.8 and 3.9, respectively.

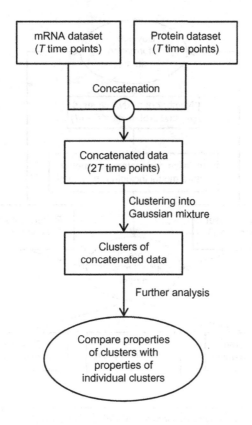

FIG. 3.9

Method for clustering after concatenation [11].

In coupled mixture modeling, the mRNA dataset was clustered as U different clusters with $p(u)(u \in 1, 2, \ldots, U)$ denoting the probability that mRNA expression of a gene belongs to the uth cluster. Similarly protein dataset was clustered as V different clusters with $p(v)(v \in 1, 2, \ldots, V)$ denoting the probability that protein expression of a gene belongs to the vth cluster. The joint probability can be described as $p(u, v) = p(u)p(v|u)$, where $p(v|u)$ is the parameter that provides the relationship between mRNA expression and protein level.

EM algorithm was used to maximize a log-likelihood function to infer the desired parameters. The number of clusters (U and V) in each dataset was derived to be $U = 15$ and $V = 20$ using Bayesian information criterion (BIC, proposed by Schwarz [65]). Fig. 3.10 demonstrates the coupled mixture model.

The values of $p(v|u)$ can unravel important information about the complexity of the relationship between mRNA and protein expressions. For example, the protein cluster $v = 4$ had a total of 19 proteins in it, 18 of those were ribosomal proteins. There were 7 mRNA clusters that had positive $p(u|v = 4)$ within the protein cluster $v = 4$. The most connected mRNA cluster with this protein cluster was the cluster

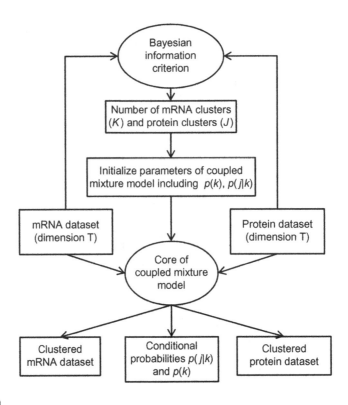

FIG. 3.10

Coupled clustering method used by Haider and Pal [11] and Rogers et al. [60].

$u = 3$ because $p(u = 3|v = 4) = 0.3653$ (which was the highest among all $p(u|v = 4)$). If we consider the other protein clusters that are related to this mRNA cluster $u = 3$, we note that there are 14 protein clusters with positive $p(v|u = 3)$.

This complex set of information conforms to the complex relationships between mRNA and protein expression. The inference of complex relationships from the conditional probabilities remains an open problem where the use of further biological knowledge about the involvement of different sets of mRNA and proteins in different biological processes might throw more light.

3.4.7 CATEGORY 7: DYNAMIC MODELING

A number of studies reported in the literature have inferred dynamic models (such as Boolean network, linear models, differential equation models, Bayesian networks, etc.) of GRNs from time series transcriptomic data alone. For example, Liang et al. [66] used REVEAL algorithm for inference of Boolean network model from time series mRNA expression data and Haider and Pal [67] proposed a BN inference algorithm from limited time series mRNA data. A basic linear modeling has been proposed by D'haeseleer [68]. GRN models consisting of differential equations was employed by Guthke et al. [69]. Validation of inference procedures of GRN was discussed by Dougherty [70]. Friedman used Bayesian networks to analyze and model gene expression data [53]. Among the existing network models, Bayesian networks can be applied to combine heterogeneous data and prior biological knowledge. For example, Nariai et al. [71] used protein-protein interaction network data for refining the Bayesian network model of the GRN produced by mRNA data alone. Zhang et al. [72] used TFBS data and gene expression data (transcriptomic) to model GRN using Bayesian network approach. Werhli and Husmeier [73] integrated multiple sources of prior biological knowledge (TF binding location) with microarray expression data to generate a Bayesian network model.

Example: Nariai et al. [71] used cell cycle microarray data [74] of *Saccharomyces cerevisiae* and 9030 protein-protein interaction data derived from MIPS database [75] to construct a Bayesian network model. The authors proved that the use of p-p interaction data had refined the estimated gene network produced by using only microarray data. The algorithm used to construct the network can be simply illustrated by the flowchart shown in Fig. 3.11. The algorithm is designated as the greedy hill-climbing algorithm.

In the greedy hill-climbing algorithm, each network was evaluated by Bayesian network and Nonparameteric Regression Criterion (BNRC) score [71]. Parents of each gene (genes that regulate a gene are called parents of that gene) were determined using this algorithm; the parents can be protein complex or other genes. PCA was used to find the protein complexes that were involved in regulating certain genes. Three gene networks were estimated: (i) by using only microarray data, (ii) by using only p-p interaction data, and (iii) by using the greedy hill-climbing algorithm. The three networks were then compared with the KEGG compiled network for evaluation. The network edges agreed with the KEGG pathway used for comparison. By using 350 chosen genes from the MIPS functional category *mitotic cell cycle and cell cycle control*, 34 protein complexes were discovered in this study (22 of these 34 complexes are listed in MIPS complex catalog). Also, incorporating phase information of cell cycle (e.g., G1/S, S, M phase) revealed biologically important relationships of several genes that are not included in the KEGG pathway.

It is well established that inferring GRNs from mRNA data alone has a huge computational complexity and often lack in accuracy. Thus a number of studies [71–73] used prior biological knowledge to reduce the complexity and/or incorporated other type(s) of data to increase the accuracy of the inference process. But to our knowledge, no such study has been conducted that uses protein abundance data along with transcriptomic data to infer a GRN which can reveal a dynamic relationship model between transcriptomic and proteomic network.

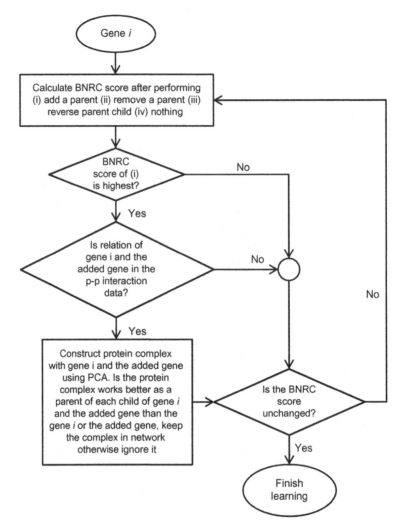

FIG. 3.11

The greedy hill-climbing algorithm for finding and modeling protein complexes and estimating a gene network [11].

REFERENCES

[1] J. Barretina, et al., The Cancer Cell Line Encyclopedia enables predictive modelling of anticancer drug sensitivity, Nature 483 (7391) (2012) 603–607, http://dx.doi.org/10.1038/nature11003.

[2] H. Zou, T. Hastie, Regularization and variable selection via the Elastic Net, J. R. Stat. Soc. Ser. B 67 (2005) 301–320.

[3] L. Breiman, Random forests, Mach. Learn. 45 (1) (2001) 5–32, ISSN 0885-6125.

[4] R. Kohavi, G. John, Wrappers for feature subset selection, Pattern Recog. Lett. 97 (1997) 273–324.

[5] A.K. Jain, D. Zongker, Feature selection-evaluation, application, and small sample performance, IEEE Trans. Pattern Anal. Machine Intell. 19 (1997) 153–158.

[6] K. Kira, L.A. Rendell, A practical approach to feature selection, in: Proceedings of the 9th International Workshop on Machine Learning, ML92, Morgan Kaufmann Publishers Inc., San Francisco, CA, ISBN 1-5586-247-X, 1992, pp. 249–256, http://dl.acm.org/citation.cfm?id=141975.142034.

[7] I. Kononenko, Estimating attributes: analysis and extensions of RELIEF, in: Proceedings of the European Conference on Machine Learning on Machine Learning, ECML-94, Springer-Verlag New York, Inc., Secaucus, NJ, ISBN 3-540-57868-4, 1994, pp. 171–182, http://dl.acm.org/citation.cfm?id=188408.188427.

[8] M.R. Šikonja, I. Kononenko, Theoretical and empirical analysis of ReliefF and RReliefF, Mach. Learn. 53 (1–2) (2003) 23–69.

[9] P. Pudil, J. Novovicova, J. Kittler, Floating search methods in feature selection, Pattern Recog. Lett. 15 (1994) 1119–1125.

[10] T.K. Moon, W.C. Stirling, Mathematical Methods and Algorithms for Signal Processing, Prentice Hall, Upper Saddle River, NJ, 2000.

[11] S. Haider, R. Pal, Integrated analysis of transcriptomic and proteomic data, Curr. Genomics 14 (2) (2013) 91–110.

[12] Y.W. Zhang, Y. Zheng, J.Z. Wang, X.X. Lu, Z. Wang, L.B. Chen, X.X. Guan, J.D. Tong, Integrated analysis of DNA methylation and mRNA expression profiling reveals candidate genes associated with cisplatin resistance in non-small cell lung cancer, Epigenetics 9 (6) (2014) 896–909.

[13] D.C. Kim, M. Kang, B. Zhang, X. Wu, C. Liu, J. Gao, Integration of DNA methylation, copy number variation, and gene expression for gene regulatory network inference and application to psychiatric disorders, in: 2014 IEEE International Conference on Bioinformatics and Bioengineering (BIBE), 2014, pp. 238–242, doi:10.1109/BIBE.2014.71.

[14] R. Louhimo, S. Hautaniemi, CNAmet: an R package for integrating copy number, methylation and expression data, Bioinformatics 27 (6) (2011) 887–888.

[15] Z. Sun, Y.W. Asmann, K.R. Kalari, B. Bot, J.E. Eckel-Passow, T.R. Baker, J.M. Carr, I. Khrebtukova, S. Luo, L. Zhang, G.P. Schroth, E.A. Perez, E.A. Thompson, Integrated analysis of gene expression, CpG island methylation, and gene copy number in breast cancer cells by deep sequencing, PLoS ONE 6 (2) (2011) e17490.

[16] S.A. Selamat, B.S. Chung, L. Girard, W. Zhang, Y. Zhang, M. Campan, K.D. Siegmund, M.N. Koss, J.A. Hagen, W.L. Lam, S. Lam, A.F. Gazdar, I.A. Laird-Offringa, Genome-scale analysis of DNA methylation in lung adenocarcinoma and integration with mRNA expression, Genome Res. 22 (7) (2012) 1197–1211.

[17] N. An, X. Yang, S. Cheng, G. Wang, K. Zhang, Developmental genes significantly afflicted by aberrant promoter methylation and somatic mutation predict overall survival of late-stage colorectal cancer, Sci. Rep. 5 (2015) 18616.

[18] G. Chen, T.G. Gharib, C.-C. Huang, J.M.G. Taylor, D.E. Misek, S.L.R. Kardia, T.J. Giordano, M.D. Iannettoni, M.B. Orringer, S.M. Hanash, D.G. Beer, Discordant protein and mRNA expression in lung adenocarcinomas, Mol. Cell. Proteomics 1 (4) (2002) 304–313.

[19] L.E. Pascal, L.D. True, D.S. Campbell, E.W. Deutsch, M. Risk, I.M. Coleman, L.J. Eichner, P.S. Nelson, A.Y. Liu, Correlation of mRNA and protein levels: cell

type-specific gene expression of cluster designation antigens in the prostate, BMC Genomics 9 (2008) 246.

[20] S.P. Gygi, Y. Rochon, B.R. Franza, R. Aebersold, Correlation between protein and mRNA abundance in yeast, Mol. Cell Biol. 19 (1999) 1720–1730.

[21] E.S. Yeung, Genome-wide correlation between mRNA and protein in a single cell, Angew. Chem. Int. Ed. 50 (3) (2011) 583–585, ISSN 1521-3773, doi: 10.1002/anie.201005969.

[22] A. Ghazalpour, B. Bennett, V.A. Petyuk, L. Orozco, R. Hagopian, I.N. Mungrue, C.R. Farber, J. Sinsheimer, H.M. Kang, N. Furlotte, C.C. Park, P.-Z. Wen, H. Brewer, K. Weitz, I.I. Camp, G. David, C. Pan, R. Yordanova, I. Neuhaus, C. Tilford, N. Siemers, P. Gargalovic, E. Eskin, T. Kirchgessner, D.J. Smith, R.D. Smith, A.J. Lusis, Comparative analysis of proteome and transcriptome variation in mouse, PLoS Genet. 7 (6) (2011) e1001393.

[23] S.B. Altenbach, W.H. Vensel, F.M. DuPont, Integration of transcriptomic and proteomic data from a single wheat cultivar provides new tools for understanding the roles of individual alpha gliadin proteins in flour quality and celiac disease, J. Cereal Sci. 52 (2) (2010) 143–151.

[24] J.P. McRedmond, S.D. Park, D.F. Reilly, J.A. Coppinger, P.B. Maguire, D.C. Shields, D.J. Fitzgerald, Integration of proteomics and genomics in platelets, Mol. Cell. Proteomics 3 (2) (2004) 133–144.

[25] N. Delmotte, C.H. Ahrens, C. Knief, E. Qeli, M. Koch, H.-M. Fischer, J.A. Vorholt, H. Hennecke, G. Pessi, An integrated proteomics and transcriptomics reference data set provides new insights into the *Bradyrhizobium japonicum* bacteroid metabolism in soybean root nodules, Proteomics 10 (2010) 1391–1400.

[26] T. Kaneko, Y. Nakamura, S. Sato, K. Minamisawa, T. Uchiumi, S. Sasamoto, A. Watanabe, K. Idesawa, M. Iriguchi, K. Kawashima, M. Kohara, M. Matsumoto, S. Shimpo, H. Tsuruoka, T. Wada, M. Yamada, S. Tabata, Complete genomic sequence of nitrogen-fixing symbiotic bacterium *Bradyrhizobium japonicum* USDA110, DNA Res. 9 (2002) 189–197.

[27] D.N. Perkins, D.J.C. Pappin, D.M. Creasy, J.S. Cottrell, Probability-based protein identification by searching sequence databases using mass spectrometry data, Electrophoresis 20 (18) (1999) 3551–3567, ISSN 1522-2683.

[28] A.D. Sarma, D.W. Emerich, Global protein expression pattern of *Bradyrhizobium japonicum* bacteroids: a prelude to functional proteomics, Proteomics 5 (16) (2005) 4170–4184.

[29] T. Maier, M. Gell, L. Serrano, Correlation of mRNA and protein in complex biological samples, FEBS Lett. 583 (24) (2009) 3966–3973.

[30] P. Perco, I. Muhlberger, G. Mayer, R. Oberbauer, A. Lukas, B. Mayer, Linking transcriptomic and proteomic data on the level of protein interaction network, Electrophoresis 31 (2010) 1780–1789.

[31] E. Com, E. Boitier, J.-P. Marchandeau, A. Brandenburg, S. Schroeder, D. Hoffmann, A. Mally, J.-C. Gautier, Integrated transcriptomic and proteomic evaluation of gentamicin nephrotoxicity in rats, Toxicol. Appl. Pharmacol. 258 (1) (2012) 124–133.

[32] H. Schmid, A. Boucherot, Y. Yasuda, A. Henger, B. Brunner, F. Eichinger, A. Nitsche, E. Kiss, M. Bleich, H.J. Grne, P.J. Nelson, D. Schlndorff, C.D. Cohen, M. Kretzler, Modular activation of nuclear factor-kappaB transcriptional programs in human diabetic nephropathy, Diabetes 55 (11) (2006) 2993–3003.

[33] H.J. Baelde, M. Eikmans, P.P. Doran, D.W.P. Lappin, E. de Heer, J.A. Bruijn, Gene expression profiling in glomeruli from human kidneys with diabetic nephropathy, Am. J. Kidney Dis. 43 (4) (2004) 636–650.

[34] M. Rudnicki, S. Eder, P. Perco, J. Enrich, K. Scheiber, C. Koppelstatter, G. Schratzberger, B. Mayer, R. Oberbauer, T.W. Meyer, G. Mayer, Gene expression profiles of human proximal tubular epithelial cells in proteinuric nephropathies, Kidney Int. 71 (2006) 325–335.

[35] A. Bairoch, B. Boeckmann, S. Ferro, E. Gasteiger, Swiss-Prot: juggling between evolution and stability, Brief. Bioinformatics 5 (1) (2004) 39–55.

[36] P.D. Thomas, A. Kejariwal, M.J. Campbell, H. Mi, K. Diemer, N. Guo, I. Ladunga, B. Ulitsky-Lazareva, A. Muruganujan, S. Rabkin, J.A. Vandergriff, O. Doremieux, PANTHER: a browsable database of gene products organized by biological function, using curated protein family and subfamily classification, Nucleic Acids Res. 31 (1) (2003) 334–341.

[37] H. Mi, Q. Dong, A. Muruganujan, P. Gaudet, S. Lewis, P.D. Thomas, PANTHER version 7: improved phylogenetic trees, orthologs and collaboration with the Gene Ontology Consortium, Nucleic Acids Res. 38 (Suppl. 1) (2010) D204–D210.

[38] D.W. Huang, B.T. Sherman, R.A. Lempicki, Systematic and integrative analysis of large gene lists using DAVID bioinformatics resources, Nat. Protoc. 4 (2008) 44–57.

[39] A. Bernthaler, I. Muhlberger, R. Fechete, P. Perco, A. Lukas, B. Mayer, A dependency graph approach for the analysis of differential gene expression profiles, Mol. BioSyst. 5 (2009) 1720–1731.

[40] V.D. Marinescu, I.S. Kohane, A. Riva, The MAPPER database: a multi-genome catalog of putative transcription factor binding sites, Nucleic Acids Res. 33 (Suppl. 1) (2005) D91–D97.

[41] E. Piruzian, S. Bruskin, A. Ishkin, R. Abdeev, S. Moshkovskii, S. Melnik, Y. Nikolsky, T. Nikolskaya, Integrated network analysis of transriptomic and proteomic data in psoriasis, BMC Syst. Biol. 4 (2010) 41.

[42] M. Imielinski, S. Cha, T. Rejtar, E.A. Richardson, B.L. Karger, D.C. Sgroi, Integrated proteomic, transcriptomic, and biological network analysis of breast carcinoma reveals molecular features of tumorigenesis and clinical relapse, Mol. Cell. Proteomics 11(6) (2012) M111.014910.

[43] Y. Nikolsky, E. Sviridov, J. Yao, D. Dosymbekov, V. Ustyansky, V. Kaznacheev, Z. Dezso, L. Mulvey, L.E. Macconaill, W. Winckler, T. Serebryiskaya, T. Nikolskaya, K. Polyak, Genome-wide functional synergy between amplified and mutated genes in human breast cancer, Cancer Res. 68 (22) (2008) 9532–9540.

[44] Z. Dezso, Y. Nikolsky, T. Nikolskaya, J. Miller, D. Cherba, C. Webb, A. Bugrim, Identifying disease-specific genes based on their topological significance in protein networks, BMC Syst. Biol. 3 (2009) 36.

[45] V. Pihur, S. Datta, S. Datta, RankAggreg, an R package for weighted rank aggregation, BMC Bioinformatics 10 (1) (2009) 62.

[46] P. Diaconis, R.L. Graham, Spearman's footrule as a measure of disarray, J. R. Stat. Soc. Ser. B Methodol. 39 (2) (1977) 262–268, ISSN 00359246, http://www.jstor.org/stable/2984804.

[47] R.Y. Rubinstein, D.P. Kroese, The Cross-Entropy Method: A Unified Approach to Combinatorial Optimization, Monte-Carlo Simulation and Machine Learning, Springer-Verlag, New York, 2004.

[48] W. Torres-Garca, W. Zhang, G.C. Runger, R.H. Johnson, D.R. Meldrum, Integrative analysis of transcriptomic and proteomic data of *Desulfovibrio vulgaris*: a non-linear model to predict abundance of undetected proteins, Bioinformatics 25 (15) (2009) 1905–1914.

[49] J.H. Friedman, Stochastic gradient boosting, Comput. Stat. Data Anal. 38 (4) (2002) 367–378.

[50] W. Torres-Garcia, S.D. Brown, R.H. Johnson, W. Zhang, G.C. Runger, D.R. Meldrum, Integrative analysis of transcriptomic and proteomic data of *Shewanella oneidensis*: missing value imputation using temporal datasets, Mol. BioSyst. 7 (2011) 1093–1104.

[51] F. Li, L. Nie, G. Wu, J. Qiao, W. Zhang, Prediction and characterization of missing proteomic data in *Desulfovibrio vulgaris*, Comp. Funct. Genomics 2011 (2011).

[52] J.H. Friedman, Greedy function approximation: a gradient boosting machine, Ann. Stat. 29 (5) (2001) 1189–1232.

[53] N. Friedman, M. Linial, I. Nachman, D. Pe'er, Bayesian networks to analyze expression data, in: Proceedings of the Fourth Annual International Conference on Computational Molecular Biology, 2000, pp. 127–135.

[54] J.F. Heidelberg, R. Seshadri, S.A. Haveman, C.L. Hemme, I.T. Paulsen, J.F. Kolonay, J.A. Eisen, N. Ward, B. Methe, L.M. Brinkac, S.C. Daugherty, R.T. Deboy, R.J. Dodson, A.S. Durkin, R. Madupu, W.C. Nelson, S.A. Sullivan, D. Fouts, D.H. Haft, J. Selengut, J.D. Peterson, T.M. Davidsen, N. Zafar, L. Zhou, D. Radune, G. Dimitrov, M. Hance, K. Tran, H. Khouri, J. Gill, T.R. Utterback, T.V. Feldblyum, J.D. Wall, G. Voordouw, C.M. Fraser, The genome sequence of the anaerobic, sulfate-reducing bacterium *Desulfovibrio vulgaris* Hildenborough, Nat. Biotech. 22 (2004) 554–559.

[55] L. Nie, G. Wu, W. Zhang, Correlation of mRNA expression and protein abundance affected by multiple sequence features related to translational efficiency in *Desulfovibrio vulgaris*: a quantitative analysis, Genet. Soc. Am. 174 (2006) 2229–2243.

[56] L. Nie, G. Wu, F.J. Brockman, W. Zhang, Integrated analysis of transcriptomic and proteomic data of *Desulfovibrio vulgaris*: zero-inflated Poisson regression models to predict abundance of undetected proteins, Bioinformatics 22 (13) (2006) 1641–1647.

[57] L. Nie, G. Wu, W. Zhang, Correlation between mRNA and protein abundance in *Desulfovibrio vulgaris*: a multiple regression to identify sources of variations, Biochem. Biophys. Res. Commun. 339 (2) (2006) 603–610.

[58] G. Lithwick, H. Margalit, Hierarchy of sequence-dependent features associated with prokaryotic translation, Genome Res. 13 (12) (2003) 2665–2673.

[59] R.G. Lomax, Statistical Concepts: A Second Course for Education and the Behavioral Sciences, Longman, New York, 1992.

[60] S. Rogers, M. Girolami, W. Kolch, K.M. Waters, T. Liu, B. Thrall, H.S. Wiley, Investigating the correspondence between transcriptomic and proteomic expression profiles using coupled cluster models, Bioinformatics 24 (2008) 2894–2900.

[61] S. Rogers, Statistical methods and models for bridging omics data levels, Methods Mol. Biol. 719 (1) (2011) 133–151.

[62] B.G. Lindsay, Mixture models: theory, geometry and applications, NSF-CBMS Reg. Conf. Ser. Probab. Stat. 5 (1995) i–iii + v–ix + 1–163, ISSN 19355920.

[63] A.P. Dempster, N.M. Laird, D.B. Rubin, Maximum likelihood from incomplete data via the EM algorithm, J. R. Stat. Soc. Ser. B Methodol. 39 (1) (1977) 1–38, ISSN 00359246.

[64] M. Meila, Comparing clusterings: an information based distance, J. Multivar. Anal. 98 (5) (2007) 873–895.

[65] G. Schwarz, Estimating the dimension of a model, Ann. Stat. 6 (2) (1978) 461–464.

[66] S. Liang, S. Fuhrman, R. Somogyi, Reveal, a general reverse engineering algorithm for inference of genetic network architectures, Pac. Symp. Biocomput. 3 (1998) 18–29.

[67] S. Haider, R. Pal, Boolean network inference from time series data incorporating prior biological knowledge, BMC Genomics 13 (Suppl. 6) (2012) S9.

[68] P. D'haeseleer, Linear modeling of mRNA expression levels during CNS development and injury, Pac. Symp. Biocomput. 4 (1999) 41–52.

[69] R. Guthke, U. Mller, M. Hoffmann, F. Thies, S. Tpfer, Dynamic network reconstruction from gene expression data applied to immune response during bacterial infection, Bioinformatics 21 (8) (2005) 1626–1634.

[70] E.R. Dougherty, Validation of inference procedures for gene regulatory networks, Curr. Genomics 8 (6) (2007) 351–359.

[71] N. Nariai, S. Kim, S. Imoto, S. Miyano, Using protein-protein interactions for refining gene networks estimated from microarray data by Bayesian networks, Pac. Symp. Biocomput. 9 (2004) 336–347.

[72] Y. Zhang, Z. Deng, H. Jiang, P. Jia, Inferring gene regulatory networks from multiple data sources via a dynamic Bayesian network with structural EM, in: Data Integration in the Life Sciences, vol. 4544, 2007, pp. 204–214.

[73] A.V. Werhli, D. Husmeier, Reconstructing gene regulatory networks with Bayesian networks by combining expression data with multiple sources of prior knowledge, Stat. Appl. Genet. Mol. Biol. 6 (2007) 15.

[74] P.T. Spellman, G. Sherlock, M.Q. Zhang, V.R. Iyer, K. Anders, M.B. Eisen, P.O. Brown, D. Botstein, B. Futcher, Comprehensive identification of cell cycleregulated genes of the yeast *Saccharomyces cerevisiae* by microarray hybridization, Mol. Biol. Cell 9 (12) (1998) 3273–3297.

[75] H.W. Mewes, D. Frishman, U. Gldener, G. Mannhaupt, K. Mayer, M. Mokrejs, B. Morgenstern, M. Mnsterktter, S. Rudd, B. Weil, MIPS: a database for genomes and protein sequences, Nucleic Acids Res. 30 (1) (2002) 31–34.

Validation methodologies

CHAPTER OUTLINE

4.1 INTRODUCTION

A regression model generated for drug sensitivity prediction has to be tested for accuracy before being applied in any clinical decision-making scenario. Additionally, there can be multiple potential predictive models generated and their predictive

Predictive Modeling of Drug Sensitivity. http://dx.doi.org/10.1016/B978-0-12-805274-7.00004-X
Copyright © 2017 Elsevier Inc. All rights reserved.

accuracy estimates can be used to select a model for further application. For validating the accuracy of a designed model, we first need to select a form of accuracy measure which can be a difference measure, such as norm or a similarity measure, such as correlation coefficient between the vector of predicted and actual experimentally derived sensitivities. The accuracy measure can be absolute, such as correlation coefficient or relative, such as Akaike Information Criterion (AIC). The various forms of commonly used fitness measures are discussed in Section 4.2. For applying the fitness measures, we need to decide on the samples to be used in estimating the accuracy of the model. For instance, we can evaluate the model fitness on the same data that has been used to estimate the model parameters which will result in an optimistic view of the model accuracy, whereas a hold out of a section of the data for testing that is significantly distinct from the training data may result in a pessimistic estimate of the model accuracy. The emphasis of the earlier chapter was on overfitting due to increased model complexity under small samples and in this chapter, we will consider the issue of overfitting caused by optimistic model accuracy estimates. The accuracy estimates obtained from the data used to train the model are typically optimistic, and thus techniques to estimate the true error by generalizing the model to predict on new samples not used for training are considered. We discuss the commonly used accuracy estimation techniques for testing sample selection in Section 4.3.

The accuracy estimates can be highly dependent on the number of samples available for model training and testing, and the estimates can change with different selection of the samples. Furthermore, a more complex model containing a larger set of parameters will require larger training data to avoid overfitting. Section 4.4 discusses the effectiveness of various error estimation techniques when the number of available samples is limited. The effectiveness is discussed in terms of *bias* and *variance* of the difference between the error estimate and the true generalization error.

The final section of this chapter discusses various in vitro and in vivo experimental models used for validating drug sensitivity predictions. We discuss the commonly used in vitro cell line models and primary cultures and compare their advantages and disadvantages. For in vivo testing, we discuss the genetically engineered mice models (GEMMs), xenograft mice models, and animals such as dogs that develop cancer spontaneously.

Fig. 4.1 provides a pictorial representation of the various topics to be discussed in this chapter.

4.1.1 MODEL EVALUATION

Before discussing the model evaluation approaches in details, it is important to point out the purpose of model fitness evaluation. *Model selection* refers to the performance evaluation of multiple models and selection of the best model [1]. Model fitness evaluation can also be considered for *Model assessment* that refers to

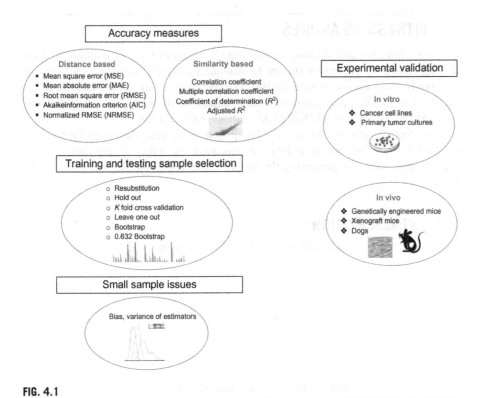

FIG. 4.1

Pictorial representation of the various topics to be discussed in this chapter.

estimating the generalization error of the chosen final model. For data-rich problems with large samples, it is usually recommended to create three random partitions of *training set*, *validation set*, and *test set* [1]. The training set is used to estimate the parameters of the models, the validation set is used to estimate the prediction error for model selection, and the test set is used for assessment of the generalization error of the final selected model. The test set is expected to be used toward the end of the analysis to estimate selected model performance. However, if the test set is repeatedly used to select the model with lowest error, the test set error of the selected model can potentially underestimate the generalization error. There is no golden rule for selecting the data percentages for training, validation, and test. With large sample sizes, numbers such as 50%, 25%, and 25% for training, validation, and test, respectively, are often used. Similarly, there are no general rules on how much training data is enough and is very much dependent on the type of model design and problem complexity. An output response linearly dependent on only one or two known features can potentially be trained on a few samples, as compared to a response dependent on a complex relationship between numerous covariates may require thousands of training samples.

4.2 FITNESS MEASURES

In this section, we will discuss various fitness measures that are commonly used in drug sensitivity prediction studies to evaluate model performance. Some of the commonly used measures to evaluate the prediction significance includes mean square error (MSE), normalized root mean square error (NRMSE), mean absolute error (MAE), correlation coefficient, R^2, Adjusted R^2, and AIC.

Note that the fitness measures can be applied to the same data used for training the model or can be applied to data that was not used in the training process. The various approaches for generating the testing and training samples are discussed in Section 4.3.

DATA REPRESENTATION

Let $Y = [y_1, y_2, \ldots, y_n]^T$ denote the actual output drug sensitivity response for n samples,

$$X = \begin{bmatrix} x_{1,1} & x_{1,2} & \cdots & x_{1,P} \\ x_{2,1} & x_{2,2} & \cdots & x_{2,P} \\ \vdots & \vdots & \ddots & \vdots \\ x_{n,1} & x_{n,2} & \cdots & x_{n,P} \end{bmatrix}$$

denote the predictor variables for n samples with P features and $\tilde{Y} = [\tilde{y}_1, \tilde{y}_2, \ldots, \tilde{y}_n]^T$ denote the predicted responses.

4.2.1 NORM-BASED FITNESS MEASURES

MAE denotes the ratio of the **1** norm of the error vector $Y - \tilde{Y}$ to the number of samples and is defined as

$$MAE = \frac{1}{n} \sum_{i=1}^{n} |y_i - \tilde{y}_i| \tag{4.1}$$

Mean bias error (MBE) captures the average bias in the prediction and is calculated as

$$MBE = \frac{1}{n} \sum_{i=1}^{n} (\tilde{y}_i - y_i) \tag{4.2}$$

MSE denotes the ratio of the square of the two norms of the error vector to the number of samples and is defined as

$$MSE = \frac{1}{n} \sum_{i=1}^{n} (y_i - \tilde{y}_i)^2 \tag{4.3}$$

Root mean square error (RMSE) denotes the square root of the MSE.

The MBE is usually not used as a measure of the model error as high individual errors in prediction can also produce a low MBE. MBE is primarily used to estimate the average bias in the model and to decide if any steps need to be taken to correct the model bias. The MBE, MAE, and RMSE are related by the following inequalities: $MBE \leq MAE \leq RMSE \leq \sqrt{n}MAE$. If any subsequent theoretical analysis is conducted on the error measure, MSE or RMSE is often preferred, as compared to MAE due to the ease of applying derivatives and other analytical measures. However, studies [2] have pointed out that RMSE is an inappropriate measure for average model performance, as it is a function of three characteristics: the variability in the error distribution, the square root of the number of error samples, and the average error magnitude (MAE). Due to the squaring portion of RMSE, larger errors will have more impact on the MSE than smaller errors. Furthermore, the upper limit of RMSE ($MAE \leq RMSE \leq \sqrt{n}MAE$) varies with $\sqrt{n}$ and can have different interpretations for different sample sizes.

Note that the above measures will have the same units as the variable to be predicted and thus cannot be compared for different variables that are scaled differently. The normalization of the error can be handled either as normalized versions described next or using measures such as correlation coefficients and R^2.

NRMSE as used in [3] is defined as

$$NRMSE = \sqrt{\frac{\sum_{i=1}^{n}(y_i - \tilde{y}_i)^2}{\sum_{i=1}^{n}(y_i - \bar{y})^2}} \tag{4.4}$$

or it can be defined as ratio of the RMSE to the range of the data or mean of the data.

4.2.2 CORRELATION COEFFICIENT

The **Pearson correlation coefficient** is defined as the ratio of covariance between two random variables to the product of standard deviations of the random variables as shown in Eq. (4.5).

$$\rho_{Y,\tilde{Y}} = \frac{Cov(\mathbf{Y}, \tilde{\mathbf{Y}})}{\sigma_Y \sigma_{\tilde{Y}}}$$

$$= \frac{\mathbb{E}(\mathbf{Y}\tilde{\mathbf{Y}}) - \mathbb{E}(\mathbf{Y})\mathbb{E}(\tilde{\mathbf{Y}})}{\sqrt{\mathbb{E}(\mathbf{Y}^2) - \mathbb{E}(\mathbf{Y})^2}\sqrt{\mathbb{E}(\tilde{\mathbf{Y}}^2) - \mathbb{E}(\tilde{\mathbf{Y}})^2}} \tag{4.5}$$

where $\mathbb{E}$ denotes the expectation operator and $\mathbf{Y}$ and $\tilde{\mathbf{Y}}$ denote the random variables representing the actual experimental and predicted drug sensitivities, respectively.

Because the exact distribution of the actual and predicted output responses are usually not known to calculate the expectation, we estimate the sample correlation coefficient. The sample Pearson correlation coefficient is usually calculated as

$$r_{Y,\tilde{Y}} = \frac{\sum_{i=1}^{n} y_i \tilde{y}_i - n\bar{y}\bar{\tilde{y}}}{\sqrt{(\sum_{i=1}^{n} y_i^2 - n\bar{y}^2)}\sqrt{(\sum_{i=1}^{n} \tilde{y}_i^2 - n\bar{\tilde{y}}^2)}} \tag{4.6}$$

where $\bar{y} = \frac{\sum_{i=1}^{n} y_i}{n}$ and $\bar{\tilde{y}} = \frac{\sum_{i=1}^{n} \tilde{y}_i}{n}$ denote the sample means.

The Pearson correlation coefficient provides a measure of linear relationship between the two variables and can fail to capture nonlinear relationships. For instance, if our prediction is $\tilde{Y} = aY + b$, Pearson correlation coefficient will capture this linear relationship and produce the maximum value of 1, whereas if our prediction is $\tilde{Y} = aY^3 + b$, the Pearson correlation coefficient will be less than 1 even when there is a one-to-one monotonic relationship between $\tilde{Y}$ and Y. Furthermore, outliers can have a significant impact on calculation of the Pearson correlation coefficient. Thus other forms of correlation coefficients are also considered, such as Spearman's rank correlation coefficient that measures the Pearson correlation coefficient between the ranks of the variables. Being based on the ranks, the **Spearman's rank correlation coefficient** provides a measure of any monotonic relationship between the two variables, whereas Pearson correlation coefficient provides a measure of the linear relationship between the variables. Another example of rank-based correlation coefficient is **Kendall Tau rank correlation coefficient** [4] that provides a measure of the number of changes required to match the ranking of one variable with the other.

Multiple correlation coefficient refers to the correlation between the actual response and the response predicted by a linear function of predictor variables.

For instance, we can be faced with a situation where we have different pipelines producing RNA-seq quantization with different scaling and thus to measure the ability of a linear combination of the pipelines to approximate the ground truth (e.g., assumed to be given by PCR experiments), we can employ the use of multiple correlation coefficient. A direct averaging of the outputs from different pipelines will not be the most informed prediction due to different scaling factors. Fig. 4.2 shows an example where we compare the multiple correlation coefficient between RNA-seq information for genes of sample A processed through Illumina platform (denoted by ILM) versus a linear combination of RNA-seq quantization based on three other platforms.

4.2.3 COEFFICIENT OF DETERMINATION R^2

The coefficient of determination R^2 is a measure of how well the regression model can explain the fraction of the variance in the data. It is defined as

$$R^2 = \frac{\sum_{i=1}^{n} (\tilde{y}_i - \bar{y})^2}{\sum_{i=1}^{n} (y_i - \bar{y})^2} = 1 - \frac{\sum_{i=1}^{n} (y_i - \tilde{y}_i)^2}{\sum_{i=1}^{n} (y_i - \bar{y})^2} \tag{4.7}$$

$$= 1 - \frac{SS_{\text{Residual}}}{SS_{\text{Total}}} \tag{4.8}$$

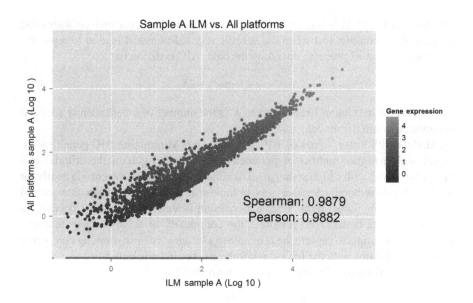

FIG. 4.2

Multiple correlation coefficient between RNA-seq data from different RNA-seq platforms.

where SS_{Residual} denotes the sum of squared residuals for the regression model and SS_{Total} denotes the total sum of squares which is proportional to the variance in the data.

An issue with R^2 is that the value of R^2 will increase when a predictor variable is added to the regression model and thus *overfitted* models with too many predictors can also have a misleading high R^2 value.

The **Adjusted** R^2 is adjusted for the number of predictors in the regression model and is defined as

$$
\begin{aligned}
R^2_{\text{adj}} &= 1 - \frac{n-1}{n-p-1}\left(1 - \frac{SS_{\text{Residual}}}{SS_{\text{Total}}}\right) \\
&= 1 - \frac{n-1}{n-p-1}(1 - R^2)
\end{aligned}
\tag{4.9}
$$

where n and p denote the number of observations and number of regression predictors, respectively.

4.2.4 AKAIKE INFORMATION CRITERION

The previous fitness methodologies provided a fitness measure in the absolute sense. For instance, a correlation coefficient of 0.95 between actual and predicted responses denote high prediction accuracy for the model. *AIC*, on the other hand, provides a relative measure of evaluating the fitness of a statistical model and assists in model selection. AIC [5] is based on information theoretic concepts and estimates the amount of information lost when the specific regression model is used to represent the actual biological process generating the data. AIC is defined as

$$AIC = -2\log(M) + 2K \tag{4.10}$$

where M is the maximum likelihood and K is the number of independently adjusted parameters within the model.

Models with lower values of AIC are preferred. Minimizing AIC penalizes the model with a higher number of parameters (the second part of the criteria) and favors the model with a higher statistical likelihood (first part of the criteria involving negative log-likelihood). Another way to look at it is that AIC incorporates model complexity in the fitness term. Since a more complicated model will always explain the training data better, AIC penalizes the complexity to achieve a parsimonious model. AIC attempts to capture the principle of *Occam's razor* that among competing hypotheses, the one with the fewest assumptions should be selected.

Similarly, **Bayesian Information Criterion** (BIC) is defined as

$$BIC = -2\log(M) + K\log(n) \tag{4.11}$$

where M is the maximum likelihood, K is the number of free parameters in the model to be estimated, and n is the number of samples. If the model is linear regression, K is the number of *regressors* including the intercept. BIC tries to select the model with the largest posterior probability and is asymptotically consistent as a selection criteria [1]. However, the performance of BIC suffers if n is not significantly larger than K.

Some additional model assessment approaches from the analytical side include *minimum description length* [6] and *Vapnik-Chervonenkis (VC) dimension* [7].

4.3 SAMPLE SELECTION TECHNIQUES FOR ACCURACY ESTIMATION

In this section, we will discuss various techniques for selecting the samples for training and testing the accuracy of the model. The main objective of these approaches is to estimate the *generalization error*, or in other words, how well a model will perform on new unseen data. Any measure of accuracy as explained in the previous section can be applied on the testing samples for calculating the model fitness.

4.3.1 RESUBSTITUTION OR TRAINING ERROR

Resubstitution is the most basic approach that uses the training samples as the testing samples. Since the testing samples have already been used to train the model, resubstitution estimates will provide an optimistic view of the model accuracy. In small sample scenarios (<100 samples), resubstitution estimates do not further reduce the number of limited available samples for training and can potentially be used for model selection. However, we need to be careful about overfitting, as training error estimates will keep improving when we increase the number of features or complexity of the model.

The pseudocode for resubstitution accuracy estimation is shown in Algorithm 4.1.

ALGORITHM 4.1　RESUBSTITUTION ACCURACY ESTIMATE

INPUT: Initial Data X of dimension $n \times D$ and Y of dimension $N \times 1$
OUTPUT: Accuracy Estimate α_{resub}
Training Data Indices $= I_{tr} = [1, 2, \ldots, n]$
Testing Data Indices $= I_{te} = [1, 2, \ldots, n]$
Train Model F using $X(I_{tr})$ and $Y(I_{tr})$
Evaluate accuracy α_{resub} of F using $X(I_{te})$ and $Y(I_{te})$

4.3.2 HOLD OUT

In Hold-Out accuracy estimate, we partition the data into distinct training and testing sets such that the accuracy estimate is based on data that has not been used during model training. The percentage of data that is hold-out for testing is dependent on the number of available samples and the specific application, but it primarily ranges from 50% to 20%. For instance, in NCI-DREAM drug sensitivity prediction challenge 2012, 18 out of 53 samples (34%) were hold out for testing the accuracy of the submissions [8,9].

The pseudocode for Hold-Out accuracy estimation with m out of n samples being used for training is shown in Algorithm 4.2.

ALGORITHM 4.2　HOLD-OUT ACCURACY ESTIMATE

INPUT: Initial Data X of dimension $n \times D$ and Y of dimension $n \times 1$
OUTPUT: Accuracy Estimate $\alpha_{HoldOut}$
Training Data Indices $= I_{tr} =$ select m distinct indices from $[1, 2, \ldots, n]$
Testing Data Indices $= I_{te} = [1, 2, \ldots, n] \setminus I_{tr}$
Train Model F using $X(I_{tr})$ and $Y(I_{tr})$
Evaluate accuracy $\alpha_{HoldOut}$ of F using $X(I_{te})$ and $Y(I_{te})$

Note that *hold-out* validation can be problematic when the data that is hold out belongs to a different population distribution than the training dataset. This can happen in drug sensitivity prediction scenarios where hold-out data can come from a different laboratory where application of different normalization and observation methodology can change the response distribution. This scenario can potentially be avoided by selecting a random set of data points from the union of available datasets for training and the remainder for testing.

4.3.3 *K*-FOLD CROSS VALIDATION

In *K*-fold Cross Validation accuracy estimate, we partition the data randomly into *K* distinct folds of roughly equivalent sizes. We train the model on $K - 1$ folds and test on the remaining fold and repeat it *K* times, each time selecting a different fold to hold out for testing. The final *K*-fold Cross Validation accuracy estimate is the average accuracy over the *K* testing folds. Normally, *K* is selected to be around 10 as was considered in this [10] drug sensitivity analysis article.

The pseudocode for *K*-fold Cross-Validation accuracy estimation is shown in Algorithm 4.3.

ALGORITHM 4.3 *K*-FOLD CROSS-VALIDATION ACCURACY ESTIMATE

INPUT: Initial Data X of dimension $n \times D$ and Y of dimension $n \times 1$
OUTPUT: Accuracy Estimate $\alpha_{K\text{-fold}}$
Randomly partition $[1, 2, \ldots, n]$ into K-folds $F_1, F_2, \ldots, F_K$ such that
$F_1 \cup F_2 \cdots \cup F_K = [1, 2, \ldots, n]$ and $F_i \cap F_j = \varnothing$ for $i \neq j \in [1, 2, \ldots, K]$
  for all $i = 1 : K$ **do**
    Testing Data Indices $= I_{te} = F_i$
    Training Data Indices $= I_{tr} = [1, 2, \ldots, n] \setminus F_i$
    Train Model F_i using $X(I_{tr})$ and $Y(I_{tr})$
    Evaluate accuracy α_i of F_i using $X(I_{te})$ and $Y(I_{te})$
  end for
  Accuracy $\alpha_{K\text{-fold}} = \frac{\sum_{i=1}^{K} \alpha_i}{K}$

Leave-One-Out accuracy estimate is a special form of *K*-fold Cross Validation where $K = n$, such as we leave one sample out for testing at each iteration.

When sample size is small, such as 50, 5- or 10-fold cross-validation might result in higher estimate of the error due to model training on a smaller set of samples. In other words, if the learning curve (actual error vs training sample size) has considerable slope at the training set size, 5- or 10-fold cross-validation will overestimate the true prediction error. Thus commonly used 5- or 10-fold cross-validation error estimates can exhibit *bias*, but usually have low variance. On the other hand, leave-one-out error estimates can have low bias, but higher variance as the training sets are usually very similar. Furthermore, computational complexity of generating leave-one-out error is significantly higher as *n* models have to be trained.

4.3.4 BOOTSTRAP

Bootstrap [11] accuracy estimation considers sampling with replacement while generating the training samples. Consider that the samples $X_1, X_2, \ldots, X_n$ belongs to a underlying distribution and we want to train a model on that distribution.

Bootstrap considers sampling from the underlying empirical distribution represented by $X_1, X_2, \ldots, X_n$. Thus n training samples are generated from the samples $X_1, X_2, \ldots, X_n$ using sampling with replacement and the samples that are not selected are used for testing the model accuracy. This procedure is repeated B number of times and average accuracy estimated. B is usually recommended to be between 25 and 200 [12].

The pseudocode for Bootstrap accuracy estimation with B repetitions is shown in Algorithm 4.4.

ALGORITHM 4.4 BOOTSTRAP ACCURACY ESTIMATE

INPUT: Initial Data X of dimension $n \times D$ and Y of dimension $n \times 1$
OUTPUT: Accuracy Estimate α
for all $i = 1 : B$ **do**

   Training Data Indices $= I_{tr} = n$ random sample indices with replacement from $[1, 2, \ldots, n]$
   Testing Data Indices $= I_{te} = [1, 2, \ldots, n] \setminus I_{tr}$
   Train Model F_i using $X(I_{tr})$ and $Y(I_{tr})$
   Evaluate accuracy α_i of F_i using $X(I_{te})$ and $Y(I_{te})$

end for
Accuracy $\alpha_{\text{boot}} = \frac{\sum_{i=1}^{B} \alpha_i}{B}$

We can calculate the asymptotic probability of a specific sample being selected for training in one bootstrap set. The probability of not selecting a sample in n independent selections is $(1 - 1/n)^n$ and thus the probability of selecting a sample is $1 - (1 - 1/n)^n$. The asymptotic behavior ($n \to \infty$) is given by

$$P_{\text{asym}} = \lim_{n \to \infty} 1 - \left(1 - \frac{1}{n}\right)^n = 1 - e^{-1} = 0.632 \tag{4.12}$$

Thus around 37% of the samples are not selected in the bootstrap training selection procedure and are used for testing.

The **Bias** of an estimator is the difference between the expected value and the true value of the parameter being estimated. The Bootstrap error estimator can be upward biased, such as the error estimate can be higher than the actual value. On the other hand, resubstitution error estimate can be downward biased, such as the error estimate can be lower than the true value. The **0.632 Bootstrap** method [12,13] considers a combination of these two estimators (one upward biased and one downward biased) to potentially arrive at an unbiased estimator. The **0.632 Bootstrap** error estimate is given by

$$\alpha_{0.632\text{Boot}} = 0.632 \cdot \alpha_{\text{boot}} + 0.368 \cdot \alpha_{\text{resub}} \tag{4.13}$$

4.3.5 CONFIDENCE INTERVAL

Jackknife-After-Bootstrap approach [11] can be used for generating the confidence intervals of the Bootstrap errors. Let N_i denote the set of bootstrap samples that do not contain sample X_i and the bootstrap estimate computed from N_i is denoted by ϵ_i. The standard error can be computed as

$$s = \sqrt{\frac{n-1}{n} \sum_{i=1}^{n} (\epsilon_i - \bar{\epsilon})^2} \tag{4.14}$$

where $\bar{\epsilon} = (1/n) \sum_{i=1}^{n} \epsilon_i$. The $100(1-\gamma)\%$ prediction intervals for the true error can then be computed as $[\bar{\epsilon} - sz_{\gamma/2}, \bar{\epsilon} + sz_{\gamma/2}]$, where z_γ is the γ quantile of the standard normal distribution.

4.4 SMALL SAMPLE ISSUES

When the number of samples are large, hold out, or cross-validation estimates can be fairly close to the true estimates. However, when the number of samples are small, the error estimates can differ significantly from the true error. While analyzing the performance of an estimator, we often study the **Bias** and **Variance** of the estimator. Let α be the true value of a parameter to be estimated (for our case, it denotes the true error of the model) and $\hat{\alpha}$ denote the estimate of the parameter α. The bias is defined as

$$Bias(\hat{\alpha}) = \mathbb{E}(\hat{\alpha}) - \alpha \tag{4.15}$$

When $Bias(\hat{\alpha}) = 0$, the estimator is defined to be unbiased. The variance of an estimator is defined as

$$Variance(\hat{\alpha}) = \mathbb{E}(\hat{\alpha} - \mathbb{E}(\hat{\alpha}))^2 \tag{4.16}$$

The MSE of the estimator $\mathbb{E}(|\hat{\alpha} - \alpha|^2)$ is related to the Bias and Variance of the estimator as follows:

$$\mathbb{E}(|\hat{\alpha} - \alpha|^2) = \mathbb{E}(\hat{\alpha} - \mathbb{E}(\hat{\alpha}))^2 + (\mathbb{E}(\hat{\alpha}) - \alpha)^2$$
$$MSE(\hat{\alpha}) = Bias^2(\hat{\alpha}) + Variance(\hat{\alpha}) \tag{4.17}$$

The earlier described estimators can be categorized as follows:

- $\hat{\alpha}_{resub}$ tends to be low biased, but extremely fast to compute and often have low variance.
- $\hat{\alpha}_{K\text{-fold}}$ tends to be high biased for low K such as 2 or 5. For higher K such as $K = n$ in case of Leave-One-Out, the estimator is close to unbiased, but can have high variance [14,15]. Computational complexity for leave-one-out error estimation is around n times higher than resubstitution error estimation.

- $\hat{\alpha}_{\text{boot}}$ tends to have lower variance than cross-validation, but high biased because around 37% of the samples are left out from training. $\hat{\alpha}_{0.632\text{Boot}}$ tends toward unbiasedness by combining with low biased $\hat{\alpha}_{\text{resub}}$. However, $\hat{\alpha}_{0.632\text{Boot}}$ is slow to compute due to multiple repeats required for an estimate.

4.4.1 SIMULATION STUDY

In this section, we consider an analysis of various error estimation methodologies using experimental drug sensitivity data. Random forest is used as the regression model and MAE is used as the fitness measure.

The first analysis is based on NCI-DREAM drug sensitivity dataset [8,9] and the second study is based on Cancer Cell Line Encyclopedia database [16]. We use the error on hold-out data to represent true error (denoted as validation error) and compare the validation error with Resubstitution, Leave-one-out, Bootstrap, and 0.632 Bootstrap estimates.

4.4.1.1 NCI-DREAM drug sensitivity dataset

The training set of NCI-DREAM challenge dataset [8,9] consists of genomic characterization and drug responses for 35 cell lines, while 18 cell lines are used as validation data. For different error estimations, we have used the following five datasets: gene expression, methylation, RNA sequence, RPPA, and copy number variation. Using these five datasets, we have generated five random forest models whose combinations (model stacking) are used to generate the integrated models to estimate the errors [9]. Initially provided response of 35 cell lines have been used to estimate Leave-one-out, Bootstrap, Resubstitution, and 0.632 Bootstrap errors. The Bootstrap errors were calculated with $B = 40$ repeats. The response to the remaining 18 cell lines that were provided later in the challenge process have been used to estimate the true error. In Fig. 4.3, different error estimations are shown for all 31 drugs of this dataset. The errors shown in Fig. 4.3 refer to MAE. We note that for the majority of the drugs, Leave-one-out error is closer to the validation error as compared to the other three estimates for this small sample example. However, there is still significant difference between the validation error and the considered error estimators.

We next calculated the confidence interval for DREAM challenge dataset using Jackknife-After-Bootstrap approach. The 80% confidence interval of the drugs for the integrated model using five datasets is shown in Fig. 4.4. Fig. 4.4 shows that the confidence intervals are relatively tight along with small 0.632 Bootstrap errors. Out of the 31 drugs, the validation or true error in 27 drugs is within 80% confidence interval.

4.4.1.2 CCLE dataset

The CCLE dataset consists of 24 drugs. For each drug, the number of cell lines with genomic characterization data and pharmacological data varies between 350 and 490. We utilize 100 cell lines to build the integrated random forest model along with

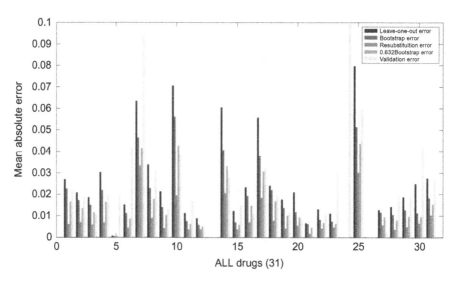

FIG. 4.3

Leave-one-out, Bootstrap, Resubstitution, 0.632 Bootstrap, and True error for all 31 drugs for NCI-DREAM challenge dataset. The prediction for drugs 13, 24, and 26 shows zero error as they contained minimal variations in sensitivity.

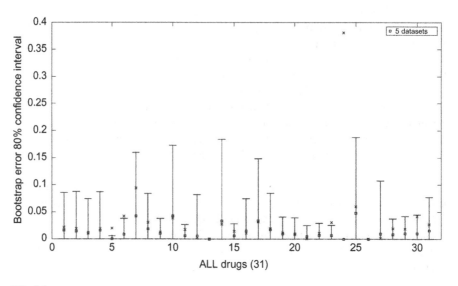

FIG. 4.4

The 80% confidence interval of 0.632 Bootstrap error for the 31 drugs in the NCI-DREAM challenge dataset. A *square in the line* indicates 0.632 Bootstrap error and *cross* indicates the validation error.

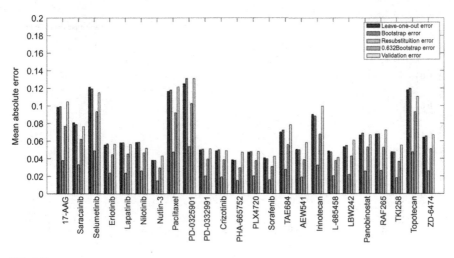

FIG. 4.5

Leave-one-out, Bootstrap, Resubstitution, 0.632 Bootstrap, and True error for all 24 drugs for 1 set of 100 training cell lines. For different runs, the cell lines selected for training can change significantly but we observed that the change in the errors is minimal.

estimation of 0.632 Bootstrap and Leave-one-out errors. The remaining cell lines of that drug are used as testing data to calculate validation or true error. Fig. 4.5 shows the Bootstrap (BSP), Resubstitution, 0.632 Bootstrap error, Leave-one-out (LOO) error, and True error for all the 24 drugs. The Bootstrap errors were calculated with $B = 40$. All the errors are in the form of MAE. We observe from Fig. 4.5 that the value of LOO error is higher than 0.632 BSP error for all the drugs, but when compared to validation/true error, LOO is closer to true error as compared to 0.632 BSP.

The 80% confidence interval of 0.632 Bootstrap error calculated using *Jackknife-After-Bootstrap* approach is shown in Fig. 4.6. The validation error is also represented in the same figure using a cross, and the validation error for all 24 drugs stays within the 80% confidence interval. Note that as compared to the DREAM challenge dataset figure for confidence interval shown in Fig. 4.4, the confidence intervals for the CCLE dataset are larger. The primary reason behind this is the integration of five datasets for prediction in DREAM challenge dataset, whereas only two datasets are used for the integrated prediction in the CCLE case. The results indicate that the confidence interval can be made significantly tighter with the use of more datasets in prediction. We will discuss these integration aspects in later chapters.

4.4.1.3 Bias correction

We observed that the different error estimation approaches displayed bias in multiple scenarios as shown in Fig. 4.7. The figure shows the distribution of the validation error estimate for 24 drugs of CCLE for different runs.

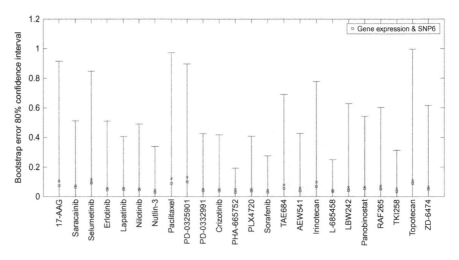

FIG. 4.6

The 80% confidence intervals of 0.632 Bootstrap error for all 24 drugs. *Square in the line* indicates the 0.632 Bootstrap error and *cross* indicates the validation error.

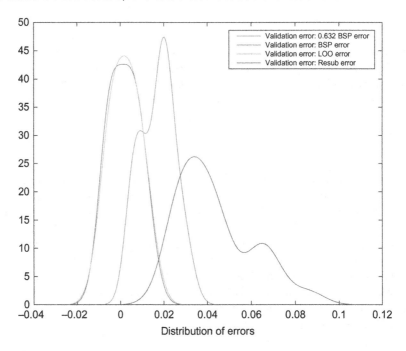

FIG. 4.7

Leave-one-out, Bootstrap, Resubstitution, and 0.632 Bootstrap error distribution for all drugs of CCLE before bias correction.

The observed bias was used to correct the error estimate in a new run. We observed reduction in bias close to zero utilizing this approach. The variance did not change while changing the bias and because $MSE = Bias^2 + Variance$, the MSE reduced following bias correction. Table 4.1 shows the mean, variance, and MSE before and after bias correction. The distribution of the errors after bias correction is shown in Fig. 4.8.

4.5 EXPERIMENTAL VALIDATION TECHNIQUES

In the previous sections, we discussed various approaches for estimating the accuracy of a predictive model using existing drug sensitivity data. In this section, we will discuss the various experimental methodologies to validate the accuracy of a predictive model. We next discuss the primary types of in vitro and in vivo techniques that are commonly used for validation.

4.5.1 IN VITRO CELL LINES

Cell Lines are the most commonly used approach to study cancer biology and test various cancer treatments. Some commonly used cancer cell line databases are NCI60, Cancer Cell Line Encyclopedia, and Genomic of Drug Sensitivity in Cancer. Cell lines usually contain cells of one type and can be genetically identical (homogenous population) or genetically diverse (heterogeneous population). The individual cells in cell lines can divide on their own and continue to grow until restricted by cell culture nutrients. The cell lines are easy to grow in a laboratory and multiple research groups can have access to the same cell line for corroborating research findings, and building a body of knowledge based on a specific genetic type of cells. However, the laboratory environment along with absence of other regular cell types in the cell line restricts the ability of a cell line to mimic in vivo cancer growth and the efficacy of anticancer drugs [17].

Cancer cell lines are expected to reflect the properties (such as genotypic and phenotypic characteristics, mutation status and gene expression, drug sensitivity) of the original cancer type from which they were cultured [18,19]. However, some of the cell lines in use may have properties distinctly different from those observed in the majority of the clinical disease, as they might have been selected from badly differentiated tumors based on their propensity to grow in controlled environments [18,19]. Following multiple differentiations of the cell line, the cell line may loose some of its original features, such as differentiation abilities. However, the cell line might become more homogeneous, as the most rapidly growing subclone will dominate. Furthermore, some studies have shown that substantial proportion of cell lines are mislabeled or replaced by cells derived from different individual, tissue, or species [20–23]. A potential approach to solve the problem has been approached through authentication of human cell lines using short tandem repeat profiling.

Table 4.1 Bias Correction

Error Type	Mean		Variance		Mean Squared Error	
	Before Bias Correction	After Bias Correction	Before Bias Correction	After Bias Correction	Before Bias Correction	After Bias Correction
Leave-one-out	0.0012	−1.99e−4	4.4e−5	4.43e−5	4.55e−5	4.43e−5
Bootstrap	7.13e−4	−7.32e−4	4.42e−5	4.08e−5	4.47e−5	4.13e−5
0.632 Bootstrap error	0.0162	−8.72e−4	7.15e−5	6.16e−5	3.34e−4	6.23e−5
Resubstitution error	0.0428	−0.0011	2.97e−4	2.66e−4	0.0021	2.67e−4

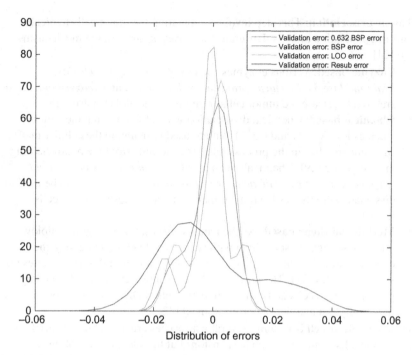

FIG. 4.8

Leave-one-out, Bootstrap, Resubstitution, and 0.632 Bootstrap error distribution for all drugs of CCLE after bias correction.

4.5.2 IN VITRO PRIMARY TUMOR CULTURES

Primary tumor cultures are the cell cultures established from the tumor biopsy of the patient and reflect the surroundings in the original tissue. As compared to cell lines, primary cell cultures are more heterogeneous and also contain a small fraction of normal stromal cells, which can be an advantage because tumor response is often influenced by an interplay of the tumor cells and the nonmalignant cells in the tumor microenvironment. Because different cells have diverse growth rates in culture conditions, the primary cell culture may loose the heterogeneity in the original tumor biopsy with time, as cells that grow rapidly in culture conditions will become predominant. Thus short-term drug sensitivity studies on the recently extracted primary tumor culture can capture the heterogeneity in the patient tumor.

Various steps involved in generating a primary tumor culture can be broadly categorized as [24]:

Tumor acquisition: This step involves tumor biopsy to extract the neoplastic mass with minimal damage to surrounding tissue. There are standard procedures for tumor resection depending on the tissue type [24].

Tumor dissociation: This step involves separating the tumor cells to create uniform population of cells. Different categories of dissociation mechanisms are [24]:

- **Enzyme based**: Various enzymes such as *trypsin, papain, elastase, hyaluronidase, collagenase, pronase, hyaluronidase*, and *deoxyribonuclease* are used to release the tumor cells from the extracellular matrix [25–28].
- **Chemical based**: Chemical dissociation works by removing the cations (such as K^+, Mg^{2+}, and Ca^{2+}) that are used to maintain the cellular matrix structure and thus in the process releasing the cells from the bonds holding them together [29]. Chemicals such as *ethylenediaminetetraacetic acid, ethylene glycol tetraacetic acid*, or *sodium tetraphenylborate* can be used to dissociate solid tissues from liver, mammary, and intestinal crypt cells [24,30].
- **Mechanical shear based**: As the name suggests, this technique employs mincing, sieving, or scratching with scissors or blades and passing the contents through a filtration mesh along with aspiration through pipettes and smaller needles [31]. The mechanical-based approach can produce a larger portion of dead cells and cells with hyperdiploid DNA (having more than the normal diploid number of chromosomes) [32].

Primary tumor cell isolation: Several approaches are available for enriching the tumor cells from single cell suspensions but the success rates of these techniques for generating primary tumor cultures are usually less than 100%. The commonly used techniques are as follows [24]:

- **Two-dimensional cultures**: The 2D cultures are monolayers that have been extensively used to study basics of cell biology. A extracellular matrix type environment is simulated on a culture plate using various components to grow primary cells from the tumor [24]. An issue with drug sensitivity studies using 2D cultures is the limited ability of the 2D culture to capture the tissue-specific characteristics of topology, intracellular contacts, differentiation, and functional behavior of tumor cells.
- **Explant cell cultures**: Explant cell cultures are created by breaking the tumor tissue into small parts and allowing them to grow on a nutrient-rich medium coated with fetal bovine serum on well plates. This technique can require multiple serial subculturing to generate the primary tumor culture, which can result in genetic variability, as compared to the original tumor. However, explant cell cultures better represent the tumor microenvironment as compared to 2D cultures. Explant cell cultures can usually be created with greater than 50% efficiency [33].
- **Precision-cut slice cultures**: This is similar to explant cell cultures, but have limitations on the thickness of the slices, which is considered optimal for growth at 160 μm [34]. The cell viability with this technique is usually above 50%, but the stringent thickness requirement restricts its use in multiple scenarios [35].

- **Three-dimensional cell cultures**: The 3D cell cultures attempt to provide an environment that is similar to what is available to in vivo tissues. This requires [35] (a) different cell types to interact and exchange growth factors and other biological effectors, (b) availability of an extracellular matrix to provide a 3D scaffolding for mechanical stability, and (c) availability of nutrient-rich interstitial fluid for cellular differentiation and growth. The 3D scaffolds are generated based on tissue type using various biocompatible material such as collagen [36], hydrogel [37], matrigel [38]. The tumor cells are seeded in a monolayer fashion between two layers of extracellular matrix (whose selection is dependent on the tumor cell type) to simulate in vivo conditions. The comparison of tumors with the 3D cultures in terms of capturing the phenotype can be accomplished by comparing transcriptional profiles, mutation characterizations, and protein expression profiles [39].

  As compared to 2D cultures, 3D cultures are better suited to capture numerous features of tumors such as proliferation capabilities, metastatic potential, immune evasion, and angiogenic properties [40–42]. However, 3D cultures can be expensive to produce and require tissue-dependent scaffold.

4.5.3 IN VIVO GENETICALLY ENGINEERED MICE MODELS

As the name suggests, GEMMs consist of mice whose genes have been modified to reflect genetic aberrations observed in a specific human tumor. The modifications can be approached by (a) injecting the desired altered DNA into a single cell of the mouse embryo [43] or (b) altering the embryonic stem cells with a DNA construct that contains the desired DNA sequences [44] and injecting the embryonic stem cells with recombinant DNA into mice blastocysts. GEMMs [45] have played a vital role in understanding tumor proliferation as they effectively capture the tumor microenvironment, along with the human tumor phenotype.

Nevertheless, GEMMs have been limited use in evaluating targeted therapies as compared to in vitro cell lines or tumor cultures or in vivo mouse xenograft models. The generation of a test bed of similar GEMMs is expensive and requires a large number of animals to be bred as compared to easier generation of synchronized xenograft models or cell lines. The timelines required to generate genetically engineered mice can be quite long and may require 12–24 months from design to initial experimental cohort [46]. Recent advances in genome editing, such as CRISPR holds the promise of generation of GEM models in as little as 4 weeks [47].

GEMMs have many advantages over xenografts [48] such as (a) knowledge of the initiating mutation; (b) tumor developing spontaneously in the normal tissue for the specific cancer type with an evolving microenvironment; and (c) intact immune system. GEMMs can play an important role in the coming years to evaluate the efficacy of cancer therapies, identify biomarkers, and analyze therapy toxicities [45]. Many types of GEMMs have been created to reflect the genetic and phenotypic traits of various forms of human cancers [48].

4.5.4 IN VIVO XENOGRAFT MICE

To create a xenograft mouse, human tumor cells are transplanted, into the organ type in which the tumor originated, in immunocompromised mice that accept human cells [49]. Various types of mice can be used for xenograft models, such as T-cell deficient nude athymic (nu) mice [50]; B- and T-cell deficient severe combined immunodeficient (SCID) mice [51], or recombination-activating gene 2 (Rag2)-killed mice [52]. The success rates for establishing mouse xenografts vary across tumors with around 40% success for rhabdomyosarcoma, 47% for osteosarcoma, and around 31% for pediatric brain tumor, rhabdoid tumors of the CNS and kidney, and Wilms tumor [52].

Some advantages of xenograft models as compared to GEM models include [49] (a) use of actual human tumor tissue incorporating the complex genetic and epigenetic aberrations that may not be captured by GEMM, (b) evaluation of therapies can be conducted in few weeks or months but GEMM can require close to a year, (c) genomic characterizations can be measured from both human biopsy and xenograft model before and after therapy, and (d) stromal cells from human tumor microenvironment can be injected into the xenograft to create the tumor microenvironment. Xenografts designed from human primary tumors, as compared to xenografts designed from cancer cell lines have shown high drug efficacy predictive capability with respect to human clinical activity [53–55]. The challenges with xenograft models include (a) the lack of replication of immune system and thus inability to model the lymphocyte-mediated response to the tumor [49] and (b) the inability to study various stages of the cancer as can be done with GEMM.

4.5.5 OTHER IN VIVO ANIMAL MODELS

Mice models are the most commonly used models for analyzing cancer therapies but some more complex organisms are also being used for evaluating therapeutic regimes. One such example is a dog which is frequently used as a model in drug discovery and development research due to its similarities to humans with respect to cardiovascular, urogenital, nervous, and musculoskeletal systems [56]. Furthermore, dogs develop spontaneous tumors (without genetic manipulations as in case of mice) which is highly desirable for cancer research. As compared to pigs [57] or nonhuman primates [58], dogs as pets are cared for until old age that is commonly associated with spontaneous cancer development. Additionally, the number of pet dogs in United States alone exceeds 70 million and 1 out of 4 dogs develops a tumor of some kind during its lifetime (American Veterinary Medical Association).

REFERENCES

[1] T. Hastie, R. Tibshirani, J. Friedman, The Elements of Statistical Learning: Data Mining, Inference and Prediction, Springer, New York, NY, USA, 2009.

[2] C.J. Willmott, K. Matsuura, Advantages of the mean absolute error (MAE) over the root mean square error (RMSE) in assessing average model performance, Clim. Res. 30 (2005) 79–82.

[3] S. Haider, R. Rahman, S. Ghosh, R. Pal, A copula based approach for design of multivariate random forests for drug sensitivity prediction, PLoS ONE 10 (12) (2015) e0144490.

[4] M. Kendall, A new measure of rank correlation, Biometrika 30 (1–2) (1938) 81–89.

[5] H. Akaike, A new look at the statistical model identification, IEEE Trans. Autom. Control 19 (6) (1974) 716–723.

[6] J. Rissanen, Modeling by shortest data description, Automatica 14 (1978) 465–471.

[7] V.N. Vapnik, Statistical Learning Theory, Wiley-Interscience, New York, NY, USA, 1998.

[8] J.C. Costello, et al., A community effort to assess and improve drug sensitivity prediction algorithms, Nat. Biotechnol. (2014), doi:10.1038/nbt.2877.

[9] Q. Wan, R. Pal, An ensemble based top performing approach for NCI-DREAM drug sensitivity prediction challenge, PLoS ONE 9 (6) (2014) e101183.

[10] N. Berlow, L.E. Davis, E.L. Cantor, B. Seguin, C. Keller, R. Pal, A new approach for prediction of tumor sensitivity to targeted drugs based on functional data, BMC Bioinform. 14 (2013) 239.

[11] B. Efron, Bootstrap methods: another look at jackknife, Ann. Statist. 7 (1979) 1–26.

[12] B. Efron, Estimating the error rate of a prediction rule: improvements on cross-validation, J. Am. Stat. Assoc. 78 (1983) 316–331.

[13] R.T.B. Efron, Improvements on cross-validation: the .632+ bootstrap method, J. Am. Stat. Assoc. 92 (438) (1997) 548–560.

[14] L. Devroye, L. Gyorfi, G. Lugosi, A Probabilistic Theory of Pattern Recognition, Springer, Berlin, 1997.

[15] U.M. Braga-Neto, E.R. Dougherty, Is cross-validation valid for small-sample microarray classification?, Bioinformatics 20 (3) (2004) 374–380.

[16] J. Barretina, et al., The Cancer Cell Line Encyclopedia enables predictive modelling of anticancer drug sensitivity, Nature 483 (7391) (2012) 603–607.

[17] J.P. Gillet, et al., Redefining the relevance of established cancer cell lines to the study of mechanisms of clinical anti-cancer drug resistance, Proc. Natl. Acad. Sci. U. S. A. 108 (46) (2011) 18708–18713.

[18] S.P. Langdon, Cancer Cell Culture: Methods and Protocols, first ed., Humana Press, Totowa, NJ, USA, 2004.

[19] J.R.W. Masters, Human cancer cell lines: fact and fantasy, Nat. Rev. Mol. Cell Biol. 1 (2000) 233–236.

[20] C. Alston-Roberts, R. Barallon, S.R. Bauer, J. Butler, A. Capes-Davis, W.G. Dirks, E. Elmore, M. Furtado, L. Kerrigan, M.C. Kline, A. Kohara, G.V. Los, R.A. MacLeod, J.R. Masters, M. Nardone, R.M. Nardone, R.W. Nims, P.J. Price, Y.A. Reid, J. Shewale, A.F. Steuer, D.R. Storts, G. Sykes, Z. Taraporewala, J. Thomson, Cell line misidentification: the beginning of the end, Nat. Rev. Cancer 10 (6) (2010) 441–448.

[21] P. Hughes, D. Marshall, Y. Reid, H. Parkes, C. Gelber, The costs of using unauthenticated, over-passaged cell lines: how much more data do we need?, BioTechniques 43 (5) (2007) 577–578.

[22] A. Capes-Davis, G. Theodosopoulos, I. Atkin, H.G. Drexler, A. Kohara, R.A. MacLeod, J.R. Masters, Y. Nakamura, Y.A. Reid, R.R. Reddel, R.I. Freshney, Check your cultures! A list of cross-contaminated or misidentified cell lines, Int. J. Cancer 127 (1) (2010) 1–8.

[23] J.R. Lorsch, F.S. Collins, J. Lippincott-Schwartz, Cell biology. Fixing problems with cell lines, Science 346 (6216) (2014) 1452–1453.

[24] A. Mitra, L. Mishra, S. Li, Technologies for deriving primary tumor cells for use in personalized cancer therapy, Trends Biotechnol. 31 (6) (2013) 347–354.

[25] W.-C. Li, K. Ralphs, D. Tosh, Isolation and culture of adult mouse hepatocytes, in: A. Ward, D. Tosh (Eds.), Mouse Cell Culture, Methods in Molecular Biology, vol. 633, Humana Press, New York, NY, USA, 2010 pp. 185–196.

[26] T. Mitaka, The current status of primary hepatocyte culture, Int. J. Exp. Pathol. 79 (6) (1998) 393–409.

[27] R. Mitra, M. Morad, A uniform enzymatic method for dissociation of myocytes from hearts and stomachs of vertebrates, Am. J. Physiol. Heart Circ. Physiol. 249 (5) (1985) H1056–H1060.

[28] R.I. Freshney, Culture of Animal Cells: A Manual of Basic Technique and Specialized Applications, John Wiley & Sons, Inc., Hoboken, NJ, USA, 2010.

[29] J.V. Castell, M.J. Gomez-Lechon, Liver cell culture techniques, Methods Mol. Biol. 481 (2009) 35–46.

[30] C.C. Harris, C.A. Leone, Some effects of EDTA and tetraphenylboron on the ultrastructure of mitochondria in mouse liver cells, J. Cell Biol. 28 (2) (1966) 405–408.

[31] R.E. Cunningham, Tissue disaggregation, Methods Mol. Biol. 588 (2010) 327–330.

[32] B.M. Ljung, B. Mayall, C. Lottich, C. Boyer, S.S. Sylvester, G.S. Leight, H.F. Siegler, H.S. Smith, Cell dissociation techniques in human breast cancer–variations in tumor cell viability and DNA ploidy, Breast Cancer Res. Treat. 13 (2) (1989) 153–159.

[33] X.F. Pei, M.S. Noble, M.A. Davoli, E. Rosfjord, M.T. Tilli, P.A. Furth, R. Russell, M.D. Johnson, R.B. Dickson, Explant-cell culture of primary mammary tumors from MMTV-c-Myc transgenic mice, In Vitro Cell. Dev. Biol. Anim. 40 (1–2) (2004) 14–21.

[34] N. Parajuli, W. Doppler, Precision-cut slice cultures of tumors from MMTV-neu mice for the study of the ex vivo response to cytokines and cytotoxic drugs, In Vitro Cell. Dev. Biol. Anim. 45 (8) (2009) 442–450.

[35] J.B. Kim, R. Stein, M.J. O'Hare, Three-dimensional in vitro tissue culture models of breast cancer—a review, Breast Cancer Res. Treat. 85 (3) (2004) 281–291.

[36] F. Berdichevsky, D. Alford, B. D'Souza, J. Taylor-Papadimitriou, Branching morphogenesis of human mammary epithelial cells in collagen gels, J. Cell. Sci. 107 (Pt 12) (1994) 3557–3568.

[37] M.M. Malinen, H. Palokangas, M. Yliperttula, A. Urtti, Peptide nanofiber hydrogel induces formation of bile canaliculi structures in three-dimensional hepatic cell culture, Tissue Eng. Part A 18 (23–24) (2012) 2418–2425.

[38] C. Fischbach, R. Chen, T. Matsumoto, T. Schmelzle, J.S. Brugge, P.J. Polverini, D.J. Mooney, Engineering tumors with 3D scaffolds, Nat. Methods 4 (10) (2007) 855–860.

[39] P.A. Kenny, G.Y. Lee, C.A. Myers, R.M. Neve, J.R. Semeiks, P.T. Spellman, K. Lorenz, E.H. Lee, M.H. Barcellos-Hoff, O.W. Petersen, J.W. Gray, M.J. Bissell, The morphologies of breast cancer cell lines in three-dimensional assays correlate with their profiles of gene expression, Mol. Oncol. 1 (1) (2007) 84–96.

[40] R. Fridman, M.C. Kibbey, L.S. Royce, M. Zain, M. Sweeney, D.L. Jicha, J.R. Yannelli, G.R. Martin, H.K. Kleinman, Enhanced tumor growth of both primary and established human and murine tumor cells in athymic mice after coinjection with Matrigel, J. Natl. Cancer Inst. 83 (11) (1991) 769–774.

[41] M. Jechlinger, K. Podsypanina, H. Varmus, Regulation of transgenes in three-dimensional cultures of primary mouse mammary cells demonstrates oncogene dependence and identifies cells that survive deinduction, Genes Dev. 23 (14) (2009) 1677–1688.

[42] S. Wang, D. Nagrath, P.C. Chen, F. Berthiaume, M.L. Yarmush, Three-dimensional primary hepatocyte culture in synthetic self-assembling peptide hydrogel, Tissue Eng. Part A 14 (2) (2008) 227–236.

[43] J.W. Gordon, G.A. Scangos, D.J. Plotkin, J.A. Barbosa, F.H. Ruddle, Genetic transformation of mouse embryos by microinjection of purified DNA, Proc. Natl. Acad. Sci. U. S. A. 77 (12) (1980) 7380–7384.

[44] K.R. Thomas, M.R. Capecchi, Site-directed mutagenesis by gene targeting in mouse embryo-derived stem cells, Cell 51 (3) (1987) 503–512.

[45] J.C. Walrath, J.J. Hawes, T. Van Dyke, K.M. Reilly, Genetically engineered mouse models in cancer research, Adv. Cancer Res. 106 (2010) 113–164.

[46] H. Lee, Genetically engineered mouse models for drug development and preclinical trials, Biomol. Ther. (Seoul) 22 (4) (2014) 267–274.

[47] H. Yang, H. Wang, R. Jaenisch, Generating genetically modified mice using CRISPR/Cas-mediated genome engineering, Nat. Protoc. 9 (8) (2014) 1956–1968.

[48] O.J. Becher, E.C. Holland, Genetically engineered models have advantages over xenografts for preclinical studies, Cancer Res. 66 (7) (2006) 3355–3358.

[49] A. Richmond, Y. Su, Mouse xenograft models vs GEM models for human cancer therapeutics, Dis. Model. Mech. 1 (2–3) (2008) 78–82.

[50] S.P. Flanagan, "Nude", a new hairless gene with pleiotropic effects in the mouse, Genet. Res. 8 (3) (1966) 295–309.

[51] M.J. Bosma, A.M. Carroll, The SCID mouse mutant: definition, characterization, and potential uses, Annu. Rev. Immunol. 9 (1991) 323–350.

[52] C.L. Morton, P.J. Houghton, Establishment of human tumor xenografts in immunodeficient mice, Nat. Protoc. 2 (2) (2007) 247–250.

[53] R.S. Kerbel, Human tumor xenografts as predictive preclinical models for anticancer drug activity in humans: better than commonly perceived-but they can be improved, Cancer Biol. Ther. 2 (4 Suppl. 1) (2003) S134–S139.

[54] J.I. Johnson, S. Decker, D. Zaharevitz, L.V. Rubinstein, J.M. Venditti, S. Schepartz, S. Kalyandrug, M. Christian, S. Arbuck, M. Hollingshead, E.A. Sausville, Relationships between drug activity in NCI preclinical in vitro and in vivo models and early clinical trials, Br. J. Cancer 84 (10) (2001) 1424–1431.

[55] C.C. Scholz, D.P. Berger, B.R. Winterhalter, H. Henss, H.H. Fiebig, Correlation of drug response in patients and in the clonogenic assay with solid human tumour xenografts, Eur. J. Cancer 26 (8) (1990) 901–905.

[56] C. Khanna, K. Lindblad-Toh, D. Vail, C. London, P. Bergman, L. Barber, M. Breen, B. Kitchell, E. McNeil, J.F. Modiano, S. Niemi, K.E. Comstock, E. Ostrander, S. Westmoreland, S. Withrow, The dog as a cancer model, Nat. Biotechnol. 24 (9) (2006) 1065–1066.

[57] L.B. Schook, T.V. Collares, W. Hu, Y. Liang, F.M. Rodrigues, L.A. Rund, K.M. Schachtschneider, F.K. Seixas, K. Singh, K.D. Wells, E.M. Walters, R.S. Prather, C.M. Counter, A genetic porcine model of cancer, PLoS ONE 10 (7) (2015) e0128864.

[58] H.J. Xia, C.S. Chen, Progress of non-human primate animal models of cancers, Zool. Res. 32 (1) (2011) 70–80.

Tumor growth models

5

5.1 INTRODUCTION

Tumor growth models consist of dynamic mathematical models, such as partial differential equations to model the tumor size data and can be used to characterize tumor growth and specific drug sensitivity response. However, tumor growth models requiring time series information for training and the inability to handle large numbers of unknown parameters are not suited for drug sensitivity modeling from high-throughput genomic or functional characterizations, where the emphasis is on deciphering genetic interconnections for modeling the steady-state behavior starting from a large set of potential features and limited model information. The purpose of this chapter is to introduce the readers to this area closely related to drug sensitivity modeling that can potentially be used to model the in vitro growth dynamics of patient tumor cultures and the effect of specific therapy on tumor volume, and for predicting tumor volumes at later time points. Furthermore, when in vivo behavior is represented by tumor growth models, they can be used to compare the response, due to therapy with the predicted untreated response. However, readers are advised that tumor growth models require a significant amount of time series data for model parameter estimation, which is usually not available for individual tumor patients and

Predictive Modeling of Drug Sensitivity. http://dx.doi.org/10.1016/B978-0-12-805274-7.00005-1
Copyright © 2017 Elsevier Inc. All rights reserved.

they are not suitable for inferring and representing numerous relationships between genetic entities. A potential application of the discussed approaches can follow once other approaches are used to narrow down the relevant features and drugs for a tumor and additional time-series experiments are subsequently conducted to decipher a tumor growth model to characterize the growth dynamics.

To understand tumor growth, let us consider the type of mutations occurring in a cell. Cell autonomous mutations are ones that provide growth advantage to the cell in which they appeared, whereas noncell-autonomous mechanisms provide growth advantage to cells that do not have that mutation [1]. The majority of mutations studied for cancer exhibit cell autonomous behavior by conferring a growth advantage to the cells where they originate. Some examples of cell autonomous behavior include KRAS mutations [2] and *PDGFRA-α* mutations [3]. Some noncell-autonomous behavior has also been observed in cancer signaling, but the suppression of those mechanisms has resulted in overtaking of the cell growth using cell-autonomous mechanisms [4,5].

Cell growth is usually controlled by noncell-autonomous mechanisms. Cancer usually begins when noncell-autonomous growth suppression mechanisms are overruled by cell autonomous driver mutations in the originating tumor cells [1]. There are two main theories for growth of cancer cells: (a) cancer stem cell theory and (b) stochastic model of cancer growth. The cancer stem cell theory [6] postulates that growth of cancer cells is similar to normal cells in the manner that it is hierarchical with cancer stem cells as the starting point. Based on this theory, cancer stem cells are assumed to posses the ability to divide and can produce cells similar to stem cells, or transit amplifying cells. The transit amplifying cells can divide for a certain number of times before differentiating to specialized tumor cells that do not divide.

In contrast to cancer stem cell model that assumes an organized structure with stem cells at the top, stochastic model of cancer growth postulates that each tumor cell has the same potential for cell division and they decide at random between self-renewal or differentiation.

To analyze cancer growth through a simplistic stochastic model, let us denote by f the fraction of the cells in a generation that will divide again, d is the probability that the cell dies, l is the probability that the cell lives without dividing and $f + d + l = 1$. To maintain cell growth in check, homeostatic mechanisms maintain $f = 0.5$ and a cancerous mutation increases the $f > 0.5$. While calculating the time required or tumor progression to a fixed number of cells, the probability of nondividing cells dying (d) or living (l) is only important in cell growth when f is close to 0.5. For instance, reaching 10^6 cells starting from 1 cell with $f = 0.52$ and $d = 0.49, l = 0$ and $d = 0, l = 0.49$ are around 340 and 285 generations, respectively. When $f = 0.7$ the generations required for $d = 0.49, l = 0$ and $d = 0, l = 0.49$ to reach 10^6 cells are 41 and 40, respectively. It shows that when the driver mutations induce higher probability of division f, the life or death of the nondividing cells is inconsequential to tumor development. This is expected when the fraction of dividing cells is significantly higher than 0.5 and thus in every new generation, the importance of nondividing cells keeps decreasing.

Our exploration of drug sensitivity modeling has primarily been considered from the perspectives of either mapping genetic and epigenetic characterizations to drug response, or mapping drug inhibition profiles to drug sensitivities. In this chapter, we will consider the modeling of the dynamics of tumor volume based on growth rates and initial tumor volume. Note that in comparison, dynamical genetic regulatory network models represent the time-dependent relationships between gene or protein expressions and not the phenotypical response. In that sense, we can categorize the tumor growth models as modeling *phenotype dynamics*, and dynamical genetic regulatory network models as modeling the *genotype dynamics*.

Tumor growth models have been studied for more than 60 years [7–9] starting from an era before the advent of genetic profiling. As with most dynamical modeling, the steps involved include (i) observing the tumor growth curves, (ii) designing appropriate models that reflect the observed data, and (iii) validating the designed models based on measures such as future growth prediction accuracy.

The observation of the temporal tumor growth curves can be achieved through in vitro assays, three-dimensional in vitro spheroids or in vivo tumor models [10]. Each approach presents its own set of advantages and drawbacks. In vitro assays are easier to maintain and it has been shown that they can maintain the in vivo three-dimensional growth properties in vitro in some scenarios [11]. However, the lack of microenvironment and multiple regeneration can change the behavior of the in vitro assays. In comparison, in vivo xenograft mice models can be expensive to design and may posses their own unique set of differences as compared to the in vivo human tumor which fails to capture important aspects of tumorigenesis including tumor-initiating events and interactions with the immune system.

Studies have shown that the tumor growth rate decreases with time [12,13] and models incorporating a deceleration of tumor growth rate with time have shown to fit observed data better than unconstrained exponential models [14]. The overall exponential growth of tumors motivated the modeling of the growth using exponential models, but the observations of change in growth rate resulted in the application of other constrained models such as the Gompertz model [15,16].

In terms of applications, tumor growth models can be applied to estimate the future course of tumor progression from a clinical or preclinical drug testing scenario. For instance, Simeoni et al. [17] considered the modeling of nontreated xenograft tumors using models with exponential growth, followed by linear growth and altered the growth rate of treated tumors based on drug concentration and available tumor cells. The model was tested on multiple anticancer drugs and was able to predict the response of tumors exposed to drugs given at different dose levels and schedules [17]. Note that the application of this form of modeling will be limited when a new drug is considered or a different patient sample is considered. Results reported in [18] using Gompertz model for untreated mice and two-compartment pharmacokinetic/pharmacodynamic model for treated mice illustrate that there is huge variability in growth rates of mice with the same form of cancer and thus patient-specific models are essential to accurately model tumor growth as compared to models representing the average behavior of the subject population.

5.2 EXPONENTIAL LINEAR MODELS

The initial exponential models were based on the belief that tumor growth will continue exponentially if sufficient nutrients and oxygen are available. These growth models assumed that the cells proliferate with constant growth rate and are represented by a basic differential equation model shown in Eq. (5.1).

$$\frac{dV}{dt} = r_0 V \tag{5.1}$$

where V denotes the tumor volume and the initial tumor volume is given by $V(t = 0) = V_0$. The solution to this equation is given by $V = V_0 e^{r_0 t}$. This model was applied for analyzing cancer growth back in 1956 [7] and they introduced the tumor doubling time $DT = \frac{\ln 2}{r_0}$ to quantify the tumor growth rate. This model has also been applied for analyzing chemotherapy induced tumor progression [19].

Since the unrestricted exponential growth is usually not observed in actual tumor growth data due to limitations on nutrients, space, and oxygen, alternative formulations have been proposed [20]. One such approach considered incorporating a linear growth after a specific time, as shown in Eq. (5.2).

$$\frac{dV}{dt} = \begin{cases} r_0 V, & \text{if } t \leq \tau \\ r_1, & \text{if } t > \tau \end{cases} \tag{5.2}$$

If we assume continuous differentiability of solution to Eq. (5.2), we arrive at the unique value of $\tau = \frac{1}{r_0} \log(\frac{r_1}{r_0 V_0})$.

5.3 LOGISTIC AND GOMPERTZ MODELS

The reduced rate of growth observed in tumor progression has also been modeled by a logistic model that incorporates a linear decrease of the relative growth rate $\frac{1}{V}\frac{dV}{dt}$ as shown in Eq. (5.3).

$$\frac{dV}{dt} = rV\left(1 - \frac{V}{K}\right) \tag{5.3}$$

where K is the maximal volume of the tumor. The model can be considered as cells competing for available finite resources. A *generalized logistic model* is given by:

$$\frac{dV}{dt} = rV\left(1 - \left(\frac{V}{K}\right)^{\beta}\right) \tag{5.4}$$

The analytical solution to the generalized logistic model is given by

$$V(t) = \frac{V_0 K}{(V_0^{\beta} + (K^{\beta} V_0^{\beta})e^{-r\beta t})^{1/\beta}} \tag{5.5}$$

Gompertz model considers the reduction in tumor growth rate based on a function of $\ln V(t)$ as shown in Eq. (5.6).

$$\frac{dV}{dt} = V(t)(\alpha - \beta \ln V(t)) \tag{5.6}$$

Note that model shown in Eq. (5.6) can be considered as $\frac{dV}{dt} = f(t)V(t)$ where $\frac{df}{dV} = -\frac{\beta}{V}$. The solution to *Gompertz* model is given by

$$V(t) = e^{\frac{\alpha}{\beta} - \left(\frac{\alpha}{\beta} - \ln V_0\right)e^{-\beta t}} \tag{5.7}$$

The carrying capacity of the *Gompertz* model is $K_G = \lim_{t \to \infty} V(t) = e^{\frac{\alpha}{\beta}}$. K_G also reflects an equilibrium point with $\frac{dV}{dt} = 0$. By equating $\frac{d^2V}{dt^2} = 0$, we arrive at the *inflection point* $V_i = e^{\frac{\alpha}{\beta} - 1}$ where maximum growth rate is achieved. In terms of biological understanding, inflection points reflect a course in the tumor development after which growth rate slows down, and the reason for the slowdown can be external mechanisms, such as nutrient availability or intrinsic self-regulation mechanisms. As compared to exponential growth models, Gompertz model captures the dependence of tumor models on nutrient, oxygen, and space [21]. As the tumor grows, the resources are depleted and tumor growth decreases until it reaches the saturation volume of $K_G = e^{\alpha/\beta}$. A disadvantage with Gompertz model is the difficulty to estimate the carrying capacity of the model based on preclinical or clinical data, as the maximum volume is often not observed in preclinical or clinical settings [21].

5.4 POWER LAW MODELS

Power law models generalize the exponential model as follows:

$$\frac{dV}{dt} = r_0 V^{\alpha} \tag{5.8}$$

$\alpha = 1$ produces the basic exponential model and the solution for general α is

$$V(t) = (V_0^{1-\alpha} + (1 - \alpha)r_0 t)^{\frac{1}{1-\alpha}} \tag{5.9}$$

If we incorporate a linear death rate, we arrive at the following power law model

$$\frac{dV}{dt} = r_0 V^{\alpha} - r_1 V \tag{5.10}$$

Eq. (5.10) with $\alpha = 2/3$ is commonly known as Bertalanffy model [22].
    Explicit solution of the model is given by:

$$V(t) = \left(\frac{r_0}{r_1} + \left(V_0^{1-\alpha} - \frac{r_0}{r_1}\right)e^{-r_1(1-\alpha)t}\right)^{\frac{1}{1-\alpha}} \tag{5.11}$$

5.5 STOCHASTIC TUMOR GROWTH MODELS

The previously described tumor growth models, such as Gompertz model describes the population size $X(t)$ at time t based on the solution to a deterministic differential equation model. However, deterministic models fail to accurately capture any external environment fluctuations and thus stochastic tumor growth models have been proposed to capture the random effects caused by external environmental conditions [23–26].

To explain the idea of stochastic tumor growth models, we consider the approach illustrated in [26]. Let us consider the discrete version of the Gompertz equation as follows:

$$v_{(n+1)\tau} - v_{n\tau} = (\alpha\tau - \beta\tau \ln v_{n\tau})v_{n\tau} \quad (n = 0, 1, \ldots) \tag{5.12}$$

In reality, tumor growth in a tissue is based on division or death of tumor cells and thus is a discrete process. To incorporate random fluctuations, the intrinsic change $\alpha\tau$ of the population during time duration $[n\tau, (n+1)\tau]$, $n = 0, 1, \ldots$, is modeled as the mean of a sequence of independent and identically distributed Bernoulli random variables $Z_0, Z_\tau, \ldots$ with probability distribution as follows:

$$Pr(Z_{n\tau} = \sigma\sqrt{\tau}) = \frac{1}{2} + \frac{\alpha\sqrt{\tau}}{2\sigma}$$

$$Pr(Z_{n\tau} = -\sigma\sqrt{\tau}) = \frac{1}{2} - \frac{\alpha\sqrt{\tau}}{2\sigma} \tag{5.13}$$

where $\sigma > 0$ is a measure of the environment fluctuations. The mean and variance of $Z_{n\tau}$ is given by

$$\mathbb{E}(Z_{n\tau}) = \alpha\tau \tag{5.14}$$

$$Var(Z_{n\tau}) = \sigma^2\tau - (\alpha\tau)^2 \tag{5.15}$$

Based on this modification, the stochastic model can be described by

$$V_{(n+1)\tau} - V_{n\tau} = (Z_{n\tau} - \beta\tau \ln V_{n\tau})V_{n\tau} \tag{5.16}$$

where $V_{n\tau}$ is a discrete space, continuous time stochastic process. Considering the continuous version of this process, we will arrive at a diffusion process with drift (mean)

$$\mu(v) = \alpha v - \beta v \ln v \tag{5.17}$$

and variance

$$Var(v) = \alpha^2 v^2 \tag{5.18}$$

Further stochastic extensions have been considered in [27] where a two-compartment growth model is proposed, consisting of a compartment with proliferating cells with nonzero growth rate and a compartment of quiescent cells with zero growth rate.

Merrill [28] considers tumor growth models in the form of density-dependent jump Markov processes. A state space-based model for incorporating stochastic aspects in the Gompertz model has been proposed in [29].

Note that the stochastic modeling of tumor growth as represented by stochastic extension of Gompertz model has similarities to stochastic modeling approaches in genetic regulatory networks. Fine-scale stochastic models of genetic regulatory networks are also modeled as stochastic master equation models [30,31] whose continuous approximations are given by stochastic differential equations [32] similar to the diffusion process considered here. The deterministic models can be considered as the mean approximations of the stochastic models. Furthermore, the stochastic modeling of tumor growth is more relevant when the tumor size is smaller [28]. For larger tumor sizes, the deterministic approximations provide results close to the stochastic models.

5.6 MODELING TUMOR SPHEROID GROWTH

In vitro tumor spheroid cultures are often considered to be closer representation of in vivo tumors as compared to single layer in vitro cell growths [33]. Experimental imaging data [34] suggest that tumor spheroids eventually develop the following three zones [35]:

- An outside layer of *proliferating cells* that forms a concentric shell at the exterior of the spheroid. These shells can convert to quiescent cells, can die, or shed into the surrounding medium. During the earlier stages of development, a large fraction of the tumor spheroid may consist of proliferating cells.
- A secondary layer of *quiescent cells* inside the layer of proliferating cells. These cells arise from proliferating cells and posses the ability to convert back to proliferating cells.
- *Necrotic cells* form the center of the spheroid. These cells are more observed in older cultures as compared to smaller spheroids where they may be absent. These cells arise when quiescent cells die.

When the growth of spheroids finally stops, a thin layer of actively proliferating cells is usually observed at the boundary [35,36]. The three layers of a tumor spheroid are shown pictorially in Fig. 5.1.

To model a tumor spheroid, Wallace and Guo [35] considered the following variables: P (number of proliferating cells), Q (number of quiescent cells), N (number of necrotic cells), and total spheroid size. The differential equations relating the variables are given by:

$$\frac{dP}{dt} = G(P)b_{P,Q}P + c_{Q,P}QF(P,Q,N)dP \tag{5.19}$$

$$\frac{dQ}{dt} = b_{P,Q}P - c_{Q,P}Qe_{Q,N}Q + H(P,Q,N) \tag{5.20}$$

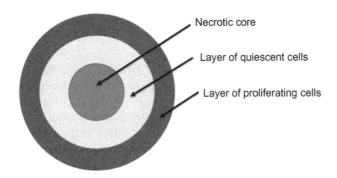

FIG. 5.1

Three layers of a tumor spheroid.

$$\frac{dN}{dt} = e_{Q,N}Q - mN \qquad (5.21)$$

The various parameters involved in the above model are explained next: (a) $G(P)$ is the growth of proliferating cells which can be modeled as a function of surface area of the spheroid $(P + Q + N)^{2/3}$, volume of proliferating cells, and a form of logistic or Gompertzian growth factor. (b) $b_{P,Q}P$ denotes the rate of transition from proliferative to quiescent cell. Similarly, $c_{Q,P}Q$ denotes the rate of transition from quiescent to proliferative and $e_{Q,N}Q$ denotes the rate of transition from quiescent to necrotic cells. (c) $F(P,Q,N)$ and $H(P,Q,N)$ denote the effect of tumor necrosis factor.

Wallace and Guo [35] considered various forms of the model parameters and showed that some of the realizations of the simple model representing the three compartments of proliferative, quiescent, and necrotic cells can successfully model a diverse set of experimentally observed in vitro tumor spheroid data. The high accuracy models have growth and death terms for proliferative cells proportional to surface area and necrotic cells, respectively.

5.7 DISCUSSION

A significant amount of research has been performed in modeling tumor cell growth and this chapter attempted to provide a basic overview without the intention of comprehensively covering all directions of the research area. For additional details, readers are referred to some recent works in mathematical modeling of tumor cell dynamics including immersed boundary cell model [37], Cellular Potts model [38–40], hybrid discrete-continuum model [41–43], partial differential equation models [44,45], image integrated mathematical models [46], and evolutionary game theory models [47,48].

In this chapter, we considered modeling of tumor growth in the form of tracking the change in tumor volume with time. The parameters of the model estimated from experimental data inform us on the severity of the tumor. For instance, the tumor doubling time (represented by $\frac{\ln 2}{r_0}$ for the exponential model) denotes the time it takes for the tumor to double in size. In terms of application of these models to evaluate cancer therapies, researchers have considered answering questions such as how do therapies alter the parameters of the model? [17–19]. Usually, new sets of experimental data are generated following application of therapies and models fitted to these new datasets for analyzing the effect of therapies. The approach imposes large experimental requirements for model parameter estimation, which can restrict the evaluation to fewer drugs and limited cell lines or cultures. Because patient tumors of the same type can harbor diverse phenotypic and genotypic behavior, parameter estimation based on population data or dissimilar patient data can restrict the use of tumor growth models for new patients. However, tumor growth models once inferred can be quite useful in modeling the in vitro growth dynamics of patient tumor cultures and the effect of specific therapy on tumor volume, and for predicting tumor volumes at later time points. Furthermore, when in vivo behavior is represented by tumor growth models, they can be used to compare the response due to therapy with the predicted untreated response.

To summarize, tumor growth models represent the tissue-specific dynamics in terms of differential equations. Tumor growth models characterize a scale different from the genetic regulatory network models, which models at the protein and cell-specific scale. We can argue that the parameters of the tumor growth model capture in a transformed space the internal drivers of tumor growth, such as gene mutations and interactions that are explicitly modeled in genetic regulatory networks. Since the number of parameters in tumor growth models is limited, their parameters are easier to estimate from experimental data and the models are computationally inexpensive to simulate, as compared to dynamic genetic regulatory network models. However, tumor growth models are limited in their applicability to new patients and restricted in predicting the effect of a new drug or drug combination.

REFERENCES

[1] A. Sidow, N. Spies, Concepts in solid tumor evolution, Trends Genet. 31 (4) (2015) 208–214.
[2] S. von Karstedt, A. Conti, M. Nobis, A. Montinaro, T. Hartwig, J. Lemke, K. Legler, F. Annewanter, A.D. Campbell, L. Taraborrelli, A. Grosse-Wilde, J.F. Coy, M.A. El-Bahrawy, F. Bergmann, R. Koschny, J. Werner, T.M. Ganten, T. Schweiger, K. Hoetzenecker, I. Kenessey, B. Hegedus, M. Bergmann, C. Hauser, J.H. Egberts, T. Becker, C. Rocken, H. Kalthoff, A. Trauzold, K.I. Anderson, O.J. Sansom, H. Walczak, Cancer cell-autonomous TRAIL-R signaling promotes KRAS-driven cancer progression, invasion, and metastasis, Cancer Cell 27 (4) (2015) 561–573.

[3] S. Hirota, A. Ohashi, T. Nishida, K. Isozaki, K. Kinoshita, Y. Shinomura, Y. Kitamura, Gain-of-function mutations of platelet-derived growth factor receptor alpha gene in gastrointestinal stromal tumors, Gastroenterology 125 (3) (2003) 660–667.

[4] C.S. Grasso, Y.M. Wu, D.R. Robinson, X. Cao, S.M. Dhanasekaran, A.P. Khan, M.J. Quist, X. Jing, R.J. Lonigro, J.C. Brenner, I.A. Asangani, B. Ateeq, S.Y. Chun, J. Siddiqui, L. Sam, M. Anstett, R. Mehra, J.R. Prensner, N. Palanisamy, G.A. Ryslik, F. Vandin, B.J. Raphael, L.P. Kunju, D.R. Rhodes, K.J. Pienta, A.M. Chinnaiyan, S.A. Tomlins, The mutational landscape of lethal castration-resistant prostate cancer, Nature 487 (7406) (2012) 239–243.

[5] L.A. Diaz, R.T. Williams, J. Wu, I. Kinde, J.R. Hecht, J. Berlin, B. Allen, I. Bozic, J.G. Reiter, M.A. Nowak, K.W. Kinzler, K.S. Oliner, B. Vogelstein, The molecular evolution of acquired resistance to targeted EGFR blockade in colorectal cancers, Nature 486 (7404) (2012) 537–540.

[6] A. Kreso, J.E. Dick, Evolution of the cancer stem cell model, Cell Stem Cell 14 (3) (2014) 275–291.

[7] V.P. Collins, R.K. Loeffler, H. Tivey, Observations on growth rates of human tumors, Am. J. Roentgenol. Radium Ther. Nucl. Med. 76 (5) (1956) 988–1000.

[8] G.G. Steel, Growth Kinetics of Tumours: Cell Population Kinetics in Relation to the Growth and Treatment of Cancer, Clarendon Press, Oxford, 1977.

[9] L. Heuser, J.S. Spratt, H.C. Polk, Growth rates of primary breast cancers, Cancer 43 (5) (1979) 1888–1894.

[10] S. Benzekry, C. Lamont, A. Beheshti, A. Tracz, J.M. Ebos, L. Hlatky, P. Hahnfeldt, Classical mathematical models for description and prediction of experimental tumor growth, PLoS Comput. Biol. 10 (8) (2014) e1003800.

[11] A.E. Freeman, R.M. Hoffman, In vivo-like growth of human tumors in vitro, Proc. Natl. Acad. Sci. U. S. A. 83 (8) (1986) 2694–2698.

[12] A.K. Laird, Dynamics of tumour growth: comparison of growth rates and extrapolation of growth curve to one cell, Br. J. Cancer 19 (1965) 278–291.

[13] G.G. Steel, L.F. Lamerton, The growth rate of human tumours, Br. J. Cancer 20 (1) (1966) 74–86.

[14] J.A. Spratt, D. von Fournier, J.S. Spratt, E.E. Weber, Decelerating growth and human breast cancer, Cancer 71 (6) (1993) 2013–2019.

[15] L. Norton, A Gompertzian model of human breast cancer growth, Cancer Res. 48 (24 Pt 1) (1988) 7067–7071.

[16] V.G. Vaidya, F.J. Alexandro, Evaluation of some mathematical models for tumor growth, Int. J. Biomed. Comput. 13 (1) (1982) 19–36.

[17] M. Simeoni, P. Magni, C. Cammia, G. De Nicolao, V. Croci, E. Pesenti, M. Germani, I. Poggesi, M. Rocchetti, Predictive pharmacokinetic-pharmacodynamic modeling of tumor growth kinetics in xenograft models after administration of anticancer agents, Cancer Res. 64 (3) (2004) 1094–1101.

[18] C. Loizides, D. Iacovides, M.M. Hadjiandreou, G. Rizki, A. Achilleos, K. Strati, G.D. Mitsis, Model-based tumor growth dynamics and therapy response in a mouse model of de novo carcinogenesis, PLoS ONE 10 (12) (2015) e0143840.

[19] S.E. Shackney, A computer model for tumor growth and chemotherapy, and its application to L1210 leukemia treated with cytosine arabinoside (NSC-63878), Cancer Chemother. Rep. 54 (6) (1970) 399–429.

[20] A. Talkington, R. Durrett, Estimating tumor growth rates in vivo, Bull. Math. Biol. 77 (2015) 1934–1954.

[21] B. Ribba, N.H. Holford, P. Magni, I. Troconiz, I. Gueorguieva, P. Girard, C. Sarr, M. Elishmereni, C. Kloft, L.E. Friberg, A review of mixed-effects models of tumor growth and effects of anticancer drug treatment used in population analysis, CPT Pharmacometrics Syst. Pharmacol. 3 (2014) e113.

[22] L. Bertalanffy, Problems of organic growth, Nature 163 (4135) (1949) 156–158.

[23] Prajneshu, Diffusion approximations for models of population growth with logarithmic interactions, Stoch. Process. Their Appl. 10 (1) (1980) 87–99.

[24] S. Sahoo, A. Sahoo, S.F.C. Shearer, Dynamics of Gompertzian tumour growth under environmental fluctuations, Phys. A Stat. Mech. Appl. 389 (6) (2010) 1197–1207.

[25] E.K. Moummou, R. Gutirrez, R. Gutirrez-Sanchez, A stochastic Gompertz model with logarithmic therapy functions: parameters estimation, Appl. Math. Comput. 219 (8) (2012) 3729–3739.

[26] G. Albano, V. Giorno, A stochastic model in tumor growth, J. Theor. Biol. 242 (2) (2006) 329–336.

[27] G. Albano, V. Giorno, P. Romn-Romn, F. Torres-Ruiz, Inference on a stochastic two-compartment model in tumor growth, Comput. Stat. Data Anal. 56 (6) (2012) 1723–1736.

[28] S.J. Merrill, Stochastic models of tumor growth and the probability of elimination by cytotoxic cells, J. Math. Biol. 20 (3) (1984) 305–320.

[29] W.-Y. Tan, W. Ke, G. Webb, A stochastic and state space model for tumour growth and applications, Comput. Math. Method Med. 10 (2) (2009) 117–138.

[30] D.T. Gillespie, Exact stochastic simulation of coupled chemical reactions, J. Phys. Chem. 81 (1977) 2340–2361.

[31] R. Pal, S. Bhattacharya, U. Caglar, Robust approaches for genetic regulatory network modeling and intervention, IEEE Signal Process. Mag. 29 (1) (2012) 66–76.

[32] X. Cai, X. Wang, Stochastic modeling and simulation of gene networks: a review of the state-of-the-art research on stochastic simulations, IEEE Signal Process. Mag. 24 (2007) 27–36.

[33] M.T. Santini, G. Rainaldi, P.L. Indovina, Apoptosis, cell adhesion and the extracellular matrix in the three-dimensional growth of multicellular tumor spheroids, Crit. Rev. Oncol. Hematol. 36 (2–3) (2000) 75–87.

[34] M.D. Sherar, M.B. Noss, F.S. Foster, Ultrasound backscatter microscopy images the internal structure of living tumour spheroids, Nature 330 (6147) (1987) 493–495.

[35] D.I. Wallace, X. Guo, Properties of tumor spheroid growth exhibited by simple mathematical models, Front. Oncol. 3 (2013) 51.

[36] J. Folkman, M. Hochberg, Self-regulation of growth in three dimensions, J. Exp. Med. 138 (4) (1973) 745–753.

[37] K.A. Rejniak, An immersed boundary framework for modelling the growth of individual cells: an application to the early tumour development, J. Theor. Biol. 247 (1) (2007) 186–204.

[38] M. Scianna, E. Bassino, L. Munaron, A cellular Potts model analyzing differentiated cell behavior during in vivo vascularization of a hypoxic tissue, Comput. Biol. Med. 63 (2015) 143–156.

[39] A. Voss-Bohme, Multi-scale modeling in morphogenesis: a critical analysis of the cellular Potts model, PLoS ONE 7 (9) (2012) e42852.

[40] A. Szabo, R.M. Merks, Cellular Potts modeling of tumor growth, tumor invasion, and tumor evolution, Front. Oncol. 3 (2013) 87.

[41] J. Jeon, V. Quaranta, P.T. Cummings, An off-lattice hybrid discrete-continuum model of tumor growth and invasion, Biophys. J. 98 (1) (2010) 37–47.

[42] A.R.A. Anderson, M.A.J. Chaplain, S. McDougall, A Hybrid Discrete-Continuum Model of Tumour Induced Angiogenesis, Springer, New York, NY, 2012, pp. 105–133.

[43] A. Araujo, D. Bastanta, Hybrid discrete-continuum cellular automaton (HCA) model of prostate to bone metastasis, bioRxiv (2016). http://dx.doi.org/10.1101/043620.

[44] P. Hinow, S.E. Wang, C.L. Arteaga, G.F. Webb, A mathematical model separates quantitatively the cytostatic and cytotoxic effects of a HER2 tyrosine kinase inhibitor, Theor. Biol. Med. Model. 4 (2007) 14.

[45] N. Hartung, S. Mollard, D. Barbolosi, A. Benabdallah, G. Chapuisat, G. Henry, S. Giacometti, A. Iliadis, J. Ciccolini, C. Faivre, F. Hubert, Mathematical modeling of tumor growth and metastatic spreading: validation in tumor-bearing mice, Cancer Res. 74 (22) (2014) 6397–6407.

[46] T.E. Yankeelov, N. Atuegwu, D. Hormuth, J.A. Weis, S.L. Barnes, M.I. Miga, E.C. Rericha, V. Quaranta, Clinically relevant modeling of tumor growth and treatment response, Sci. Transl. Med. 5 (187) (2013) 187ps9.

[47] P.M. Altrock, L.L. Liu, F. Michor, The mathematics of cancer: integrating quantitative models, Nat. Rev. Cancer 15 (12) (2015) 730–745.

[48] J.M. Pacheco, F.C. Santos, D. Dingli, The ecology of cancer from an evolutionary game theory perspective, Interface Focus 4 (4) (2014) 20140019.

Overview of predictive modeling based on genomic characterizations

6.1 INTRODUCTION

A variety of methodologies have been proposed for drug sensitivity prediction based on genomic characterizations. A common approach is to consider a training set of cell lines with experimentally measured genomic characterizations (RNA expression, protein expression, methylation, SNPs, etc.) and responses to different drugs, and design supervised predictive models for each individual drug based on one or more genomic characterizations. The modeling techniques employed have been diverse involving commonly used traditional regression approaches, such as linear regression with regularization or problem-specific new developments that incorporate prior biological knowledge of pathways. The majority of developments have considered steady-state models due to limited availability of time-series data following drug perturbations. The performance of similar class of models can also vary significantly based on the type of data preprocessing, feature selection, data integration, and model parameter selection applied during the design process. Some of the key challenges of drug sensitivity model inference and analysis include (C_1) extremely large dimension of features as compared to number of samples, (C_2) limited model accuracy and precision, (C_3) data generation and measurement noises, and (C_4) lack of interpretability of models in terms of biological relevance of selected features. Whenever feasible throughout this chapter, we will discuss the model performance in terms of these key challenges.

The chapter outlines the commonly used predictive modeling techniques applied to drug sensitivity prediction based on genomic characterizations. We broadly categorize the approaches into groups of linear regression techniques, nonlinear regression methods, kernel approaches, ensemble methods, and dynamical models. A final section is devoted to application of these techniques on NCI-DREAM drug sensitivity prediction challenge.

6.2 PREDICTIVE MODELING TECHNIQUES

6.2.1 LINEAR REGRESSION MODELS

Linear regression models are one of the most commonly applied models that have been studied extensively and the statistical properties of the parameter estimations for linear models are easier to determine as compared to nonlinear scenarios.

To explain a linear model, consider $x_{tr}(i,j)$ and $y(i)$ $(i = 1, \ldots, n; j = 1, \ldots, M)$ to denote the training predictor features and output response samples, respectively. A linear regression model considers generation of regression coefficients $\beta_1, \beta_2, \ldots, \beta_M$ such that $y(i) - \sum_{j=1}^{M} \beta_j x_{tr}(i,j) = \epsilon_i$ is minimized. The minimization of the error term ϵ_i is usually considered in the sum of squared error sense, that is, minimize $\sum_{i=1}^{n} \epsilon_i^2$. For linear regression, we also generally incorporate a constant term β_{M+1} commonly known as *intercept*, that is, $\epsilon_i = y(i) - \sum_{j=1}^{M} \beta_j x_{tr}(i,j) - \beta_{M+1}$. In terms of matrix notation, we can write it as

$$\mathbf{Y} = \mathbf{X}\beta + \epsilon \tag{6.1}$$

$$\mathbf{Y} = \begin{bmatrix} y(1) \\ y(2) \\ \cdot \\ \cdot \\ y(n) \end{bmatrix} \quad \mathbf{X} = \begin{bmatrix} x_{tr}(1,1) & x_{tr}(1,2) & \cdots & x_{tr}(1,M) & 1 \\ x_{tr}(2,1) & x_{tr}(2,2) & \cdots & x_{tr}(2,M) & 1 \\ \cdot & \cdot & \cdots & \cdot & \cdot \\ \cdot & \cdot & \cdots & \cdot & \cdot \\ x_{tr}(n,1) & x_{tr}(n,2) & \cdots & x_{tr}(n,M) & 1 \end{bmatrix} \quad \beta = \begin{bmatrix} \beta_1 \\ \beta_2 \\ \cdot \\ \cdot \\ \beta_{M+1} \end{bmatrix} \quad \epsilon = \begin{bmatrix} \epsilon_1 \\ \epsilon_2 \\ \cdot \\ \cdot \\ \epsilon_n \end{bmatrix}$$

The vector $\mathbf{Y}$ consists of the *response or dependent variables*, $\mathbf{X}$ consists of the *regressors, predictor*, or *independent variables*, vector β is the *parameter* vector that needs to be estimated, and ϵ denotes the *error or disturbance* terms.

Note that the development of the linear regression framework consists of a linear combination of parameters β_i and predictor variables. Transforming the predictor variables to higher orders can allow us to utilize the framework in modeling with nonlinear terms, which is known as polynomial regression [1].

6.2.1.1 Ordinary least square

Ordinary least square is the simplest form of linear regression where our goal is to select β that minimizes $\sum_{i=1}^{n} \epsilon_i^2 = (\mathbf{Y} - \mathbf{X}\beta)^T(\mathbf{Y} - \mathbf{X}\beta)$. Differentiating $(\mathbf{Y} - \mathbf{X}\beta)^T(\mathbf{Y} - \mathbf{X}\beta)$ with respect to the parameter vector β and equating it to zero results in

$$\mathbf{X}^T\mathbf{X}\hat{\beta} = \mathbf{X}^T\mathbf{Y} \tag{6.2}$$

If $\mathbf{X}^T\mathbf{X}$ is invertible, the parameter vector $\hat{\beta}$ is given by $\hat{\beta} = (\mathbf{X}^T\mathbf{X})^{-1}\mathbf{X}^T\mathbf{Y}$.

Statistical properties: We can analyze the statistical properties of the above estimator such as *bias*. Bias of an estimator denotes the difference between the expected value of the estimator and the true value of the parameter being estimated. For analyzing bias, we have to consider the assumptions on the data distribution being modeled. Linear regression considers the following assumptions:

- Linear summation of products of parameters and features: $\mathbf{Y} = \mathbf{X}\beta + \epsilon$
- No correlation between independent variables and errors: $\mathbb{E}(\epsilon_i | \mathbf{X}(j, 1 : M)) = 0$
- Homoscedasticity and uncorrelated: Error ϵ_i has finite constant variance $Var(\epsilon_i) = \sigma^2$ and uncorrelated to each other.
- Full column rank of $\mathbf{X}$ is required for nonsingularity of $\mathbf{X}^T\mathbf{X}$ and that restricts multicolinearity of predictor variables.

To analyze the *Bias* of estimator $\hat{\beta}$, consider that

$$\hat{\beta} = (\mathbf{X}^T\mathbf{X})^{-1}\mathbf{X}^T\mathbf{Y} = (\mathbf{X}^T\mathbf{X})^{-1}\mathbf{X}^T(\mathbf{X}\beta + \epsilon) = \beta + (\mathbf{X}^T\mathbf{X})^{-1}\mathbf{X}^T\epsilon \tag{6.3}$$

And thus taking expectations, we have

$$\mathbb{E}(\hat{\beta}|\mathbf{X}) = \beta + \mathbb{E}((\mathbf{X}^T\mathbf{X})^{-1}\mathbf{X}^T\epsilon|\mathbf{X}) \tag{6.4}$$
$$= \beta \tag{6.5}$$

Using iterated expectations, we have

$$\mathbb{E}(\hat{\beta}) = \mathbb{E}_{\mathbf{X}}(\mathbb{E}(\hat{\beta}|\mathbf{X})) = \mathbb{E}_{\mathbf{X}}(\beta) = \beta \tag{6.6}$$

Thus the least square estimator is an unbiased estimator under the considered assumptions.

Under the considered assumptions, ordinary least square estimate $\hat{\beta}$ is also *consistent*, that is, when the number of samples is increased to ∞, the estimator converges in probability to the true value [1].

Ordinary least squares-based linear regression models are simple to understand and quite easy to interpret in terms of feature (reflecting genomic entities) weights and their positive or negative influences and thus can tackle challenge C_4. However, it is not suitable to solve the large feature dimension problem with limited samples (challenge C_1) and can frequently overfit. A number of the assumptions of ordinary least squares are also not satisfied in drug sensitivity studies, such as output response being a linear summation of covariates, and thus ordinary least square is not strong in solving the accuracy challenge C_2. In terms of robustness to measurement noises (C_3), ordinary least square-based linear regressions have decent performance. Thus ordinary least squares are suitable for scenarios where the number of features have already been reduced to a smaller set and we are interested in an easily interpretable and robust model rather than a highly accurate model.

6.2.1.2 Regularized least square regression

Ordinary least square regression can lead to overfitting and thus regularization attempts to solve this issue by adding a regularization term in the cost function $((\mathbf{Y} - \mathbf{X}\beta)^T(\mathbf{Y} - \mathbf{X}\beta)$ for ordinary least square) that is being minimized. Regularization usually consists of adding a penalty on the complexity of the model, which in regression is often included in the form of $\|\beta\|_1$ or $\|\beta\|_2$ norms. For regularization, we usually do not consider the intercept term and thus, for this section, $\mathbf{X}$ will represent an $n \times M$ matrix and β will represent an M length vector. For application of regularized regression methods, it is common practice to center the data (subtracting the mean) as an initial step.

Ridge regression: Ridge regression is similar to ordinary least square with an additional penalty term consisting of the square of l_2 norm of β that shrinks the parameter estimates [2].

$$\hat{\beta}^{\text{ridge}} = \arg\min_{\beta}(\mathbf{Y} - \mathbf{X}\beta)^T(\mathbf{Y} - \mathbf{X}\beta) + \lambda \sum_{i=1}^{M} \beta(i)^2 \tag{6.7}$$

$$= \arg\min_{\beta} \|\mathbf{Y} - \mathbf{X}\beta\|_2^2 + \lambda\|\beta\|_2^2 \tag{6.8}$$

$\lambda \geq 0$ is a tuning parameter that provides a control on fitting a linear model and shrinking the parameter coefficients. The solution to the Ridge regression problem is given by

$$\hat{\beta}^{\text{ridge}} = (\mathbf{X}^T\mathbf{X} + \lambda\mathbf{I})^{-1}\mathbf{X}^T\mathbf{Y} \tag{6.9}$$

$\lambda = 0$ provides the ordinary least square estimate $\hat{\beta}^{\text{ridge}} = \hat{\beta}$ and $\lambda = \infty$ shrinks the coefficients to $\hat{\beta}^{\text{ridge}} = 0$. Note that a nonzero tuning parameter λ can make $(\mathbf{X}^T\mathbf{X} + \lambda\mathbf{I})^{-1}$ invertible, even when $\mathbf{X}^T\mathbf{X}$ is singular. Because each value of λ can produce a different solution for the parameter, the tuning parameter is usually selected based on cross-validation.

Ridge regression estimator is biased [3] because

$$\mathbb{E}(\hat{\beta}^{\text{ridge}}) = \mathbb{E}[(\mathbf{I} + \lambda(\mathbf{X}^T\mathbf{X})^{-1})\hat{\beta}] \tag{6.10}$$

$$= (\mathbf{I} + \lambda(\mathbf{X}^T\mathbf{X})^{-1})\beta \tag{6.11}$$

$$\neq \beta \quad (\text{if } \lambda \neq 0) \tag{6.12}$$

Lasso: In *Least Absolute Selection and Shrinkage Operator* (*Lasso*), the penalty term consists of the l_1 norm of the parameter vector.

$$\hat{\beta}^{lasso} = \arg\min_{\beta}(\mathbf{Y} - \mathbf{X}\beta)^T(\mathbf{Y} - \mathbf{X}\beta) + \lambda\sum_{i=1}^{M}|\beta(i)| \qquad (6.13)$$

$$= \arg\min_{\beta}\|\mathbf{Y} - \mathbf{X}\beta\|_2^2 + \lambda\|\beta\|_1 \qquad (6.14)$$

Similar to Ridge regression, $\hat{\beta}^{lasso} = \hat{\beta}$ when $\lambda = 0$ and $\hat{\beta}^{lasso} = 0$ when $\lambda = \infty$. However, Ridge regression continues shrinking of all parameters with an increase of λ and cannot be utilized for variable selection. Whereas the $|l|_1$ norm as penalty in *Lasso* enables variable selection by converting more coefficients of β to zero and shrinking the remaining nonzero coefficients with increase of λ. The bias increases and variance decreases with increase in λ [4].

Lasso estimator $\hat{\beta}^{lasso}$ has no closed form solution similar to *ridge regression* and can be solved using quadratic programming techniques. Some packages to solve **lasso** include

- *lars* package in **R**: Least Angle Regression, Lasso, and Forward Stagewise [5]
- *lasso* function in Matlab Statistics and Machine Learning Toolbox
- *L1Packv2* package in Mathematica [6]

Elastic Net regularization: Elastic Net regularization attempts to overcome the limitations of LASSO [7], such as (a) when n is smaller than M, LASSO selects at most n features before it saturates, (b) LASSO usually selects one variable from a group of highly correlated variables, ignoring the others. Elastic Net combines the penalties of ridge regression and LASSO to arrive at

$$\hat{\beta}^{EN} = \arg\min_{\beta}\|\mathbf{Y} - \mathbf{X}\beta\|_2^2 + \lambda_1\|\beta\|_1 + \lambda_2\|\beta\|_2^2 \qquad (6.15)$$

Note that $\lambda_1 = 0, \lambda_2 = \lambda$ results in $\hat{\beta}^{EN} = \hat{\beta}^{ridge}$ and $\lambda_1 = \lambda, \lambda_2 = 0$ produces $\hat{\beta}^{EN} = \hat{\beta}^{lasso}$.

It has been shown that the Elastic Net regression can be formulated as a support vector (SV) machine problem and resources applied to SV machine solutions can be applied to Elastic Net as well [8]. Elastic Net regression can also be solved using Matlab *lasso* function and *elasticnet* package in R [9].

Similar to ordinary least square regression, regularized least square regression is easily interpretable in terms of selecting the significant features for the specific problem and handles challenge C_4. Due to penalization of model complexity, regularized regression is able to handle challenge C_1 of large number of features. Regularized regression has robustness properties that can handle C_3. However, the linear combination of selected features limits the accuracy of the model (C_2).

6.2.1.3 Logistic regression

Logistic regression refers to a regression model where the response or dependent variable is categorical [10,11]. The model estimates the probability of the categorical dependent variable based on the predictor variables, using logistic function $(g(t) = \frac{1}{1+e^{-t}})$. To explain logistic regression, consider a binary outcome Y and we want to model $Pr(Y = 1|X = x)$. The logistic regression model based on logit transformation (inverse of sigmoidal or logistic function) of the probability (range is now unbounded) and equating it to linear combination of predictors will be

$$\ln \frac{p(x)}{1 - p(x)} = \beta_0 + x\beta \tag{6.16}$$

and $p(x)$ can be computed as

$$p(x) = \frac{1}{1 + e^{-(\beta_0 + x\beta)}} \tag{6.17}$$

6.2.1.4 Principal component regression

As the name suggests, principal component regression (PCR) is based on principal component analysis. We calculate the principal components $\mathbf{Z}$ of the dependent matrix $\mathbf{X}$ and using ordinary least squares, fit the regression of $\mathbf{Y}$ on $\mathbf{Z}$. The regression coefficients in the transformed space are mapped back to the original space to arrive at the PCR coefficients. Thus the main difference with ordinary least squares is that the regression coefficients are computed based on the principle components of the data. When a subset of the principal components is used for generating the regression coefficients, it acts similar to a regularization process. PCR can also overcome the problem of multicollinearity in ordinary least squares [12].

Partial least squares (PLS) regression: Geladi and Kowalski [13] have similarities to PCR, but projects both $\mathbf{X}$ and $\mathbf{Y}$ to new spaces before fitting a regression model. The idea is that the model might have few underlying factors that are important and using these latent factors in prediction can avoid overfitting [14].

Some packages with *PCR* and *PLS Regression* functionality include

- *pls* package in **R**: PLS and PCR [15]
- *plsregress* and *pca* functions in Matlab Statistics and Machine Learning Toolbox

PCR and PLS regression utilize derived input directions and select a subset of the derived directions for modeling. PCR keeps the high variance directions and discards the low variance ones, whereas PLS also tends to shrink the low variance directions, but can inflate the high variance regions [16]. Both PCR and PLS have similarities with Ridge regression that shrinks in all directions and more so in low variance directions. In terms of predictive performance, Ridge regression is usually preferable to PLS and PCR [17].

6.2.2 KERNEL METHODS

Various kernel-based approaches are popular for regression modeling. Next, we will discuss the most common approach of SV regression. Other forms of kernel-based approach include Bayesian Kernel-Based Multi-Task Learning [18] which was the best performer in NCI-DREAM drug sensitivity prediction challenge.

6.2.2.1 Support vector regression

SV regression is based on the statistical learning theory, or *Vapnik-Chervonenkis* (VC) theory developed by Vladimir Vapnik and Alexey Chervonenkis during the 1960s [19–21] that attempts to provide the statistical underpinnings of learning algorithms and their generalization abilities. Originating from research at Bell Labs [22], the development of SV machines had a strong application component part to it and the initial application consisted of SV classifiers for optical character recognition. SV classifiers and subsequently SV regression showed high accuracy in comparison to existing approaches and became a commonly used tool in classification and regression modeling [22–25].

Let $x_{tr}(i,j)$ and $y(i)$ $(i = 1, \ldots, n; j = 1, \ldots, M)$ denote the training predictor features and output response samples, respectively. In terms of SV regression, our goal is to find a function $f(x_{tr}(i, 1 : M))$ that has at most deviation ϵ from actual values $y(i)$ [22] and a characteristic of the function is minimized (we will discuss those characteristics in next paragraph). Thus we are restricting the deviation of the function approximator f from the actual values while minimizing some characteristic of the approximator. For explanation purposes, we will discuss the problem based on a linear function:

$$f(x_{tr}(i, 1 : M)) = \sum_{j=1}^{M} w(j) * x_{tr}(i,j) + b \tag{6.18}$$

Here the characteristic will be the norm of w. The problem can be posed as an optimization problem as follows:

$$\text{minimize } \frac{1}{2}\|w\|^2$$

$$\text{subject to } \left| y_i - \sum_{j=1}^{M} w(j)x_{tr}(i,j) - b \right| \le \epsilon \tag{6.19}$$

To consider cases where the above solution may not be feasible, we can relax the constraints as follows:

$$\text{minimize } \frac{1}{2}\|w\|^2 + C\sum_{i=1}^{N}(\delta_i^r + \delta_i^l) \tag{6.20}$$

$$\text{subject to } \begin{cases} y_i - \sum_{j=1}^{M} w(j)x_{tr}(i,j) - b & \le \epsilon + \delta_i^r \\ \sum_{j=1}^{M} w(j) * x_{tr}(i,j) + b - y_i & \le \epsilon + \delta_i^l \\ \delta_i^r, \delta_i^l & \ge 0 \end{cases} \tag{6.21}$$

where $C > 0$ provides a trade-off between deviation from ϵ and minimization of weight vector.

The above formulation can be converted to a dual problem using Lagrange multipliers [26]:

$$
\text{maximize} \quad \begin{cases} -\frac{1}{2} \sum_{i=1}^{N} \sum_{j=1}^{M} (\alpha_i^r - \alpha_i^l)(\alpha_j^r - \alpha_j^l) \sum_{k=1}^{M} x_{\text{tr}}(i,k) x_{\text{tr}}(j,k) \\ -\epsilon \sum_{i=1}^{N} (\delta_i^r + \delta_i^l) + \sum_{i=1}^{N} y(i)(\delta_i^r - \delta_i^l) \end{cases} \tag{6.22}
$$

$$
\text{subject to} \quad \sum_{i=1}^{N} (\delta_i^r - \delta_i^l) = 0 \text{ and } \delta_i^r, \delta_i^l \in [0, C] \tag{6.23}
$$

where δ_i^r, δ_i^l are Lagrange multipliers and the function approximator and weight vector are given by:

$$
f(x) = \sum_{i=1}^{N} (\delta_i^r - \delta_i^l) \langle x, x_{\text{tr}}(i) \rangle + b \tag{6.24}
$$

$$
w = \sum_{i=1}^{N} (\delta_i^r - \delta_i^l) x_{\text{tr}}(i, :) \tag{6.25}
$$

Eq. (6.24) shows that SV solution can be computed using dot product between inputs.

To extend the SV algorithm to nonlinear scenario, we will consider the use of Kernels. To explain a Kernel, let us look at an example provided by Vapnik [27]: Let us consider the map $\Phi \colon \mathbb{R}^2 \to \mathbb{R}^3$ with $\Phi(x_1, x_2) = (x_1^2, \sqrt{2}x_1x_2, x_2^2)$. Training a linear SV machine on the mapped features will result in a quadratic function. However, continuing in this fashion may become computationally intractable when the number of input features and polynomial degree increases. For instance, if we consider 20 features and a polynomial term of order 4, we will have $\binom{20+4-1}{4} = 8855$ monomial features in the transformed space. However, the observation that

$$
\langle (x_1^2, \sqrt{2}x_1x_2, x_2^2), (x_1'^2, \sqrt{2}x_1'x_2', x_2'^2) \rangle = \langle x, x' \rangle^2 \tag{6.26}
$$

where $\langle \cdot \rangle$ denotes dot product and because SV depends on dot products between patterns x_i, it suffices us to know $k(x, x') = \langle \Phi(x), \Phi(x') \rangle$ rather than $\Phi(x)$ explicitly.

The next question is then figuring out kernel functions $k(x, x)$ that correspond to dot product in some feature space. Some commonly examples of kernels are [22]:

- Homogenous polynomial kernels k with $p \in \mathbb{N}$ and

$$
k(x, x') = \langle x, x' \rangle^p \tag{6.27}
$$

- Inhomogenous polynomial kernels k with $p \in \mathbb{N}$, $c > 0$ and

$$
k(x, x') = (\langle x, x' \rangle + c)^p \tag{6.28}
$$

- Radial Basis Kernels

$$k(x, x') = e^{-\frac{\|x-x'\|^2}{2\sigma^2}} \qquad (6.29)$$

SV machines can handle nonlinearity based on the use of kernels and tend to have better predictive performance compared to regularized regression approaches [28]. However, the direct biological interpretability of the generated model is limited. SV machines do strike a balance on generalization error performance and can handle various noise models. For further details on SV regression models, readers are referred to a tutorial on SVMs by Smola and Schölkopf [22].

Multiple implementations are currently available to infer an SV regression model for application purposes. Some commonly available software tools to design an SV regression model are listed next (Note that the list is a small fraction of the available tools and is far from being comprehensive).

- *Matlab Support Vector Machine Regression*: Matlab provides a set of functions to train an SV regression model. They are included in *Statistics and Machine Learning Toolbox → Regression → Support Vector Machine Regression*. *Fitrsvm* function supports Linear, Gaussian, RBF, and Polynomial Kernel functions. The optimization can be solved using one of three algorithms: (a) Sequential Minimal Optimization [29]; (b) Iterative Single Data Algorithm [30]; and (c) L1 soft-margin minimization by quadratic programming [31].
- *IBM ILOG CPLEX Optimization Studio*: The CPLEX studio provides state-of-the-art linear and quadratic programming solver that can be used to solve for the optimization problem in SV regression.
- *Support Vector Machines in* **R**: *kernlab* [32] contains multiple kernel-based algorithms in **R** including SV implementations.
- *libsvm* implementation in **C++** [33] https://www.csie.ntu.edu.tw/~cjlin/libsvm/.

6.2.3 ENSEMBLE METHODS

This section will discuss various methods where multiple models are combined to arrive at the final prediction. The combination can consist of picking one model out of multiple regression models based on cross-validation error or a function of the individual regression predictions.

6.2.3.1 Boosting

Boosting attempts to arrive at a strong learning algorithm by combining an ensemble of weak learners. The most popular form, *adaboost* (adaptive boosting), was developed to combine weak classifiers to arrive at a high accuracy classifier [34], where later weak classifiers are adaptively designed to correctly classify harder to classify samples.

Adaboost functionality is available in multiple software packages such as *adabag* package in R [35] and *fitensemble* functions in Matlab Statistics and Machine Learning Toolbox.

6.2.3.2 Bagging

Bootstrap aggregating (or *bagging*) was introduced [36] to improve classification accuracy by combining classifiers trained on randomly generated training sets. As the name suggests, the training sets are generated using *Bootstrap* approach, or in other words, generating sets of n samples using samples with replacement from the original training set. A specified number of such training sets are generated and the response of models trained on these training sets are then averaged to arrive at the bagging response.

Bagging functionality is available in multiple software packages, such as *adabag* package in R [37] and *fitensemble* function in Matlab Statistics and Machine Learning Toolbox.

6.2.3.3 Stacking

Stacked generalization (*Stacking*) [38,39] is based on training a learning algorithm on individual model predictions. One common approach is to use a linear regression model to combine the individual model predictions [40].

6.2.3.4 Random Forests

Random Forest Regression [41] has become a commonly used tool in multiple prediction scenarios [42–50] due to their high accuracy and ability to handle large features with small samples. Random Forest [41] combines the two concepts of *Bagging* and *Random Selection of Features* [51–53] by generating a set of T regression trees where the training set for each tree is selected using Bootstrap sampling from the original sample set and the features considered for partitioning at each node is a random subset of the original set of features. Regression tree is a form of nonlinear regression model where samples are partitioned at each node of a binary tree based on the value of one selected input feature [54]. The bootstrap sampling for each regression tree generation and the random selection of features considered for partitioning at each node reduces the correlation between the generated regression trees and thus averaging their prediction responses is expected to reduce the variance of the error. We will discuss Random Forests in greater detail in the next chapter.

Random Forests tend to have high accuracy prediction (challenge C_2) and can handle large numbers of features (C_1) due to the embedded feature selection in the model generation process. Note that when the number of features is large, it is preferable to use a higher number of regression trees. Random Forests are sufficiently robust to noise (C_3), but the biological interpretability of Random Forests is limited (C_4).

6.2.4 DYNAMICAL MODELS

The previously described model types are primarily suited for steady-state modeling to map input features to output responses. To reflect the time dependencies, we utilize dynamical models to model the behavior of genetic interactions and their time-dependent behavior. A significant number of studies have been conducted for dynamical modeling of genetic regulatory networks (GRNs) in the last two decades, but their application to drug sensitivity modeling has remained limited due to (a) enormous data requirements for model parameter estimation. Most current sensitivity studies are conducted for steady-state responses and there are limited time series drug perturbation data available. Since GRNs are more complex than ordinary regression models, smaller samples can cause overfitting and bring in more uncertainties in their prediction power. (b) Due to their complexity, GRN models are usually suited for modeling fewer genes rather than a large number of genes and proteins as considered in drug sensitivity studies. Recently, researchers have tried to solve this problem by combining various types of GRN models, where each represents a subnetwork or process in the cell. A whole cell computational model of a cell has been generated in [55] by integrating mathematical models of 28 separate modules.

In this section, we will briefly review the model types considered in modeling GRNs. We will also touch upon the limited number of studies where dynamical models have been applied for drug sensitivity prediction. We expect that in the future, advances in technology will enable fast measurement of time series drug perturbation data, which will in turn, facilitate the use of dynamical models in drug sensitivity predictions.

GRNs represent the interconnections between genomic entities that govern the regulation of gene expression. Since biological regulatory networks are extremely detailed with numerous interactions, it is difficult to arrive at a single model representing the whole biological regulatory system. Depending on the purpose of modeling, the mathematical model representing the GRN brings in a level of abstraction. The focus of the modeling can be capturing interactions between RNA expressions, protein-protein interactions, or metabolite interaction networks. Usually, only parts of the regulome (genes, proteins, and metabolites involved in gene regulation), such as transcription factors, enhancers, microRNA, etc., are made explicit in a mathematical model of a GRN. To provide further background on GRN modeling, we present broad classifications of the models that are commonly used to capture the relationships between genomic entities. GRN models usually belong to either static or dynamical models. Static models are concerned with providing the topology of the network (such as the connectivity of various genomic entities) without attempting to model the dynamics of the interactions. Dynamic models, on the other hand, not only capture the connectivity of the network entities, but also provide the time-dependent functional relationships that govern the interactions. Commonly used pathway models as presented in databases such as KEGG [56] and StringDB [57] belong to the category of static models, as they primarily capture the connectivity structure without the network dynamics. Examples of dynamic models include ordinary differential equation (DE) models to model the average concentration of mRNA expressions [58] or Dynamic Bayesian Network [59] models. Based on the model complexity involving inference of connectivity and time-dependent relationships, the experimental data requirements for elucidating dynamic models are significantly higher as compared to static models.

To further explore various levels of time-dependent behavior of genes or proteins, dynamic network models can be essentially categorized as discrete where the amount of genomic entity is modeled as discrete values, or continuous where the genomic entities are models as continuous variables. Since the dynamics of biological entities consist of discrete changes based on the number of molecules created or destroyed, a fine-scale discrete model can potentially faithfully capture the dynamics of the genomic entities. Note that we have categorized the dynamic models as discrete or continuous based on the representation of the genomic entity, but a similar categorization can be applied based on discrete or continuous representation of time. Some models, such as stochastic master equations, incorporate time as a continuous quantity, whereas models like Boolean Networks (BNs) update states at discrete time intervals. The next level of characterizations is based on whether the models considered are stochastic or deterministic. A number of studies have shown that the generation of mRNA or protein expression is stochastic in nature [60–67] and thus a stochastic model is suitable to capture the true change in state probability distribution with time. Thus stochastic and discrete fine-scale models commonly known as *stochastic master equation* models have been proposed for modeling GRNs [60,61,68].

However, available experimental data may limit the inference of the parameters of a stochastic model. For instance, common genomic characterizations using microarrays provide an average observation of the mRNA expressions and are not suitable for inferring the parameters of a detailed stochastic model.

Thus deterministic models, such as DE models, are often used to represent the average behavior of the biological system [69], or in other words, modeling moments of the distribution rather than the detailed distribution. *Nonlinear ordinary differential equations* [70–73] and *piecewise linear differential equations* [74–76] have been proposed as continuous fine-scale deterministic models for GRNs. DE models assume that species concentration vary continuously and deterministically.

Dynamic models can be further classified as detailed fine-scale models or coarse-scale models where the detail refers to the level of discretization involved in modeling the gene or protein expression. For instance, our interest may be in modeling the up- or downregulation of genes and thus a coarse-scale BN [77,78] with ON and OFF states for each gene will be adequate. However, if our interest is in capturing the individual mRNA molecule generation or degradation, a detailed fine-scale model such as stochastic master equations [68,79] with an enormous state space will be required.

A coarse-scale deterministic model for GRNs has been proposed in the form of the *BN* model [74,77,78]. The BN model has yielded insights into the overall behavior of large genetic networks and can be used to model many biologically meaningful phenomena [80–87]. In the Boolean model, the assumption of a single transition rule for each gene can be problematic with respect to inference. The data are typically noisy, the number of samples is small relative to the number of parameters to be estimated, there can be unobserved (latent) variables external to the model network, and there is fundamental stochasticity of the genes in their mechanisms of action. Owing to these considerations, the *Probabilistic Boolean Network* (PBN) [88,89] was proposed as a coarse-scale stochastic model for GRNs. The probabilistic structure of the PBN can be modeled as a Markov chain. Relative to their Markovian structure, PBNs are related to Bayesian networks [59,90,91]; specifically, the transition probability structure of a PBN can be represented as a dynamic Bayesian network and every dynamic Bayesian network can be represented as a PBN (actually, a class of PBNs) [92]. The inverse problem of identifying the model from the data has been studied extensively for DE models [93–97], Dynamic Bayesian Network Models [90,91,98–100], BN models [101–106], and PBN models [88,107–113].

The above categorizations are presented in a graphical form in Fig. 6.1.

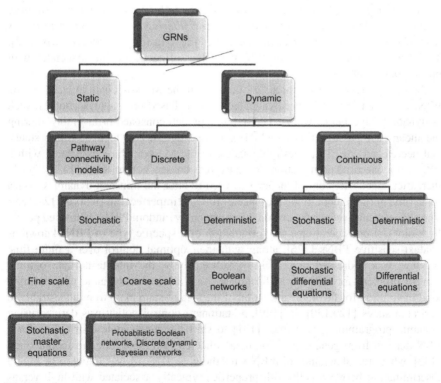

FIG. 6.1

Categorization of commonly used models for GRN modeling.

6.2.4.1 GRN intervention

In systems biology, control theory concepts have been mainly applied to better analyze and understand the biological system [58,114–118]. Most of the drug intervention studies related to translational genomics have focused on coarse-scale BN and PBN models, except for a few studies [119,120] using detailed ordinary and stochastic DE models. The problems with designing and analyzing intervention strategies in fine-scale stochastic models are twofold: (a) there is huge computational complexity involved in simulation of the fine-scale models and evaluation of intervention strategies and (b) extremely large data requirements for inference of the parameters of the fine-scale models. Consequently, a reasonable approach is to construct intervention strategies for coarse-scale models with the assumption of capturing the overall effects of intervention manifested at the phenotypic (observational) level. We next provide a brief description of the several intervention approaches that have been designed to date for PBNs.

The motivation behind application of control theory for Markovian models like PBN is to devise optimal policies for manipulating control variables that affect the transition probabilities of the network and can, therefore, be used to desirably affect its dynamic evolution. In practice, intervention is expected to be achieved by (a) *targeted small molecule kinase inhibitors (Imatinib, Gefitinib, Erlotinib, Sunitinib, etc.* [121]); (b) *Monoclonal antibodies altering the protein concentrations (Cetuximab, Alemtuzumab, Trastuzumab, etc.* [121]); or (c) *gene knockdowns* [122]. The state desirability is determined by the values of genes/proteins associated with phenotypes of interest.

Since the intervention approaches depend on the Markov chain induced by the PBN, it is extendable to other dynamical systems based on graphical models, such as dynamic Bayesian networks. The initial approach consisted of favorably altering the mean first passage times (MFPTs), increasing MFPTs to undesirable states, and decreasing MFPTs to desirable states, in the Markov chain associated with a PBN via a one-time perturbation of the expression of a control gene [123]. Shortly thereafter, stochastic control theory was used to alter the transient dynamics over a finite time horizon, both in the presence of full and imperfect information [124,125]. The above-mentioned works treated instantaneously random PBNs. Motivated purely by biological considerations, the formulation of a specific type of PBNs known as context-sensitive PBNs [126], along with their optimal control over a finite time window, was addressed in [127,128]. Subsequently, the infinite horizon optimal control problem for GRNs was formulated that allowed a dynamic programming kind of solution for favorable alteration of the steady-state mass from undesirable to desirable states [129,130]. In [129], a stationary control policy was derived using dynamic programming principles [131] to shift the steady-state distribution for a PBN derived from gene expression data collected in a metastatic melanoma study [132], where the abundance of mRNA for the gene *WNT5A* was found to be highly discriminating between cells with properties typically associated with high versus low metastatic competence, a relationship that suggests an intervention scheme to reduce the *WNT5A* gene's action. A seven-gene network containing the genes *WNT5A*, *pirin*, *S100P*, *RET1*, *MART1*, *HADHB*, and *STC2* was considered and the control objective was downregulating the *WNT5A* gene using another gene in the network as the control. Fig. 6.2 shows the steady-state distribution obtained using the proposed stationary policy with pirin as the control; both the controlled (Fig. 6.2A) and original steady-state distributions (Fig. 6.2B) are shown. States 0–63 have downregulated *WNT5A* (desirable states) and the states 64–127 have upregulated *WNT5A* (undesirable states). The steady-state distributions in Fig. 6.2 corroborate the fact that the intervention policy has shifted the probability mass to states with lower metastatic competence.

Further research in this area has considered robustness of the intervention [133] in terms of worst case [134,135] and Bayesian scenario [136,137]. Researchers have also considered constraining the number of treatments [138], intervention based on mean-first-passage-time [139], long-run behavior-based intervention [140], intervention for a family of Markovian networks [141,142], and maximal phenotype alteration [143]. Recent approaches have also considered mapping static pathways to dynamical BNs for analyzing intervention methodologies [144–146]. The effect of application of intervention designed using reduced order models on fine-scale models was analyzed in [79,147].

Dynamical regulatory networks have large data requirements and can handle few genes in the models and thus cannot tackle challenge C_1. Furthermore, GRNs are sensitive to noises and latent variables (C_3). However, the biological interpretability of GRNs is excellent and they can provide numerous insights on the relationship between genomic entities and drug response (C_4). Various types of GRN models have different levels of accuracy that are highly dependent on the initial selection of relevant features.

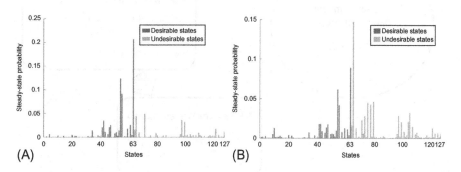

FIG. 6.2

(A) Steady-state distribution after intervention. (B) Original steady state [129].

6.3 APPLICATIONS

The predictive modeling tools described in this chapter provide an idea of the commonly used tools in regression modeling. There are numerous other predictive modeling approaches that have been developed (sometimes specific to application scenarios) and it is beyond the scope of this book to review all such possibilities. For application purposes, there is usually no single tool that always works best and often the associated steps of data preprocessing, feature selection, and error estimation play significant roles along with model selection in the final model performance. The various steps involved in predictive model generation can be broadly categorized as (A) Data Preprocessing, (B) Feature Selection, (C) Model Selection, and (D) Performance Evaluation, as shown in Fig. 6.3.

Step A involves processing the genetic and epigenetic data to numerical or categorical form that can be utilized to train the predictive model. These steps have been described in Chapter 2. A specific form of model may require a particular form of data processing; for instance, a logistic regression model will require the output response to be categorical. Normalization can allow use of data collected from different platforms to be used in the model training. Step B involves feature selection, which has been discussed in detail in the chapter on *Feature Selection*. This step

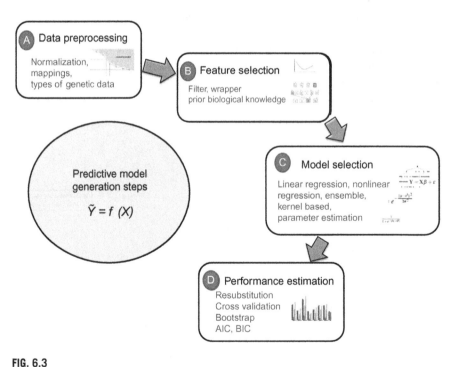

FIG. 6.3

Steps involved in predictive model generation.

may require the use of the specific model in step C if wrapper or embedded form of feature selection is used. Often, prior biological knowledge can provide information on selecting the important features. Step C involves the selection and training of the specific predictive model. Data type can sometimes guide the selection of the model class, such as when we are certain that the output response is linearly related to the input features, a linear regression form of model will likely be a good choice. Note that a complex model-to-model data that can be represented by a linear relationship might result in overfitting and will be going against the idea of *Occam's Razor*. Step D involves performance evaluation which was discussed in detail in Chapter 4. For performance evaluation, we select a performance measure, such as mean square error and the approach, such as cross-validation to measure it based on the available samples. If we are planning to select a model from competing models in part C, step D becomes even more crucial. The error estimation techniques are not guaranteed to generalize well over the real population distribution and the estimated errors may be significantly different from the true error. The final performance of the generated model is dependent on multiple factors in steps A–D, and care should be taken to select the appropriate method in each step. Studies of applications of different models on the same dataset have shown that not only the model type, but the parameters and associated steps involved in the model generation play a significant role in the accuracy of the final model. We will illustrate this behavior next based on the NCI-DREAM drug sensitivity challenge where multiple models were applied to the same drug sensitivity data and the performance evaluated.

6.3.1 NCI-DREAM CHALLENGE: COMMUNITY ASSESSMENT OF DRUG SENSITIVITY PREDICTION ALGORITHMS

DREAM challenges are organized in an attempt to solve systems biology problems based on community efforts working on a problem with access to same dataset. One of the challenges that was highly relevant to the topics covered in this book was NCI-DREAM Drug Sensitivity prediction subchallenge 1 in 2012 [148]. Genomic characterizations were provided for 53 cell lines (48 breast cancer cell lines and 5 nonmalignant breast cell lines) that were exposed to 28 therapeutic compounds at a concentration required to inhibit proliferation by 50% after 72 h (GI50) and responses to these 28 drugs were provided for 35 of these 53 cell lines [42,43]. The drug response of the remaining 18 cell lines was considered later to assess the performance of submissions. Multiple types of genomic and epigenomic data (copy number variation, methylation, gene expression through microarray, RNA sequencing, exome sequencing, and protein abundance) were generated before exposure of the cells to the drugs for each of the 53 cell lines. The dataset is represented pictorially in Fig. 6.4.

Challenge participants were asked to utilize the GI50 and genomic characterization data provided for 35 cell lines to train predictive models and predict the ranks of the remaining 18 cell line sensitivities based on the genomic characterizations of the cell lines. Forty-four teams participated in the challenge with various predictive

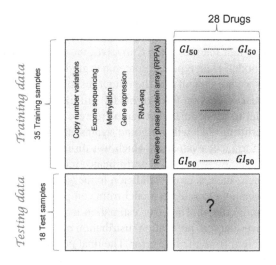

FIG. 6.4

NCI-DREAM challenge dataset.

models being implemented on the dataset. The details of the models used by the 44 submissions are available at [42]. The submissions were characterized into six groups [42]:

- **Kernel-based techniques** (four submissions) including Bayesian Multitask Multiple Kernel Learning and Support Vector Regression with Radial Basis. There were three submissions based on SV Regression that differed on feature selection approaches, such as use of correlation coefficients, bidirectional search, or separate normalizations on each genomic characterization dataset. The best performing model in this group utilized Bayesian Multiple Kernel Learning technique [18].
- **Nonlinear regression techniques** (11 submissions) including 8 based on Random Forests, 2 based on gradient boosting on regression trees, and the final 1 using Spearman's correlation. The Random Forest-based submissions differed on feature selection approaches and techniques for missing value imputation, along with the manner of utilization of the various genomic characterizations to design the composite score from individual models. For the best performing model in this group, Random Forests were designed for each genomic characterization and the combined model was generated based on a linear regression model over the individual model predictions [43].
- **Sparse linear regression techniques** (14 submissions) including *Lasso Regression* (4 submissions), *Elastic Net* (4 submissions), Linear Regression (1 submission), *spike and slab regression* (3 submissions), *regression with log-penalty* (1 submission), and *Ridge regression* (1 submission). For the best performing model in this group, features were simultaneously selected and a ranking model built for each drug by *lasso* regression.

- **Partial least square or PCR-based techniques** (four submissions): The differences were in the selection of features using correlation, principal components, or *lasso*. The top performing approach in this group considered removal of low variance features before applying correlation to drug response to select features, and designed multiple PLS models that were combined for final prediction.
- **Ensemble or model selection approaches** (five submissions) considered model selection in three cases and aggregation in the remaining two cases. The model selection was based on cross-validation error from different models, such as Lasso Regression, Ridge Regression, SV Regression, and Random Forest. The top performing approach in this group considered principal component analysis and regularization for dimensionality reduction and generated multiple regression models, and selected the top performing cross-validated model for final predictions.
- **Other approaches** (six submissions) consisted of techniques that do not fall in any of the five broad categories described earlier. These techniques considered prediction based on weighted correlation, hierarchical clustering, nearest neighbors, cox regression, and integrated voting.

Among the 44 submissions, the top 2 performers were based on Bayesian Kernel Multitask learning [18,42] and Random Forest Regression, combined with stacking [42,43]. The 44 submissions showed a diverse set of predictive tools being implemented on the same dataset, along with different ways to implement feature selection and model integration. Quite often, the preprocessing and feature selection of the data, rather than the specific predictive modeling technique, can have significant effect on predictive accuracy. For this particular challenge, an important issue was the limited number of samples (35 training samples) used to train the model. Small samples can result in significant overfitting, considering that the number of available input features was in the thousands. A potential approach to avoid the overfitting is to use feature selection or feature extraction to reduce the dimensionality of the input space. The majority of the teams used a filter-based approach (such as correlation with output response) to feature selection. For limited samples, filter-based approach can work better in terms of generalization error, as compared to wrapper feature selection approach, as incorporating the model in feature selection can accentuate model overfitting. Some of the teams in the NCI-DREAM challenge also incorporated prior biological knowledge from pathways for feature selection. Pathway information can be considered as additional information in this training problem with limited samples. If drug targets are known, the targets, along with their closely associated genes, can be used as input features for predictive model training. Some other key observations from this challenge submission include:

(a) Nonlinear models performed better than linear sparse regression models. This is not surprising when we consider the fact that the majority of biological interactions are nonlinear and drug sensitivity response is likely to be a nonlinear function of multiple features.

(b) Multiple genomic characterizations can carry complementary information and integrating them can improve prediction performance [149]. The models integrating multiple forms of genomic characterizations could improve their predictive performance, as compared to models based on single genomic characterization. The integration can be achieved by designing individual models for each dataset and combining the model predictions or using an integrated set of features from all data characterizations to train a model. Gene expression and RPPA appeared to be the highly predictive data sources. In terms of complementary behavior, methylation with gene expression data, or exome sequencing with gene expression data provided improved performance by providing complementary information. Considering the limited number of samples, it is hard to agree on which genomic or epigenomic characterization datasets are significantly better than others for drug sensitivity prediction.

(c) Combining prediction from multiple model submissions improved performance, which is expected to hold when the model predictions have complementary information. We can intuitively explain this behavior by observing that the various methodologies differed in their feature selection techniques, approaches to heterogeneous data type integration, type of model used, and model combination resulting in a diverse set of predictions with limited correlations in their differences from true sensitivities. The idea is similar to ensemble learning where multiple weak learners are combined to form a strong learner and can be termed *wisdom of crowds approach* [150], where robustness of the prediction is improved by combining predictions from multiple methodologies.

REFERENCES

[1] D.G. Kleinbaum, L.L. Kupper, K.E. Muller (Eds.), Applied Regression Analysis and Other Multivariable Methods, PWS Publishing Co., Boston, MA, 1988.

[2] A.N. Tikhonov, V.Y. Arsenin, Solutions of Ill-Posed Problems, V.H. Winston & Sons, Washington, DC; John Wiley & Sons, New York, 1977.

[3] G. James, D. Witten, T. Hastie, R. Tibshirani, An Introduction to Statistical Learning: With Applications in R, Springer Publishing Company, Inc., New York, NY, USA, 2014.

[4] R. Tibshirani, Regression shrinkage and selection via the Lasso, J. R. Stat. Soc. Ser. B 58 (1994) 267–288.

[5] T. Hastie, B. Efron, lars: Least Angle Regression, Lasso and Forward Stagewise, R package version 1.2, 2013.

[6] I. Loris, L1Packv2: a Mathematica package for minimizing an -penalized functional, Comput. Phys. Commun. 179 (12) (2008) 895–902.

[7] H. Zou, T. Hastie, Regularization and variable selection via the Elastic Net, J. R. Stat. Soc. Ser. B 67 (2005) 301–320.

[8] Q. Zhou, W. Chen, S. Song, J. Gardner, K. Weinberger, Y. Chen, A reduction of the elastic net to support vector machines with an application to GPU computing, in: AAAI Conference on Artificial Intelligence, 2015.

[9] H. Zou, T. Hastie, elasticnet: Elastic-Net for Sparse Estimation and Sparse PCA, R package version 1.1, 2012.

[10] S.H. Walker, D.B. Duncan, Estimation of the probability of an event as a function of several independent variables, Biometrika 54 (1/2) (1967) 167–179.

[11] D. Freedman, Statistical Models: Theory and Practice, Cambridge University Press, New York, NY, USA, 2005.

[12] T. Ns, B.-H. Mevik, Understanding the collinearity problem in regression and discriminant analysis, J. Chemometrics 15 (4) (2001) 413–426.

[13] P. Geladi, B.R. Kowalski, Partial least-squares regression: a tutorial, Anal. Chim. Acta 185 (1986) 1–17.

[14] T. Dijkstra, Some comments on maximum likelihood and partial least squares methods, J. Econom. 22 (1–2) (1983) 67–90.

[15] R. Wehrens, B.-H. Mevik, K.H. Liland, pls: Partial Least Squares and Principal Component Regression, R package version 2.5, 2015.

[16] T. Hastie, R. Tibshirani, J. Friedman, The Elements of Statistical Learning, Springer Series in Statistics, Springer New York Inc., New York, NY, 2001.

[17] J.H.F. Ildiko, E. Frank, A statistical view of some chemometrics regression tools, Technometrics 35 (2) (1993) 109–135.

[18] M. Gonen, A.A. Margolin, Drug susceptibility prediction against a panel of drugs using kernelized Bayesian multitask learning, Bioinformatics 30 (17) (2014) i556–i563.

[19] V. Vapnik, A. Lerner, Pattern recognition using generalized portrait method, Autom. Remote Control 24 (1963) 774–780.

[20] V. Vapnik, A. Chervonenkis, A note on one class of perceptrons, Autom. Remote Control 25 (1964) 821–837.

[21] V.N. Vapnik, Statistical Learning Theory, Wiley-Interscience, New York, NY, USA, 1998.

[22] A.J. Smola, B. Schölkopf, A tutorial on support vector regression, Stat. Comput. 14 (3) (2004) 199–222.

[23] K.-R. Mller, A.J. Smola, G. Rtsch, B. Schlkopf, J. Kohlmorgen, V. Vapnik, Predicting time series with support vector machines, in: W. Gerstner, A. Germond, M. Hasler, J.-D. Nicoud (Eds.), Artificial Neural Networks ICANN'97, Lecture Notes in Computer Science, vol. 1327, Springer, Berlin, Heidelberg, 1997, pp. 999–1004.

[24] D. Mattera, S. Haykin, Advances in kernel methods, in: Support Vector Machines for Dynamic Reconstruction of a Chaotic System, MIT Press, Cambridge, MA, 1999, pp. 211–241.

[25] N. Cristianini, J. Shawe-Taylor, An Introduction to Support Vector Machines: And Other Kernel-based Learning Methods, Cambridge University Press, New York, NY, 2000.

[26] R. Fletcher, Practical Methods of Optimization, second ed., Wiley-Interscience, New York, NY, 1987.

[27] V.N. Vapnik, The Nature of Statistical Learning Theory, Springer, New York, NY, USA, 1995.

[28] V. Cherkassky, Y. Ma, Comparison of model selection for regression, Neural Comput. 15 (7) (2003) 1691–1714.

[29] J.C. Platt, Fast training of support vector machines using sequential minimal optimization, in: Advances in Kernel Methods—Support Vector Learning, MIT Press, Cambridge, MA, USA, 1998.

[30] V. Kecman, T.-M. Huang, M. Vogt, Iterative single data algorithm for training kernel machines from huge data sets: theory and performance, in: L. Wang (Ed.), Support Vector Machines: Theory and Applications, Studies in Fuzziness and Soft Computing, vol. 177, Springer, Berlin, Heidelberg, 2005, pp. 255–274.

[31] B. Scholkopf, A.J. Smola, Learning With Kernels: Support Vector Machines, Regularization, Optimization, and Beyond, MIT Press, Cambridge, MA, 2001.

[32] A. Karatzoglou, A. Smola, K. Hornik, A. Zeileis, kernlab—an S4 package for kernel methods in R, J. Stat. Softw. 11 (9) (2004) 1–20.

[33] C.-C. Chang, C.-J. Lin, LIBSVM: a library for support vector machines, ACM Trans. Intell. Syst. Technol. 2 (2011) 27:1–27:27, software available at http://www.csie.ntu.edu.tw/~cjlin/libsvm.

[34] Y. Freund, R.E. Schapire, A decision-theoretic generalization of on-line learning and an application to boosting, J. Comput. Syst. Sci. 55 (1) (1997) 119–139.

[35] E. Alfaro, M. Gamez, N. Garcia, L. Guo, adabag: Applies Multiclass AdaBoost.M1, SAMME and Bagging, R package version 4.1, 2015.

[36] L. Breiman, Bagging predictors, Mach. Learn. 24 (2) (1996) 123–140.

[37] A. Peters, T. Hothorn, B.D. Ripley, T. Therneau, B. Atkinson, ipred: Improved Predictive Models by Indirect Classification and Bagging for Classification, Regression and Survival Problems as well as Resampling Based Estimators of Prediction Error, R package version 0.9.5, 2015.

[38] D.H. Wolpert, Stacked generalization, Neural Netw. 5 (1992) 241–259.

[39] B. Clarke, Comparing Bayes model averaging and stacking when model approximation error cannot be ignored, J. Mach. Learn. Res. 4 (2003) 683–712.

[40] L. Breiman, Stacked regressions, Mach. Learn. 24 (1) (1996) 49–64.

[41] L. Breiman, Random forests, Mach. Learn. 45 (1) (2001) 5–32.

[42] J.C. Costello, et al., A community effort to assess and improve drug sensitivity prediction algorithms, Nat. Biotechnol. (2014), doi:10.1038/nbt.2877.

[43] Q. Wan, R. Pal, An ensemble based top performing approach for NCI-DREAM drug sensitivity prediction challenge, PLoS ONE 9 (6) (2014) e101183.

[44] G. Riddick, H. Song, S. Ahn, J. Walling, D. Borges-Rivera, W. Zhang, H.A. Fine, Predicting in vitro drug sensitivity using Random Forests, Bioinformatics 27 (2) (2011) 220–224.

[45] K.L. Lunetta, L.B. Hayward, J. Segal, P. Van Eerdewegh, Screening large-scale association study data: exploiting interactions using random forests, BMC Genet. 5 (2004) 32.

[46] A. Bureau, J. Dupuis, K. Falls, K.L. Lunetta, B. Hayward, T.P. Keith, P. Van Eerdewegh, Identifying SNPs predictive of phenotype using random forests, Genet. Epidemiol. 28 (2) (2005) 171–182.

[47] R. Diaz-Uriarte, S.A. de Andres, Gene selection and classification of microarray data using random forest, BMC Bioinform. 7 (2006) 3.

[48] D. Yao, J. Yang, X. Zhan, X. Zhan, Z. Xie, A novel random forests-based feature selection method for microarray expression data analysis, Int. J. Data Min. Bioinform. 13 (1) (2015) 84–101.

[49] D.J. Yu, Y. Li, J. Hu, X. Yang, J.Y. Yang, H.B. Shen, Disulfide connectivity prediction based on modelled protein 3D structural information and random forest regression, IEEE/ACM Trans. Comput. Biol. Bioinform. 12 (3) (2015) 611–621.

[50] Y. Qi, Z. Bar-Joseph, J. Klein-Seetharaman, Evaluation of different biological data and computational classification methods for use in protein interaction prediction, Proteins 63 (3) (2006) 490–500.

[51] T.K. Ho, Random decision forests, in: Proceedings of the Third International Conference on Document Analysis and Recognition, ICDAR'95, vol. 1, 1995, p. 278.

[52] T.K. Ho, The random subspace method for constructing decision forests, IEEE Trans. Pattern Anal. Mach. Intell. 20 (8) (1998) 832–844.

[53] Y. Amit, D. Geman, Shape quantization and recognition with randomized trees, Neural Comput. 9 (7) (1997) 1545–1588.

[54] L. Breiman, J. Friedman, R. Olshen, C. Stone, Classification and Regression Trees, Wadsworth and Brooks, Monterey, CA, 1984.

[55] J.R. Karr, J.C. Sanghvi, D.N. Macklin, M.V. Gutschow, J.M. Jacobs, B. Bolival, N. Assad-Garcia, J.I. Glass, M.W. Covert, A whole-cell computational model predicts phenotype from genotype, Cell 150 (2) (2012) 389–401.

[56] M. Kanehisa, S. Goto, Y. Sato, M. Furumichi, M. Tanabe, KEGG for integration and interpretation of large-scale molecular datasets, Nucleic Acids Res. 40 (2012) D109–D114.

[57] String-db, interconnection of genes, http://string-db.org.

[58] Z. Szallasi, J. Stelling, V. Periwal, System Modeling in Cell Biology From Concepts to Nuts and Bolts, MIT Press, Cambridge, MA, 2006.

[59] K. Murphy, S. Mian, Modelling gene expression data using dynamic Bayesian networks, Technical Report, Computer Science Division, University of California, Berkeley, CA, 1999.

[60] H.H. McAdams, A. Arkin, Stochastic mechanisms in gene expression, Proc. Natl. Acad. Sci. U. S. A. 94 (1997) 814–819.

[61] A. Arkin, J. Ross, H.H. McAdams, Stochastic kinetic analysis of developmental pathway bifurcation in phage-infected *Escherichia coli* cells, Genetics 149 (1998) 1633–1648.

[62] M. Gibson, E. Mjolsness, Modeling the activity of single genes, in: Computational Methods in Molecular Biology, MIT Press, Cambridge, MA, USA, 2001.

[63] H.H. McAdams, A. Arkin, It's a noisy business! Genetic regulation at the nanomolar scale, Trends Genet. 15 (2) (1999) 65–69.

[64] G. Nicolis, I. Prigogine, Self-Organization in Nonequilibrium Systems: From Dissipative Structures to Order Through Fluctuations, Wiley, New York, NY, 1977.

[65] Z. Szallasi, Genetic network analysis in light of massively parallel biological data acquisition, in: Proceedings of Pacific Symposium in Biocomputing (PSB'99), 1999, pp. 5–16.

[66] D.T. Gillespie, Exact stochastic simulation of coupled chemical reactions, J. Phys. Chem. 81 (1977) 2340–2361.

[67] I. Golding, J. Paulsson, S.M. Zawilski, E.C. Cox, Real-time kinetics of gene activity in individual bacteria, Cell 123 (2005) 1025–1036.

[68] D.T. Gillespie, A rigorous derivation of the chemical master equation, Phys. A 188 (1992) 404–425.

[69] H. De Jong, Modeling and simulation of genetic regulatory systems: a literature review, J. Comput. Biol. 9 (2001) 67–103.

[70] J.M. Bower, H. Bolouri, Computational Modeling of Genetic and Biochemical Networks, first ed., MIT Press, Boston, MA, 2001.

[71] R. Heinrich, S. Schuster, The Regulation of Cellular Systems, Chapman and Hall, New York, NY, 1996.

[72] E.O. Voit, Computational Analysis of Biochemical Systems: A Practical Guide for Biochemists and Molecular Biologists, Cambridge University Press, Cambridge, 2000.

[73] P. Smolen, D. Baxter, J. Byrne, Mathematical modeling of gene networks, Neuron 26 (2000) 567–580.

[74] L. Glass, S.A. Kauffman, The logical analysis of continuous, nonlinear biochemical control networks, J. Theor. Biol. 39 (1973) 103–129.

[75] T. Mestl, E. Plahte, S.W. Omholt, A mathematical framework for describing and analysing gene regulatory networks, J. Theor Biol. 176 (1995) 291–300.

[76] R. Thomas, R. D'Ari, Biological Feedback, CRC Press, Boca Raton, FL, 1990.

[77] S.A. Kauffman, Metabolic stability and epigenesis in randomly constructed genetic nets, Theor. Biol. 22 (1969) 437–467.

[78] S.A. Kauffman, The Origins of Order: Self-Organization and Selection in Evolution, Oxford University Press, New York, NY, 1993.

[79] R. Pal, S. Bhattacharya, Characterizing the effect of coarse-scale PBN modeling on dynamics and intervention performance of genetic regulatory networks represented by Stochastic Master Equation models, IEEE Trans. Signal Process. 58 (2010) 3341–3351.

[80] R. Somogyi, C. Sniegoski, Modeling the complexity of gene networks: understanding multigenic and pleiotropic regulation, Complexity 1 (1996) 45–63.

[81] Z. Szallasi, S. Liang, Modeling the normal and neoplastic cell cycle with realistic Boolean genetic networks: their application for understanding carcinogenesis and assessing therapeutic strategies, Pac. Symp. Biocomput. 3 (1998) 66–76.

[82] R. Thomas, D. Thieffry, M. Kaufman, Dynamical behavior of genetic regulatory networks I. Biological role of feedback loops and practical use of the concept of the loop-characteristic state, Bull. Math. Biol. 57 (1995) 247–276.

[83] A. Wuensche, Genomic regulation modeled as a network with basins of attraction, Pac. Symp. Biocomput 3 (1998) 89–102.

[84] A. Faure, A. Naldi, C. Chaouiya, D. Theiffry, Dynamical analysis of a generic Boolean model for the control of the mammalian cell cycle, Bioinformatics 22 (2006) 124–131.

[85] J.C. Sible, J.J. Tyson, Mathematical modeling as a tool for investigating cell cycle control networks, Methods 41 (2007) 238–247.

[86] J.J. Tyson, A. Csikasz-Nagy, B. Novak, The dynamics of cell cycle regulation, Bioessays 24 (2002) 1095–1109.

[87] S. Huang, Gene expression profiling, genetic networks, and cellular states: an integrating concept for tumorigenesis and drug discovery, Mol. Med. 77 (1999) 469–480.

[88] I. Shmulevich, E.R. Dougherty, S. Kim, W. Zhang, Probabilistic Boolean networks: a rule-based uncertainty model for gene regulatory networks, Bioinformatics 18 (2002) 261–274.

[89] I. Shmulevich, E.R. Dougherty, W. Zhang, From Boolean to probabilistic Boolean networks as models of genetic regulatory networks, Proc. IEEE 90 (2002) 1778–1792.

[90] N. Friedman, M. Linial, I. Nachman, D. Pe'er, Bayesian networks to analyze expression data, in: Proceedings of the Fourth Annual International Conference on Computational Molecular Biology, 2000, pp. 127–135.

[91] A. Hartemink, D. Gifford, T. Jaakkola, R. Young, Maximum likelihood estimation of optimal scaling factors for expression array normalization, in: SPIE BIOS, 2001.

[92] H. Lahdesmaki, S. Hautaniemi, I. Shmulevich, O. Yli-Harja, Relationships between probabilistic Boolean networks and dynamic Bayesian networks as models of gene regulatory networks, Signal Process. 86 (2006) 814–834.

[93]   L. Qian, H. Wang, E. Dougherty, Inference of noisy nonlinear differential equation models for gene regulatory networks using genetic programming and Kalman filtering, IEEE Trans. Signal Process. 56 (2008) 3327–3339.

[94]   M. de Hoon, S. Imoto, S. Miyano, Inferring Gene Regulatory Networks From Time-Ordered Gene Expression Data Using Differential Equations, Springer, Berlin, 2002.

[95]   J. Goutsias, S. Kim, Stochastic transcriptional regulatory systems with time delays: a mean-field approximation, J. Comput. Biol. 13 (2006) 1049–1076.

[96]   J. Goutsias, S. Kim, A nonlinear discrete dynamical model for transcriptional regulation: construction and properties, Biophys. J. 86 (2004) 1922–1945.

[97]   A. Margolin, I. Nemenman, K. Basso, U. Klein, C. Wiggins, G. Stolovitzky, R.D. Favera, A. Califano, ARACNE: an algorithm for reconstruction of genetic networks in a mammalian cellular context, BMC Bioinform. 7 (2006).

[98]   M. Zou, S.D. Conzen, A new dynamic Bayesian network (DBN) approach for identifying gene regulatory networks from time course microarray data, Bioinformatics 21 (2005) 71–79.

[99]   D. Husmeier, Reverse engineering of genetic networks with Bayesian networks, Biochem. Soc. Trans. 31 (2003) 1516–1518.

[100] I. Lee, S.V. Date, A.T. Adai, E.M. Marcotte, A probabilistic functional network of yeast genes, Science 306 (2004) 1555–1558.

[101] T. Akutsu, S. Kuhara, O. Maruyama, S. Miyano, Identification of gene regulatory networks by strategic disruptions and gene overexpressions, in: Proceedings of the 9th Annual ACM-SIAM Symposium on Discrete Algorithms (SODA'98), 1998, pp. 695–702.

[102] T. Akutsu, S. Miyano, S. Kuhara, Identification of genetic networks from a small number of gene expression patterns under the Boolean network model, Pac. Symp. Biocomput. 4 (1999) 17–28.

[103] T. Akutsu, S. Miyano, S. Kuhara, Inferring qualitative relations in genetic networks and metabolic pathways, Bioinformatics 16 (2000) 727–743.

[104] P. D'Haeseleer, S. Liang, R. Somogyi, Genetic network inference: from co-expression clustering to reverse engineering, Bioinformatics 16 (2000) 707–726.

[105] S. Liang, S. Fuhrman, R. Somogyi, Reveal a general reverse engineering algorithm for inference of genetic network architectures, Pac. Symp. Biocomput. 3 (1998) 18–29.

[106] I. Shmulevich, A. Saarinen, O. Yli-Harja, J. Astola, Inference of genetic regulatory networks via best-fit extensions, in: W. Zhang, I. Shmulevich (Eds.), Kluwer Academic Publishers, Boston, MA, 2002, pp. 197–210.

[107] E.R. Dougherty, S. Kim, Y. Chen, Coefficient of determination in nonlinear signal processing, Signal Process. 80 (2000) 2219–2235.

[108] S. Kim, E.R. Dougherty, M.L. Bittner, Y. Chen, K. Sivakumar, P. Meltzer, J.M. Trent, General nonlinear framework for the analysis of gene interaction via multivariate expression arrays, J. Biomed. Opt. 5 (2000) 411–424.

[109] X. Zhou, X. Wang, E.R. Dougherty, Construction of genomic networks using mutual-information clustering and reversible-Jump Markov-Chain-Monte-Carlo predictor design, Signal Process. 83 (2003) 745–761.

[110] X. Zhou, X. Wang, R. Pal, I. Ivanov, M.L. Bittner, E.R. Dougherty, A Bayesian connectivity-based approach to constructing probabilistic gene regulatory networks, Bioinformatics 20 (2004) 2918–2927.

[111] R. Pal, I. Ivanov, A. Datta, M.L. Bittner, E.R. Dougherty, Generating Boolean networks with a prescribed attractor structure, Bioinformatics 21 (2005) 4021–4025.

[112] Y. Xiao, E.R. Dougherty, Optimizing consistency-based design of context-sensitive gene regulatory networks, IEEE Trans. Circ. Syst. I 53 (2006) 2431–2437.

[113] W. Zhao, E. Serpedin, E.R. Dougherty, Inferring connectivity of genetic networks using information theoretic criteria, IEEE/ACM Trans. Comput. Biol. Bioinform. 5 (2008) 262–274.

[114] D. Lauffenburger, Cell signaling pathways as control modules: complexity for simplicity?, PNAS 97 (10) (2000) 5031–5033.

[115] T.-M. Yi, Y. Huang, M.I. Simon, J. Doyle, Robust perfect adaptation in bacterial chemotaxis through integral feedback control, PNAS 97 (9) (2000) 4649–4653.

[116] C.J. Tomlin, J.D. Axelrod, Understanding biology by reverse engineering the control, Proc. Natl. Acad. Sci. U. S. A. 102 (12) (2005) 4219–4220.

[117] H. El-Samad, H. Kurata, J.C. Doyle, C.A. Gross, M. Khammash, Surviving heat shock: control strategies for robustness and performance, Proc. Natl. Acad. Sci. U. S. A. 102 (2005) 2736–2741.

[118] L. Yang, P.A. Iglesias, Positive feedback may cause the biphasic response observed in the chemoattractant-induced response of Dictyostelium cells, Syst. Control Lett. 55 (4) (2006) 329–337.

[119] A.A. Julius, A. Halasz, M.S. Sakar, H. Rubin, V. Kumar, G.J. Pappas, Stochastic modeling and control of biological systems: the lactose regulation system of *Escherichia coli*, IEEE Trans. Autom. Control 53 (2008) 51–65.

[120] A. Aswani, C. Tomlin, Topology based control of biological genetic networks, in: 47th IEEE Conference on Decision and Control, 2008, pp. 3205–3210.

[121] J.W. Clark, Molecular targeted drugs, in: Harrison's Manual of Oncology, chapter 1.10, McGraw-Hill Professional, New York, NY, USA, 2007, pp. 67–75.

[122] Roche Applied Science, Integrated solutions for gene knockdown, Biochemica (4) (2004), http://www.roche-applied-science.com/PROD_INF/BIOCHEMI/no4_04/PDF/p18.pdf.

[123] I. Shmulevich, E.R. Dougherty, W. Zhang, Gene perturbation and intervention in probabilistic Boolean networks, Bioinformatics 18 (2002) 1319–1331.

[124] A. Datta, A. Choudhary, M.L. Bittner, E.R. Dougherty, External control in Markovian genetic regulatory networks, Mach. Learn. 52 (2003) 169–191.

[125] A. Datta, A. Choudhary, M.L. Bittner, E.R. Dougherty, External control in Markovian genetic regulatory networks: the imperfect information case, Bioinformatics 20 (2004) 924–930.

[126] R. Pal, Context-sensitive probabilistic Boolean networks: steady state properties, reduction and steady state approximation, IEEE Trans. Signal Process. 58 (2010) 879–890.

[127] R. Pal, A. Datta, M.L. Bittner, E.R. Dougherty, Intervention in context-sensitive probabilistic Boolean networks, Bioinformatics 21 (2005) 1211–1218.

[128] R. Pal, A. Datta, M.L. Bittner, E.R. Dougherty, External control in a special class of probabilistic Boolean networks, in: American Control Conference, June, 2005, pp. 411–416.

[129] R. Pal, A. Datta, E.R. Dougherty, Optimal infinite horizon control for probabilistic Boolean networks, IEEE Trans. Signal Process. 54 (2006) 2375–2387.

[130] R. Pal, A. Datta, E.R. Dougherty, Optimal infinite horizon control for probabilistic Boolean networks, in: American Control Conference, 2006, pp. 668–673.

[131] D.P. Bertsekas, Dynamic Programming and Optimal Control, second ed., Athena Scientific, Nashua, NH, USA, 2001.

[132] M. Bittner, P. Meltzer, Y. Chen, Y. Jiang, E. Seftor, M. Hendrix, M. Radmacher, R. Simon, Z. Yakhini, A. Ben-Dor, N. Sampas, E. Dougherty, E. Wang, F. Marincola, C. Gooden, J. Lueders, A. Glatfelter, P. Pollock, J. Carpten, E. Gillanders, D. Leja, K. Dietrich, C. Beaudry, M. Berens, D. Alberts, V. Sondak, Molecular classification of cutaneous malignant melanoma by gene expression profiling, Nature 406 (2000) 536–540.

[133] R. Pal, S. Bhattacharya, U. Caglar, Robust approaches for genetic regulatory network modeling and intervention, IEEE Signal Process. Mag. 29 (1) (2012) 66–76.

[134] R. Pal, A. Datta, E.R. Dougherty, Robust intervention in probabilistic Boolean networks, IEEE Trans. Signal Process. 56 (2008) 1280–1294.

[135] R. Pal, A. Datta, E.R. Dougherty, Robust intervention in probabilistic Boolean networks, in: Proceedings of the American Control Conference, 2007, pp. 2405–2410.

[136] R. Pal, A. Datta, E.R. Dougherty, Bayesian Robustness in the control of gene regulatory networks, IEEE Trans. Signal Process. 57 (2009) 3667–3678.

[137] R. Pal, A. Datta, E.R. Dougherty, Bayesian robustness in the control of gene regulatory networks, in: IEEE Statistical Signal Processing Workshop, 2007, pp. 31–35.

[138] B. Faryabi, G. Vahedi, J.F. Chamberland, A. Datta, E.R. Dougherty, Optimal constrained stationary intervention in gene regulatory networks, EURASIP J. Bioinform. Syst. Biol. (2008) 620767.

[139] G. Vahedi, B. Faryabi, J.F. Chamberland, A. Datta, E.R. Dougherty, Intervention in gene regulatory networks via a stationary mean-first-passage-time control policy, IEEE Trans. Biomed. Eng. 55 (10) (2008) 2319–2331.

[140] X. Qian, E.R. Dougherty, Intervention in gene regulatory networks via phenotypically constrained control policies based on long-run behavior, IEEE/ACM Trans. Comput. Biol. Bioinform. 9 (1) (2012) 123–136.

[141] N. Berlow, R. Pal, Generation of intervention strategy for a genetic regulatory network represented by a family of Markov Chains, Conf. Proc. IEEE Eng. Med. Biol. Soc. 2011 (2011) 7610–7613.

[142] N. Berlow, R. Pal, Generation of stationary control policies with best expected performance for a family of Markov Chains, J. Biol. Syst. 20 (4) (2012) 423–440.

[143] M.R. Yousefi, E.R. Dougherty, Intervention in gene regulatory networks with maximal phenotype alteration, Bioinformatics 29 (14) (2013) 1758–1767.

[144] R.K. Layek, A. Datta, E.R. Dougherty, From biological pathways to regulatory networks, Mol. Biosyst. 7 (3) (2011) 843–851.

[145] R. Layek, A. Datta, M. Bittner, E.R. Dougherty, Cancer therapy design based on pathway logic, Bioinformatics 27 (4) (2011) 548–555.

[146] S. Sridharan, R. Layek, A. Datta, J. Venkatraj, Boolean modeling and fault diagnosis in oxidative stress response, BMC Genomics 13 (Suppl. 6) (2012) S4.

[147] R. Pal, S. Bhattacharya, Transient dynamics of reduced order models of genetic regulatory networks, IEEE Trans. Comput. Biol. Bioinform. 4 (2012) 1230–1244.

[148] D. Project, NCI-DREAM Drug Sensitivity Prediction Challenge, 2012, http://www.the-dream-project.org/challenges/nci-dream-drug-sensitivity-prediction-challenge.

[149] S. Chakradhar, Analyses that combine "omics" show cell targets in detail, Nat. Med. 21 (9) (2015) 965–966.

[150] D. Marbach, J.C. Costello, R. Kuffner, N.M. Vega, R.J. Prill, D.M. Camacho, K.R. Allison, M. Kellis, J.J. Collins, G. Stolovitzky, A. Aderhold, K.R. Allison, R. Bonneau,

D.M. Camacho, Y. Chen, J.J. Collins, F. Cordero, J.C. Costello, M. Crane, F. Dondelinger, M. Drton, R. Esposito, R. Foygel, A. de la Fuente, J. Gertheiss, P. Geurts, A. Greenfield, M. Grzegorczyk, A.C. Haury, B. Holmes, T. Hothorn, D. Husmeier, V.A. Huynh-Thu, A. Irrthum, M. Kellis, G. Karlebach, R. Kuffner, S. Lebre, V. De Leo, A. Madar, S. Mani, D. Marbach, F. Mordelet, H. Ostrer, Z. Ouyang, R. Pandya, T. Petri, A. Pinna, C.S. Poultney, R.J. Prill, S. Rezny, H.J. Ruskin, Y. Saeys, R. Shamir, A. Sirbu, M. Song, N. Soranzo, A. Statnikov, G. Stolovitzky, N. Vega, P. Vera-Licona, J.P. Vert, A. Visconti, H. Wang, L. Wehenkel, L. Windhager, Y. Zhang, R. Zimmer, Wisdom of crowds for robust gene network inference, Nat. Methods 9 (8) (2012) 796–804.

Predictive modeling based on random forests

7

CHAPTER OUTLINE

7.1 INTRODUCTION

As discussed in the previous chapter, ensemble approaches such as Random Forests have shown to perform well in drug sensitivity prediction problems. For instance, a methodology based on individual Random Forest models for each genomic characterization and integrating the predictions from various genomic characterizations was a top performer in the NCI-DREAM drug sensitivity prediction challenge [1,2]. Another instance of the use of Random Forests in drug sensitivity prediction includes [3] where a Random Forest-based ensemble approach on gene expression data was

Predictive Modeling of Drug Sensitivity. http://dx.doi.org/10.1016/B978-0-12-805274-7.00007-5
Copyright © 2017 Elsevier Inc. All rights reserved.

used for prediction of drug sensitivity and achieved a R^2 value of 0.39 between the predicted IC_{50}s and experimental IC_{50}s for NCI-60 cell lines. Random Forests have also been applied in multiple other bioinformatics' applications, such as large-scale association studies of genetic diseases [4,5], analyzing gene expression data [6,7], gene selection [8,9], identifying differential gene expression [10], connectivity prediction [11,12], protein interactions [13], and toxicity prediction [14,15]. In this chapter, we will discuss the idea behind Random Forests along with the effect of selecting various parameters, followed by probabilistic representation of Random Forests in the form of ensemble of probabilistic regression trees.

The idea of Random Forest Regression was primarily extended from Random Forest Classification, which was derived from concepts of *Bagging* [16] and *random selection of features* [17–19]. Bootstrap aggregating (or *bagging*) was introduced by Breiman [16] to improve classification accuracy by combining classifiers trained on randomly generated training sets. As the name suggests, the training sets are generated using a *Bootstrap* approach or, in other words, generating sets of *n* samples using samples with replacement from the original training set. A specified number of such training sets are generated and the response of models trained on these training sets are then averaged to arrive at the bagging response. *Bagging* is a form of ensemble learning and there can be various other types used for model aggregation, such as *boosting* (Adaptive Boosting [20]), *Bayesian Model Combination* [21], and *Stacking* [22]. Later in the chapter, we will be using stacked generalization (*Stacking*) to combine Random Forest models generated from individual genomic characterizations.

The other concept of random selection of features used in Random Forests can be considered as a form of attribute bagging where classifiers are generated on a random subspace of the original space and then combined [18]. This approach is expected to increase independence between the classifiers and the combination of such classifiers can potentially reduce the variance of the integrated classifier and increase generalization accuracy.

Random Forest [23] (henceforth Random Forest will denote Random Forest Regression, rather than Random Forest Classification) combines the two concepts of *Bagging* and *Random Subspace* by generating a set of *T* regression trees where the training set for each tree is selected using Bootstrap sampling from the original sample set, and the features considered for partitioning at each node is a random subset of the original set of features. The bootstrap sampling for each regression tree generation and the random selection of features considered for partitioning at each node reduces the correlation between the generated regression trees and thus averaging their predictive responses is expected to reduce the variance of the error.

Random Forests have gained popularity due to their high accuracy prediction even with limited samples and a large number of features. The random subspace selection incorporates an embedded feature selection in the model generation process. Random Forests can lower the prediction variance of individual regression trees by averaging over multiple regression trees that are uncorrelated.

7.2 **RANDOM FOREST REGRESSION**

Random Forest (RF) regression refers to ensembles of regression trees [23] where a set of T un-pruned regression trees are generated based on bootstrap sampling from the original training data. For each node, the optimal node splitting feature is selected from a set of m features that are picked randomly from the total M features. For $m \ll M$, the selection of the node splitting feature from a random set of features decreases the correlation between different trees and thus the average response of multiple regression trees is expected to have lower variance than individual regression trees. Larger m can improve the predictive capability of individual trees, but can also increase the correlation between trees and void any gains from averaging multiple predictions. The bootstrap resampling of the data for training each tree also increases the independence between the trees.

7.2.1 **PROCESS OF SPLITTING A NODE**

Let $x_{tr}(i, j)$ and $y(i)$ $(i = 1, \ldots, n; j = 1, \ldots, M)$ denote the training predictor features and output response samples, respectively. At any node η_P, we aim to select a feature j_s from a random set of m features and a threshold z to partition the node into two child nodes η_L (left node with samples satisfying $x_{tr}(I \in \eta_P, j_s) \leq z$) and η_R (right node with samples satisfying $x_{tr}(i \in \eta_P, j_s) > z$).

We consider the node cost as sum of square differences:

$$D(\eta_P) = \sum_{i \in \eta_P} (y(i) - \mu(\eta_P))^2 \tag{7.1}$$

where $\mu(\eta_P)$ is the expected value of $y(i)$ in node η_P. Thus the reduction in cost for partition γ at node η_P is

$$C(\gamma, \eta_P) = D(\eta_P) - D(\eta_L) - D(\eta_R) \tag{7.2}$$

The partition $\gamma*$ that maximizes $C(\gamma, \eta_P)$ for all possible partitions is selected for node η_P. Note that for a continuous feature with n samples, a total of n partitions need to be checked. Thus the computational complexity of each node split is $O(mn)$. During the tree generation process, a node with less than n_{size} training samples is not partitioned any further.

7.2.2 **FOREST PREDICTION**

Using the randomized feature selection process, we fit the tree based on the bootstrap sample $\{(\mathbf{X}_1, Y_1), \ldots, (\mathbf{X}_n, Y_n)\}$ generated from the training data.

Let us consider the prediction based on a test sample $\mathbf{x}$ for the tree Θ. Let $\eta(\mathbf{x}, \Theta)$ be the partition containing $\mathbf{x}$, the tree response takes the form [23–25]:

$$y(\mathbf{x}, \Theta) = \sum_{i=1}^{n} w_i(\mathbf{x}, \Theta) y(i) \tag{7.3}$$

where the weights $w_i(\mathbf{x}, \Theta)$ are given by

$$w_i(\mathbf{x}, \Theta) = \frac{\mathbf{1}_{\{\mathbf{x}_{tr}(i) \in \eta(\mathbf{x}, \Theta)\}}}{\#\{r : \mathbf{x}_{tr}(i) \in \eta(\mathbf{x}_{tr}(r), \Theta)\}} \tag{7.4}$$

Let the T trees of the Random Forest be denoted by $\Theta_1, \ldots, \Theta_T$ and let $w_i(\mathbf{x})$ denote the average weights over the forest, that is

$$w_i(\mathbf{x}) = \frac{1}{T} \sum_{j=1}^{T} w_i(\mathbf{x}, \Theta_j). \tag{7.5}$$

The Random Forest prediction for the test sample $\mathbf{x}$ is then given by

$$\bar{y}(\mathbf{x}) = \sum_{i=1}^{n} w_i(\mathbf{x}) y(i) \tag{7.6}$$

The above process of generating a Random Forest is represented as a step diagram in Fig. 7.1.

7.2.3 VARIABLE IMPORTANCE

Albeit Random Forest models can provide high accuracy predictions, the biological interpretability of the generated model is not straightforward. An ensemble of regression trees is hard to analyze from a biological topology or directional connectivity perspective. In comparison, linear regression models can provide a direct interpretability of individual features in terms of regression coefficients. To address this challenge, additional analysis of generated Random Forest models can be conducted to evaluate the importance of individual features. The variable (feature) importance can be assessed in multiple ways:

- **Naive approach**: This basic approach consists of assigning variable importance based on the count of the number of times the variable x_m has been selected by all the trees in the ensemble. This is simple to implement, but can be problematic because it does not incorporate the effect of the variable selection for each tree (in terms of reduction in cost or increase in overall accuracy).
- **Mean decrease impurity**: Breiman [23] proposed to evaluate the importance of variable x_m by averaging the weighted reduction in cost $w(\eta_p)C(\gamma_{\eta_p}^*, \eta_P)$ for all nodes η_P where x_m is selected, over all T trees in the Forest:

$$Imp(x_m) = \frac{1}{T} \sum_{tree} \sum_{\eta_p \in tree, \gamma_{\eta_p}^* = x_m} w(\eta_p) C(\gamma_{\eta_p}^*, \eta_P) \tag{7.7}$$

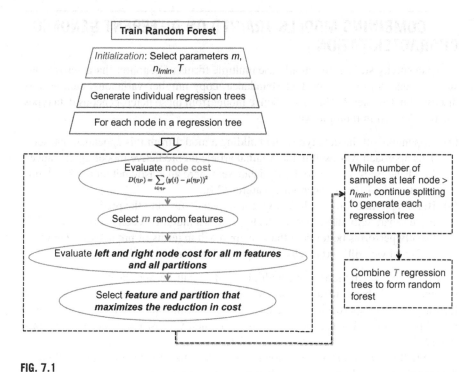

FIG. 7.1

Steps involved in Random Forest generation.

where $w(\eta_p)$ is the proportion $\frac{N_{\eta p}}{N}$ of samples reaching η_p and $\gamma^*_{\eta_P}$ is the variable used to split node η_P. If *Gini impurity* is used as the cost function, the variable importance measure is commonly known as *Gini importance*.

- **Permutation importance**: Breiman also proposed another measure based on measuring the decrease in prediction accuracy when the values of x_m are randomly permuted in the out-of-bag samples. Permutation importance performs better than the naive and mean decrease impurity approaches, but can be computationally extremely expensive.

The most commonly used variable importance measures for Random Forests are the Mean Decrease Impurity and Permutation Importance and both of these techniques can favor predictor variables with more categories [26]. However, for continuous predictor variables or variables with the same number of categories, the Mean Decrease Impurity and Permutation Importance measures do not show such biases. To avoid the bias in variable selection, Strobl et al. [26] propose the use of subsampling without replacement and forest approach where the individual trees are generated based on a conditional inference framework [27].

7.3 COMBINING MODELS TRAINED ON DIFFERENT GENOMIC CHARACTERIZATIONS

Drug sensitivity studies commonly use multiple forms of genomic characterizations, such as gene expression, protein abundance, copy number variations, etc., as was discussed in Chapter 2. Thus a predictive model from these heterogeneous data types can be designed in multiple ways:

(a) Appending all the data types and building a model from this appended dataset. This approach was used in the original article on predictive modeling of Cancer Cell Line Encyclopedia (CCLE) database, where an elastic net model was built on more than 100,000 genomic features [28].
(b) Building a model for each data type and then combining the predictions of these separate models. Our results have shown that this second approach usually performs better than the first approach in terms of predictive accuracy [1,2] and we will next discuss this approach in further detail.

Combining predictions from multiple models can be approached from various perspectives. A basic idea can be simple averaging, similar to regular ensemble learning. Another approach is based on training a learning algorithm on the individual model predictions. The second approach is commonly known as stacked generalization [29].

We will utilize a linear regression model as the stacked generalization technique for combining Random Forest Predictions from different genetic characterizations. This approach was a top performer in the NCI-DREAM drug sensitivity prediction challenge [2].

7.3.1 STACKED GENERALIZATION

Once we have the random forests built for individual datasets, we can integrate the predictions from different datasets using linear least square regression. Note that other approaches such as linear regression with regularization or regression trees can also be applied for this integration, but our simulation studies have shown that linear least square regression provides robust performance over multiple scenarios.

Least square regression: Let $\bar{y}(\mathbf{x}_{i,j})$ denote the prediction obtained by Random Forest approach for genomic characterization dataset G_i and cell line j. To utilize the biological information in different datasets for prediction, we consider a linearly weighted combination model. We use linear least square regression to estimate the weights for each dataset G_i by minimizing

$$\sum_{j=1}^{n} \left(Y_j - \sum_{i=1}^{d} \alpha_i \bar{y}(\mathbf{x}_{i,j}) \right)^2 \tag{7.8}$$

where Y_j is the experimental drug response for cell line j, α_i is the corresponding weight of dataset G_i, and d is the number of genetic characterization datasets used for integrated model generation. The **Matlab** function *fitlm* or **R** function *lm* can be used to generate the linear regression model.

Considering D genetic characterization datasets, we can produce $2^D - 1$ different nonempty combinations of these datasets. Thus $2^D - 1$ integrated regression models are evaluated for selecting the optimal integrated model.

To evaluate the integrated models, we require an error estimation approach. Before estimating the error, we normalize the samples and decide on the type of error. Some common forms of error estimation approaches that are used are Leave-one-out (LOO) error estimation and 0.632 Bootstrap error estimation.

The final integrated model is selected based on the dataset combination producing the lowest error. As discussed earlier, either LOO or 0.632 Bootstrap can be used to estimate the integrated model error. The final integrated model for application to new samples is stored as a collection of d random forests $RF(i)$ for $i = 1, \ldots, d$ whose combined prediction is given by a linear regression model with coefficients $\alpha_1, \ldots, \alpha_d$. Each Random Forest $RF(i)$ is a collection of T regression trees $RT(i, j)$ for $j = 1, \ldots, T$. Each regression tree $RT(i, j)$ is an ordered set of $\eta_{i,j,k}$ nodes where each node contains information on the node split feature and the threshold for splitting.

Note that if we had to combine a large number of predictions instead of combining model predictions from two to six genomic characterizations, regularization of the linear regression model for stacking would be desirable to avoid overfitting.

7.3.2 ANALYSIS OF NCI-DREAM CHALLENGE DATASETS

For the NCI-DREAM drug sensitivity prediction subchallenge 1, genomic characterizations were provided for 53 cell lines (48 breast cancer cell lines and 5 nonmalignant breast cell lines) that were exposed to 31 therapeutic compounds at a concentration required to inhibit proliferation by 50% after 72 h (GI50) and responses to these 31 drugs were provided for 35 of these 53 cell lines [1,2]. The drug response of the remaining 18 cell lines were provided later and used as validation data to generate the validation or true error. Multiple types of genomic and epigenetic data (copy number variation, Methylation, Gene Expression through micro-array, RNA sequencing, Exome sequencing, and protein abundance) were generated before exposure of the cells to the drugs for each of the 53 cell lines. Details of the genomic characterization datasets are provided in Table 7.1. From Table 7.1, we note that the genomic characterizations were not available for all the 53 cell lines and each dataset had missing information for some of the cell lines (the number of such cell lines is denoted by Missing cell lines in Table 7.1). The last column denotes whether the genomic dataset had some missing values for the cell lines containing that specific genomic characterization. While within the training data, all the drug responses or genomic characterization could not be reliably measured due to technical reasons. These data were provided by Prof. Joe Gray from Oregon Health and Science University.

In this section, we present results using five datasets for prediction and utilizing multiple forms of error estimation for estimating the prediction error. We consider

Table 7.1 Description of Genomic Datasets for NCI-DREAM Drug Sensitivity Challenge

Data Type	Dimension	Missing Cell Lines	Missing Values (Y/N)
Gene Expression	46 × 18,632	7	N
Methylation	41 × 27,551	12	N
RNA-Seq	44 × 36,953	9	Y
RPPA	42 × 131	11	N
SNP6	47 × 27,234	6	Y

Notes: Out of 53 cell lines, 35 cell lines are used for training and 18 for testing the prediction accuracy.

the following five types of genomic characterizations: Gene Expression, Methylation, RNA-Seq, SNP6, and RPPA. For genetic mutation information, we used the SNP6 data and did not consider the additional exome sequencing data for prediction analysis. The confidence intervals (CIs) for our estimated errors are generated using Jackknife-After-Bootstrap approach [30,31]. Since we considered mean absolute error (MAE), the lower limit of the CI is kept at $\max(0, MAE - s * z)$ where s is the standard error estimated using Jackknife-After-Bootstrap approach and z is the specific quantile of the standard normal distribution. Based on the five datasets, there are $2^5 - 1 = 31$ nonnull possible combinations of the datasets. Fig. 7.2 shows the different error estimations for the 31 feasible combinations. We note that the errors are decreasing with the integration of multiple datasets. The datasets are denoted by binary digits with the following order: Gene Expression (most significant bit), Methylation, RNA-Seq, RPPA, and SNP6 (least significant bit). For instance, 01100 denotes Methylation and RNA-Seq data combination. Fig. 7.2 shows that the 0.632 Bootstrap error for Drug 10 reduces from 0.08 for Gene Expression to 0.046 for all datasets combined. Similar behavior is also observed for other error estimation approaches.

Fig. 7.3 shows the 80% CI for drug 15 and it illustrates that the addition of datasets decreases the CI. With a single dataset, the CI is highest (leftmost combinations) and the CI decreases gradually with integration of multiple genomic characterizations (going right). The rightmost bar represents the CI of 0.632 Bootstrap error with all five datasets, which has the lowest CI among all combinations. In all the cases, the validation error denoted by cross is within 80% CI. Fig. 7.3 shows that the CI is reduced more than four times by using all the datasets as compared to gene expression alone.

The coefficient of determination R^2 between predicted and experimental sensitivities using bootstrap samples for drugs 1–31, while considering all five datasets for prediction is shown in Fig. 7.4. We should note that the R^2 values are above 0.82 for all the drugs, denoting good prediction accuracy while using five genomic characterization datasets.

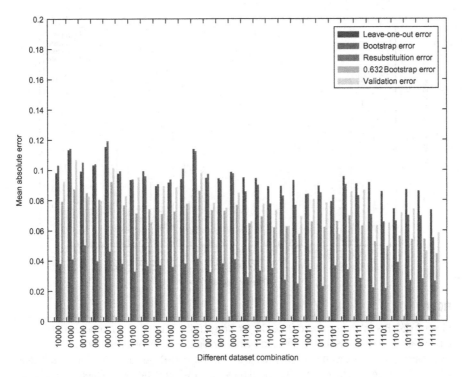

FIG. 7.2

Leave-one-out, Bootstrap, Resubstitution, 0.632 Bootstrap, and True error for Drug 10 for different dataset combinations. The datasets are denoted by binary digits with the following order: Gene Expression (most significant bit), Methylation, RNA-Seq, RPPA, and SNP6 (least significant bit). For instance, 01100 denotes Methylation and RNA-Seq data combination.

Selecting important features and appending datasets

For the results presented until now, we have used all the datasets along with all the features. Utilizing all the features may not provide the best performance, as some of the features may lack importance and thus an approach to select significant features before model generation can potentially improve performance. Thus we considered filter feature selection with RELIEFF approach [32,33]. In this section, we also report results of appending datasets before generating the predictive model, which is in contrast to earlier results where we generated individual models for each dataset and integrated the predictions from these individual models. Fig. 7.5 shows the Mean Absolute Validation Error for an integrated model with or without feature selection and with or without dataset appending for Drug-28. The five ways of generating the models are as follows:

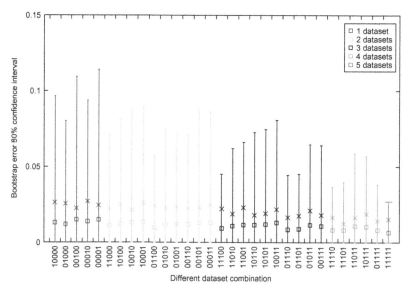

FIG. 7.3

Mean 0.632 Bootstrap error and 80% CIs for Drug 15 for $31 (= 2^5 - 1)$ different dataset combinations. The datasets are denoted by binary digits with the following order: Gene Expression (most significant bit), Methylation, RNA-Seq, RPPA, and SNP6 (least significant bit). For instance, 01100 denotes Methylation and RNA-Seq data combination [2].

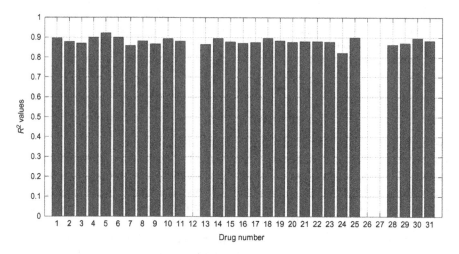

FIG. 7.4

The coefficient of determination R^2 between predicted and experimental sensitivities using bootstrap samples for Drugs 1–31, while considering all five datasets for prediction. The prediction for drugs 12, 26, and 27 was not considered as they contained minimal variations in sensitivity [2].

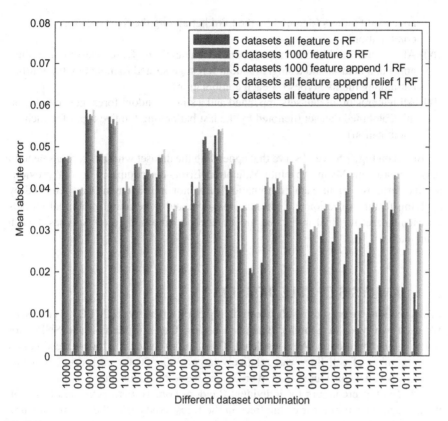

FIG. 7.5

MAE for validation set for Drug-28 with different integrated models. These models were built using different combination of datasets along with the datasets being changed in the following ways: (i) All features of the dataset are used and a random forest for each dataset being generated. (ii) Feature selection in each dataset and a random forest for each dataset being generated based on the selected features. (iii) Feature selection in each dataset followed by appending of the selected features and a single random forest being generated from the appended features. (iv) All features of all the datasets appended and applying feature selection on the appended dataset and a single random forest generated. (v) All features of all datasets appended and a single random forest generated from this combined dataset.

(i) All features of the dataset are used and a random forest for each dataset being generated (similar to the approach in previous section and denoted by the leftmost bars for each combination).

(ii) Feature selection in each dataset is conducted and a random forest for each dataset generated based on the selected features (denoted by the second bar among the five types for each combination).

(iii) Feature selection in each dataset conducted followed by appending of the selected features and a single random forest being generated from the appended features (denoted by the third bar among the five types for each combination).

(iv) All features of all the datasets appended and applying feature selection on the appended dataset and a single random forest generated (denoted by the fourth bar among the five types for each combination).

(v) All features of all datasets appended and a single random forest generated from this combined dataset (denoted by the last bar among the five types for each combination).

Based on Fig. 7.5, we observe that appending the dataset without feature selection may increase the Mean Absolute Validation Error. As compared to our previous results (leftmost bar in each combination), a minor improvement was achieved by applying feature selection in individual datasets and generating individual models for each dataset for final integrated model (second bar among the five types in each combination).

7.3.3 ANALYSIS OF CCLE DATASETS

In this section, we present results for applying our framework to CCLE [34] datasets. We used two types of genomic characterizations (Gene Expression and SNP6) for our integrated prediction. We considered 10-fold cross-validation for error estimation as the number of cell lines was relatively large (300–500 for each drug). For this analysis, we also considered the top predictors based on the integrated random forest model. The top predictors for individual Random Forests were generated based on the average bootstrap error of the trees in the forest containing the specific feature. The combined model regression coefficients were used to generate the top predictors of the joint model from the predictor weights of the individual models. Some of the top predictors were then validated by comparing with the top targets of the drugs that can be obtained from drug target inhibition profiles [35,36]. Note that not all top predictors will be targets of a drug, as the expression of upstream or downstream proteins of the drug targets may predict the efficacy of a drug without being actual targets of the drug.

For the current analysis, the Gene Expression and SNP6 datasets consist of 18,988 and 21,217 genes, respectively, and drug responses for around 500 cell lines are available for the following 24 drugs *17AAG, AEW541, AZD0530 (Saracatinib), AZD6244 (Selumetinib), Erlotinib, Irinotecan, L685458, Lapatinib, LBW242, Nilotinib, Nutlin3, Paclitaxel, Panobinostat, PD0325901, PD0332991, PF2341066 (Crizotinib), PHA665752, PLX4720, RAF265, Sorafenib, TAE684, TKI258 (dovitinib), Topotecan, and ZD6474 (Vandetanib)*. Because the majority of the applied drugs target the human kinome (the set of protein kinases), we also consider a smaller set of features containing around 400 kinase producing genes. The 400 kinases are selected from drug target inhibition profile studies [35,36]. We report our prediction results in the form of correlation coefficients between predicted and experimental sensitivities to directly compare our results with the recently published CCLE study [28]. We have also used 10-fold cross-validation similar to [28]. The CCLE study, however, used multiple other data types (for a total of 150,000 features), but we have used only Gene Expression and SNP6. The results of our prediction approach are shown in Table 7.2. We note that the average correlation coefficient across the 24 drugs was 0.421 for the CCLE Elastic net study [28], but the RF approach based on only 400 features produced a higher average correlation coefficient of 0.454 (we will term this approach *CRF-400*), an increase of 7.8%. When we used all the features (18,988

Table 7.2 CCLE Drug Sensitivity Prediction Results in the Form of Correlation Coefficients Between Experimental and Predicted Sensitivities

Drug Name	Correlation Coefficients		
	Elastic Net [28]	CRF-400	CRF-20,000
17AAG	0.43	0.4116	0.4397
AEW541	0.33	0.4037	0.3934
AZD0530 (Saracatinib)	0.18	0.2855	0.2747
AZD6244 (Selumetinib)	0.58	0.516	0.5909
Erlotinib	0.3	0.4034	0.4333
Irinotecan	0.68	0.6214	0.6776
L685458	0.47	0.5351	0.5423
Lapatinib	0.45	0.5488	0.5263
LBW242	0.08	0.184	0.1400
Nilotinib	0.76	0.5458	0.5476
Nutlin3	0.1	0.2892	0.3096
Paclitaxel	0.6	0.5453	0.5531
Panobinostat	0.65	0.616	0.6503
PD0325901	0.64	0.5837	0.6471
PD0332991	0.58	0.5077	0.5141
PF2341066 (Crizotinib)	0.36	0.5121	0.5055
PHA665752	0.27	0.3393	0.3437
PLX4720	0.55	0.4459	0.4768
RAF265	0.35	0.4378	0.4394
Sorafenib	0.27	0.4099	0.4685
TAE684	0.35	0.4073	0.4453
TKI258 (dovitinib)	0.3	0.4463	0.4611
Topotecan	0.58	0.5619	0.6226
ZD6474 (Vandetanib)	0.24	0.3331	0.3494
Average	0.421	0.454	0.473

Notes: Elastic Net denotes the approach applied in [28] for predicting sensitivity using 10-fold cross-validation from CCLE database. CRF-400 denotes our proposed combined Random Forest approach using Gene Expression and SNP6 data of only 400 genes. CRF-20000 denotes our proposed combined Random Forest approach using 18,988 Gene Expression and 21,217 SNP6 features [2].

Gene Expression features and 21,217 SNP6 features), we were able to increase our average correlation coefficient to 0.473 (we will term this approach *CRF-20,000*), an increase of 12.4%. We next applied our framework to select the top predictors for each drug for both CRF-400 and CRF-20,000. The top 20 predictors are then compared with the experimentally validated targets of the drugs, based on earlier studies [35,36]. For instance, the top predictor for the drug Erlotinib based on both CRF-400 and CRF-20,000 is EGFR, and EGFR is known to be the primary target of Erlotinib with an IC_{50} of 0.2 nM [35]. IC_{50} denotes the drug concentration required to inhibit the target protein expression by half. If we consider the drug Lapatinib, the top predictor selected was ERBB2 for both *CRF-400* and *CRF-20,000*. ERBB2 is the most potent target of Lapatinib (IC_{50} of 9.2 nM [36]) followed by EGFR (IC_{50} of 10.8 nM [36]). However, EGFR was selected as the 34th top predictor for *CRF-400* and was not picked up by *CRF-20,000*. The analysis of the top predictors shows that *CRF-400* have better chances of selecting the experimentally validated primary targets of a drug in the top 20 predictors as compared to *CRF-20,000*. However, the prediction accuracy is decreased on an average by 4% by using the smaller set of 400 features. Our ability to select the top drug targets using *CRF-20,000* might be increased if we use a larger number of trees so that each feature appears in a large number of trees in the forest. Furthermore, the inclusion of primary drug targets in the top predictors of the model will also depend on the dataset. For instance, if we consider individual Random Forest (400 features) models for Gene Expression and SNP6 for Erlotinib, EGFR is selected as the top predictor for Gene Expression but PHKG1 is selected as the top predictor for SNP6. But since the regression weights in the combined models are 0.89 for Gene Expression and 0.11 for SNP6, the top predictor for the combination model is EGFR. The analysis of the CCLE database illustrates the predictive capability of the integrated random forest approach along with the ability to generate top predictors for drug sensitivity.

7.4 PROBABILISTIC RANDOM FORESTS

In this section, we consider an extension of the RF methodology by representing a regression tree in the form of a probabilistic tree and analyzing the effect of tree weights on variance of the prediction. The probabilistic tree representation allows for analytical computation of CIs and the tree weight optimization is expected to provide stricter CIs with comparable performance in mean error. We approached the ensemble of probabilistic trees prediction from the perspectives of a mixture distribution and as a weighted sum of correlated random variables. The ensemble of probabilistic regression trees will also enable us to model the idea of *heteroscedasticity*, which implies that the variability of prediction can differ based on different values of the predictor (input) variables. Regular regression approaches consider *homoscedasticity*; that is, the predicted random variable has same finite variance.

The ensemble of probabilistic regression trees can be considered from two different perspectives.

Firstly, we can consider the ensemble as a mixture distribution for each prediction sample X_i. Consider an ensemble of T trees where the tree j produces the predicted output probability density function (pdf) of $P(Y_j|X_i)$. The pdf $P(Y|X_i)$ of the ensemble of the T regression trees with weights $\alpha_1, \ldots, \alpha_T$ is then given by $P(Y|X_i) = \sum_{j=1}^{T} \alpha_j P(Y_j|X_i)$. This approach considers that based on the weights α_j, a tree k will be selected and the prediction will be decided based on the pdf $P(Y_k|X_i)$.

For the second perspective, we can consider the output of the ensemble to be random variable Z where $Z = \sum_{j=1}^{T} \alpha_j Z_j$ is a weighted sum of T random variables $Z_1, \ldots, Z_T$ with Z_j denoting a random variable with pdf $P(Z_j|X_i)$ based on tree j. This scenario is equivalent to analyzing a weighted sum of random variables with different pdfs; that is, we model the weighted sum of the realizations of the random variables rather than the weighted sum of their distributions as was considered in the first case.

Note that the use of equal weights (i.e., $\alpha_j = \frac{1}{T}$ for $j = 1, \ldots, T$) for the regression trees is supposed to work well in terms of reducing variance of prediction when the generated trees are uncorrelated. However, some of the generated trees can often be correlated to each other and in such a scenario, we can potentially optimize the weights of the trees to reduce the variance of the ensemble prediction. Based on this idea, we analyze the variance of the prediction of a weighted sum of random variables for different forms of between tree covariance matrices. For the mixture distribution scenario, we can use maximum likelihood estimation (MLE) to generate the weights of the regression trees and analyze how the obtained weights alter the mean and variance of the error distributions.

7.4.1 PROBABILISTIC REGRESSION TREES

Let us consider the generation of regression trees from a probabilistic perspective, which will allow us to utilize well-known concepts of parameter estimations for statistical models. Estimation of regression trees using probability models has been explored in [37,38]. For a regression tree, our goal is to generate the conditional density of the form $P(\mathbf{y}|\mathbf{x}, \phi)$, where $\mathbf{y}$ and $\mathbf{x}$ refer to the output and input responses, respectively, and ϕ denotes the collection of parameters for the tree. The tree splits can be modeled by probabilistic decisions that are conditional on the input $\mathbf{x}$ and previous node decisions. As an example, consider the two-level tree shown in Fig. 7.6.

The first decision is based on the probability $P(\omega_1|\mathbf{x}, \eta)$, where ω_1 is the event signifying the left node of the tree root node and η denotes a parameter vector $\eta = [\eta_1 \tau_1]$. Note that if we consider

$$P(\omega_1|\mathbf{x}, \eta) = \frac{e^{\eta_1^T \tau_1}}{e^{\eta_1^T \mathbf{x}} + e^{\eta_1^T \tau_1}} + \frac{1}{1 + e^{\eta_1^T (\mathbf{x} - \tau_1)}} \tag{7.9}$$

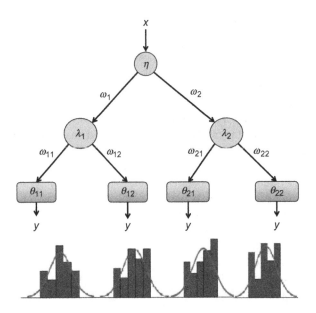

FIG. 7.6

Example probabilistic decision tree [39].

with large η_1, the split will be close to a sharp linear decision boundary similar to a regression tree. Similarly, we will have

$$P(\omega_2|\mathbf{x}, \eta) = 1 - P(\omega_1|\mathbf{x}, \eta) = \frac{1}{1 + e^{\eta_1^T(\tau_1 - \mathbf{x})}} \qquad (7.10)$$

If we consider all the branches of the tree as shown in Fig. 7.6, the corresponding distribution of y conditional on $\mathbf{x}$ and tree parameters $\phi = \{\eta, \lambda_1, \lambda_2, \theta_{11}, \theta_{12}, \theta_{21}, \theta_{22}\}$ will be given by Eq. (7.11).

$$P(y|\mathbf{x}, \phi) = \sum_{i=1}^{2} \sum_{j=1}^{2} P(y|\mathbf{x}, \omega_{ij}, \lambda_i, \omega_i, \eta) P(\omega_{ij}|\lambda_i, \omega_i, \eta, \mathbf{x}) P(\omega_i|\eta, \mathbf{x}) \qquad (7.11)$$

For larger number of branches in the tree, the above technique can be extended to obtain $P(y|\mathbf{x}, \phi)$ for a tree with parameter set ϕ. For later sections, we consider that the tree parameters ϕ are generated based on the standard random forest node generation criteria given in Eq. (7.2). The probability distribution at any leaf node is approximated by a Gaussian distribution with mean and variance equal to the mean and variance of the samples at the leaf node. For example, output response y_1 of leaf θ_{11} can have a distribution as the one shown in Fig. 7.6.

Consequently, an ensemble of T trees generated by RF regression can be represented by the T tree parameters $\phi_1, \phi_2, \ldots, \phi_T$ with each producing the conditional distribution $P(y|\mathbf{x}, \phi_i)$ for $i = 1, \ldots, T$.

7.4.2 MIXTURE DISTRIBUTION FOR PROBABILISTIC RANDOM FORESTS

As discussed earlier, we consider the ensemble as a mixture distribution for each prediction sample X_i. Consider an ensemble of T trees where the tree j produces the predicted output pdf of $P(Y_j|X_i)$. The predicted distribution for each tree is based on the estimated probabilistic regression tree model described in the previous section. The pdf $P(Y|X_i)$ of the forest of T regression trees with weights $\alpha_1, \ldots, \alpha_T$ with $\alpha_j \geq 0$ and $\sum_{j=1}^{T} \alpha_j = 1$ is then given by $P(Y|X_i) = \sum_{j=1}^{T} \alpha_j P(Y_j|X_i)$. This approach considers that based on the weights α_j, a tree k will be selected and the prediction will be decided based on the pdf $P(Y_k|X_i)$.

The mean of the mixture distribution (μ) will be the weighted sum of the means (μ_i) of the distribution of the trees, similar to the way the output of a testing sample is calculated in regular Random Forest (weighted sum of the tree predictions).

$$E[Y] = \mu = \sum_{i=1}^{T} \alpha_i \mu_i \qquad (7.12)$$

The variance of the mixture distribution (σ^2) is given by Eq. (7.13).

$$E[(Y - \mu)^2] = \sigma^2 = \sum_{i=1}^{T} \alpha_i((\mu_i - \mu)^2 + \sigma_i^2) \qquad (7.13)$$

7.4.3 WEIGHTED SUM OF RANDOM VARIABLES FOR PROBABILISTIC RANDOM FORESTS

The mixture distribution approach selects a tree based on the weights and then selects according to the distribution of the tree. Another potential approach is to consider the weighted sum of realizations from each tree. As discussed earlier, this will be equivalent to considering the output of the forest to be a random variable Z where $Z = \sum_{j=1}^{T} \alpha_j Z_j$ is a weighted sum of T random variables $Z_1, \ldots, Z_T$ with Z_j denoting a random variable with pdf $P(Z_j|X_i)$ based on tree j. The distribution of a sum of random variables can be computed as the convolution of the individual distributions. This scenario is equivalent to analyzing a weighted sum of random variables with different pdfs; that is, we model the weighted sum of the realizations of the random variables rather than the weighted sum of their distributions as was considered in the first case.

Example for weighted sum of two uncorrelated Gaussian distributions: Consider two independent random variables X_1 and X_2 that are normally distributed with pdfs $\mathbb{N}(\mu_{X_1}, \sigma_{X_1}^2)$ and $\mathbb{N}(\mu_{X_2}, \sigma_{X_2}^2)$, respectively. The distributions of $X_{1\alpha} = \alpha_1 X_1$ and $X_{2\alpha} = \alpha_2 X_2$ are given by Eq. (7.14):

$$f_{X_{1\alpha}}(x_{1\alpha}) = \frac{1}{\sqrt{2\pi \sigma_{X_1}^2 \alpha_1^2}} \exp^{-\frac{(x_{1\alpha} - \alpha_1 \mu_{X_1})^2}{2\alpha_1^2 \sigma_{X_1}^2}} \tag{7.14}$$

$$f_{X_{2\alpha}}(x_{2\alpha}) = \frac{1}{\sqrt{2\pi \sigma_{X_2}^2 \alpha_2^2}} \exp^{-\frac{(x_{2\alpha} - \alpha_2 \mu_{X_2})^2}{2\alpha_2^2 \sigma_{X_2}^2}}$$

Based on the idea of derived distributions, the pdf of random variable $Z = \alpha_1 X_1 + \alpha_2 X_2 = X_{1\alpha} + X_{2\alpha}$ is equal to the convolution of the pdfs of $X_{1\alpha}$ and $X_{2\alpha}$ and is given by Eq. (7.15):

$$f_Z(z) = \int_{-\infty}^{\infty} f_{X_{1\alpha}}(z - x) f_{X_{2\alpha}}(x) dx \tag{7.15}$$

Substituting Eq. (7.14) in Eq. (7.15), we arrive at:

$$f_Z(z) = \int_{-\infty}^{\infty} \frac{1}{\sqrt{2\pi \sigma_{X_1}^2 \alpha_1^2}} \exp^{-\frac{(z - x_{1\alpha_1} - \alpha_1 \mu_{X_1})^2}{2\sigma_{X_1}^2 \alpha_1^2}} \frac{1}{\sqrt{2\pi \sigma_{X_2}^2 \alpha_2^2}} \exp^{-\frac{(x_{2\alpha_2} - \alpha_2 \mu_{X_2})^2}{2\sigma_{X_2}^2 \alpha_2^2}} dx \tag{7.16}$$

$$= \frac{1}{\sqrt{2\pi (\sigma_{X_1}^2 \alpha_1^2 + \sigma_{X_2}^2 \alpha_2^2)}} \exp\left[-\frac{(z - (\alpha_1 \mu_{X_1} + \alpha_2 \mu_{X_2}))^2}{2(\sigma_{X_1}^2 \alpha_1^2 + \sigma_{X_2}^2 \alpha_2^2)} \right] \tag{7.17}$$

Eq. (7.17) represents the sum of 2 independent Gaussian random variables. For T independent Gaussian Random variables $X_1, \ldots, X_T$ with pdfs $\mathbb{N}(\mu_1, \sigma_1), \ldots, \mathbb{N}(\mu_T, \sigma_T)$ representing the distribution at the T leaf nodes of the forest, the distribution of the random variable Z representing their weighted sum with weights $\alpha_1, \ldots, \alpha_T$ is given by Eq. (7.18) (derived based on multiple convolutions).

$$f_Z(z) = \frac{1}{\sqrt{2\pi \left(\sum_{i=1}^{T} \sigma_{X_i}^2 \alpha_i^2 \right)}} \exp\left[-\frac{\left(z - \left(\sum_{i=1}^{T} \alpha_i \mu_{X_i} \right) \right)^2}{2 \left(\sum_{i=1}^{T} \sigma_{X_i}^2 \alpha_i^2 \right)} \right] \tag{7.18}$$

Thus Z has a normal distribution with *mean* $= \sum_{i=1}^{T} \alpha_i \mu_{X_i}$ and *Variance* $= \sum_{i=1}^{T} \alpha_i^2 \sigma_{X_i}^2$.

However, if the random variables are correlated, that is, the covariance between different tree outputs are nonzero, the mean and variance of Z are given as follows:

$$Mean\left(\sum_{i=1}^{T}\alpha_i X_i\right) = \sum_{i=1}^{T}\alpha_i \mu_{X_i} \tag{7.19}$$

$$Var\left(\sum_{i=1}^{T}\alpha_i X_i\right) = \sum_{i=1}^{T}\sum_{i=1}^{T} Cov(\alpha_i X_i, \alpha_j X_j)$$

$$= \sum_{i=1}^{T}\alpha_i^2 Var(X_i) + 2\sum_{1\leq i}\sum_{<j\leq T}\alpha_i \alpha_j Cov(X_i, X_j) \tag{7.20}$$

If the vector $C = [\alpha_1, \ldots, \alpha_T]'$ represents the weight vector and Σ represents the $T \times T$ covariance matrix, the variance of Z can be represented concisely as

$$Var(Z) = C'\Sigma C \tag{7.21}$$

where C' represents the transpose of C.

Note that the mean of Z denotes the weighted sum of the means of each tree in the forest and the prediction is the same as regular Random Forest when the tree weights are equal. The mean of Z remains the same irrespective of whether the trees are correlated or not, whereas the variance of Z is directly related to the covariance of the trees using Eq. (7.21). In the next sections, we will try to estimate the covariance among the trees in a forest and analyze how the change in C alters the variance of Z.

7.4.3.1 Empirical measure of correlation between probabilistic trees

The covariance between the trees will be estimated using empirical approaches to arrive at the covariance matrix Σ. The i,j position element of Σ denotes the covariance between the predictions of ith and jth tree represented by random variables Y_i and Y_j, that is, $\Sigma(i,j) = \mathbb{E}[(Y_i - \mathbb{E}(Y_i))(Y_j - \mathbb{E}(Y_j))]$.

For each input sample X_i and a tree j, the tree will produce a pdf $P(Y_j|X_i)$ which will be used to select a output prediction realization y_j. We will do this for all the other trees to arrive at one joint realization of the trees for sample X_i. We will do this for N input training samples to produce N joint realizations of the random variables $Y_1, \ldots, Y_T$ which will be used to calculate the sample covariance matrix shown in Eq. (7.22).

$$V = \begin{bmatrix} \frac{1}{N-1}\sum_{i=1}^{N}(Y_1(i)-\mathbb{E}(Y_1))^2 & \cdots & \frac{1}{N-1}\sum_{i=1}^{N}(Y_1(i)-\mathbb{E}(Y_1))(Y_T(i)-\mathbb{E}(Y_T)) \\ \vdots & \ddots & \vdots \\ \frac{1}{N-1}\sum_{i=1}^{N}(Y_T(i)-\mathbb{E}(Y_T))(Y_1(i)-\mathbb{E}(Y_1)) & \cdots & \frac{1}{N-1}\sum_{i=1}^{N}(Y_T(i)-\mathbb{E}(Y_T))^2 \end{bmatrix} \tag{7.22}$$

7.4.3.2 Effect of tree weight on variance

In this section, we will attempt to generate the lower and upper bounds on $C'VC$ where $C = [\alpha_1, \ldots, \alpha_T]'$ represents the tree weight vector. Assuming V is a Hermitian positive definite matrix (note that covariance matrix V is always positive semidefinite [40]), we can have the Cholesky decomposition [41] of $V = LL^T$, where

L is a lower triangular matrix with real and positive diagonal entries. Let the variance of the prediction of a specific forest be given by the function $f(C)$. We have

$$
\begin{aligned}
f(C) &= C'VC \\
&= C'LL'C \\
&= (L'C)'L'C \\
&= A'A
\end{aligned}
\tag{7.23}
$$

where $A = L'C$.

Let us analyze the minimum and maximum value of $f(C)$.

$$
f(C) = A'A = \|A\|_2^2 = \|L'C\|_2^2 \le \|L'\|_2^2\|C\|_2^2
\tag{7.24}
$$

Since

$$
\|(L')^{-1}L'C\|_2^2 \le \|(L')^{-1}\|_2^2\|L'C\|_2^2
\tag{7.25}
$$

and

$$
\|L'\|_2^2 = \text{Maximum eigenvalue of } LL' = maxeig(V)
\tag{7.26}
$$

$$
\|(L')^{-1}\|_2^2 = \frac{1}{\text{Minimum eigenvalue of } (LL')} = \frac{1}{mineig(V)}
\tag{7.27}
$$

We have

$$
mineig(V)\|C\|_2^2 \le f(C) \le maxeig(V)\|C\|_2^2
\tag{7.28}
$$

The minimum for $f(C)$ under the constraint $C'e = 1$ where $e = [1, 1, \ldots, 1]'$ is given by $f(C) = \frac{1}{e'V^{-1}e}$. Note that this does not preclude solutions with entries of C being less than zero. The details of the derivation using Langrange multipliers are included next.

If we incorporate the constraint $\sum_{i=1}^{T} C = 1$ in $f(c)$ and let $\Sigma = V$ (otherwise sign of *variance*(Σ) and *Sum*(Σ) will be confusing), then based on Lagrange Multiplier approach

$$
\begin{aligned}
f(c) &= C^T VC + \lambda \left(\sum_{i=1}^{T} C - 1\right) \\
&= \sum_{i=1}^{T}\sum_{j=1}^{T} C_i V_{ij} C_j + \lambda \left(\sum_{i=1}^{T} C - 1\right)
\end{aligned}
\tag{7.29}
$$

where $C_i = C_j$. Taking the derivative of Eq. (7.29) with respect to C_k

$$\frac{\partial f}{\partial C_k} = \sum_{j=1}^{T} 2C_j V_{kj} + \lambda \tag{7.30}$$

where $V_{kj} = V_{jk}$ and $\frac{\partial f}{\partial C_k} = 0$.

$$-\frac{\lambda}{2} = \sum_{j=1}^{T} C_j V_{kj} \tag{7.31}$$

$$\begin{bmatrix} C_1 \\ C_2 \\ \vdots \\ C_T \end{bmatrix} = -\frac{\lambda}{2} V^{-1} \begin{bmatrix} 1 \\ 1 \\ \vdots \\ 1 \end{bmatrix} \tag{7.32}$$

By solving Eq. (7.32), the value of λ can be generated and substituted in Eq. (7.29) to arrive at the value for $f(c)$. The weight vector C achieving the minimum is $C = \frac{V^{-1}e}{e'V^{-1}e}$.

Example 7.1 (Diagonal Elements of Covariance Matrix Unequal). Now consider the case where covariance matrix is a diagonal matrix and variances of the trees are unequal. Then the covariance matrix can be represented by Eq. (7.33).

$$V = \begin{bmatrix} \sigma_1^2 & 0 & \cdots & 0 \\ 0 & \sigma_2^2 & \cdots & 0 \\ \vdots & \vdots & \ddots & \vdots \\ 0 & 0 & \cdots & \sigma_T^2 \end{bmatrix} \tag{7.33}$$

where σ_i^2 is the variance of the ith tree. Without Loss of Generality, assume $\sigma_1^2 \leq \sigma_2^2 \leq \cdots \leq \sigma_T^2$. Then from Eq. (7.29), we have

$$f(c) = \sum_{i=1}^{T} \sigma_i^2 C_i^2 + \lambda \left(\sum_{i=1}^{T} C_i - 1 \right) \tag{7.34}$$

Differentiating $f(c)$ with respect to C_i,

$$\frac{\partial f(c)}{\partial C_i} = 2C_i \sigma_i^2 + \lambda = 0 \tag{7.35}$$

$$\lambda = -2C_i \sigma_i^2 \tag{7.36}$$

$f(c)$ will be minimum when $C_1 \sigma_1^2 = C_2 \sigma_2^2 = \cdots = C_T \sigma_T^2 = \gamma$, where γ is some constant.

$$f(c)_{\min} = \sum_{i=1}^{T} \sigma_i^2 C_i^2$$

$$= \sum_{i=1}^{T} C_i \gamma$$

$$= \gamma \tag{7.37}$$

as $\sum_{i=1}^{T} C_i = 1$. Now $C_i = \dfrac{\gamma}{\sigma_i^2}$, so

$$\sum_{i=1}^{T} C_i = \gamma \sum_{j=1}^{T} \frac{1}{\sigma_j^2} = 1 \tag{7.38}$$

$$\gamma = \frac{1}{\sum_{j=1}^{T} \sigma_j^2} \tag{7.39}$$

Substituting the value of γ in $C_i = \dfrac{\gamma}{\sigma_i^2}$

$$C_i = \frac{1}{\sum_{j=1}^{T} \dfrac{\sigma_i^2}{\sigma_j^2}} \tag{7.40}$$

Eq. (7.40) provides the weight of the trees that produces the lowest variance for the forest.

Example 7.2 (Random Forest With Correlated Trees). If there are correlations between the trees of a random forest, then covariance between two trees will likely be comparable to the variance of a single tree. This will require considering the nondiagonal elements of the covariance matrix. For such forests, placing higher weights on uncorrelated trees will likely result in lower variance. Let us consider a case where first two trees are correlated among themselves, while there are minimal correlations among other trees (we will assume them to be zero), and variance of all trees are considered to be same. In such a case, covariance matrix (V) will be represented as:

$$V = \begin{bmatrix} \sigma^2 & \rho & \cdots & 0 \\ \rho & \sigma^2 & \cdots & 0 \\ \vdots & \vdots & \ddots & \vdots \\ 0 & 0 & \cdots & \sigma^2 \end{bmatrix} \tag{7.41}$$

If the weight vector is C, then

$$C^T V C = \sigma^2 \sum_{i=1}^{T} C_i^2 + 2\rho C_1 C_2 \tag{7.42}$$

Subsequently

$$f(c) = \sigma^2 \sum_{i=1}^{T} C_i^2 + 2\rho C_1 C_2 + \lambda \left(\sum_{i=1}^{T} C_i - 1 \right) \tag{7.43}$$

where λ is the Lagrange multiplier. C_1 and C_2 are the weight of the trees that are correlated, while C_is for $i = 3, \ldots, T$ are the weights of all other trees that are not correlated. Differentiating Eq. (7.43) with respect to C_1, C_2, and C_i

$$\frac{\partial f}{\partial C_1} = 2C_1\sigma^2 + 2\rho C_2 + \lambda = 0 \tag{7.44}$$

$$\frac{\partial f}{\partial C_2} = 2C_2\sigma^2 + 2\rho C_1 + \lambda = 0 \tag{7.45}$$

$$\frac{\partial f}{\partial C_i} = 2C_i\sigma^2 + \lambda = 0; \quad i > 2 \tag{7.46}$$

Rearranging Eqs. (7.44), (7.45), (7.46),

$$C_1 = \frac{-\lambda}{2(\sigma^2 + \rho)} = C_2 \tag{7.47}$$

$$C_i = \frac{-\lambda}{2\sigma^2}; \quad i > 2 \tag{7.48}$$

$$\sum_{i=1}^{T} C_i = \frac{-\lambda}{2} \left[\frac{T-2}{\sigma^2} + \frac{2}{\sigma^2 + \rho} \right] = 1 \tag{7.49}$$

From Eq. (7.49),

$$- \lambda = \frac{2\sigma^2(\sigma^2 + \rho)}{T\sigma^2 + \rho(T-2)} \tag{7.50}$$

Substituting the value of Lagrange multiplier in Eqs. (7.47), (7.48),

$$C_1 = \frac{\sigma^2}{T\sigma^2 + \rho(T-2)} = C_2 \tag{7.51}$$

$$C_i = \frac{\sigma^2 + \rho}{T\sigma^2 + \rho(T-2)}; \quad i > 2 \tag{7.52}$$

The above equations can be utilized to generate the tree weights that will produce minimum variance when the between tree covariance can be represented in the form of Eq. (7.41). For example, if $\rho = \sigma^2$ and number of trees are 3, then $C_1 = C_2 = 1/4$ and $C_3 = 1/2$. For a general case with covariance matrix similar to Eq. (7.41) and $\rho = \sigma^2$, $C_1 = C_2 = \frac{1}{2(T-1)}$ and $C_i = \frac{1}{(T-1)}$ where $i > 2$.

Let us consider another canonical case, where the first two and last two trees of the forest are correlated among themselves, while there are minimal correlations among other trees (we will assume them to be zero) and variance of all the trees are equal. In that scenario, covariance matrix (V) will be given by

$$V = \begin{bmatrix} \sigma^2 & \rho_1 & \cdots & 0 & 0 \\ \rho_1 & \sigma^2 & \cdots & 0 & 0 \\ \vdots & \vdots & \ddots & \vdots & \vdots \\ 0 & 0 & \cdots & \sigma^2 & \rho_2 \\ 0 & 0 & \cdots & \rho_2 & \sigma^2 \end{bmatrix} \tag{7.53}$$

If the weight vector is C, then

$$C^T V C = \sigma^2 \sum_{i=1}^{T} C_i^2 + 2\rho_1 C_1 C_2 + 2\rho_2 C_{T-1} C_T \tag{7.54}$$

While incorporating the constraint of the weight vector $\left(\sum_{i=1}^{T} C_i = 1 \right)$ in Lagrange equation, our objective function will be given by:

$$f(c) = \sigma^2 \sum_{i=1}^{T} C_i^2 + 2\rho_1 C_1 C_2 + 2\rho_2 C_{T-1} C_T + \lambda \left(\sum_{i=1}^{T} C_i - 1 \right) \tag{7.55}$$

where λ is the Lagrange multiplier. Now, C_1 and C_2 and C_{T-1} and C_T are the weights of the first two and last two trees of the forest that are correlated, while C_is for $2 < i < T - 1$ denote the weight of all other trees that are not correlated. Differentiating Eq. (7.55) with respect to C_1, C_2, C_{T-1}, C_T, and C_i

$$\frac{\partial f}{\partial C_1} = 2C_1 \sigma^2 + 2\rho_1 C_2 + \lambda = 0 \tag{7.56}$$

$$\frac{\partial f}{\partial C_2} = 2C_2 \sigma^2 + 2\rho_1 C_1 + \lambda = 0 \tag{7.57}$$

$$\frac{\partial f}{\partial C_{T-1}} = 2C_{T-1} \sigma^2 + 2\rho_2 C_T + \lambda = 0 \tag{7.58}$$

$$\frac{\partial f}{\partial C_T} = 2C_T \sigma^2 + 2\rho_2 C_{T-1} + \lambda = 0 \tag{7.59}$$

$$\frac{\partial f}{\partial C_i} = 2C_i \sigma^2 + \lambda = 0; \quad i \neq 1, 2, T-1, T \tag{7.60}$$

In Eqs. (7.56), (7.57), if $\sigma^2 \neq \rho_1$, then $C_1 = C_2$. While in Eqs. (7.58), (7.59), if $\sigma^2 \neq \rho_2$, then $C_{T-1} = C_T$. Applying these results in the above equation, we arrive at

$$C_1 = \frac{-\lambda}{2(\sigma^2 + \rho_1)} = C_2 \tag{7.61}$$

$$C_{T-1} = \frac{-\lambda}{2(\sigma^2 + \rho_2)} = C_T \tag{7.62}$$

$$C_i = \frac{-\lambda}{2\sigma^2}; \quad i \neq 1, 2, T-1, T \tag{7.63}$$

$$\sum_{i=1}^{T} C_i = -\lambda \left[\frac{T-4}{2\sigma^2} + \frac{1}{\sigma^2 + \rho_1} + \frac{1}{\sigma^2 + \rho_2} \right] = 1 \tag{7.64}$$

If $\rho_1 = \rho_2 = \rho$, we arrive at the value of λ by solving Eq. (7.64).

$$-\lambda = \frac{2\sigma^2(\sigma^2 + \rho)}{(T-4)(\sigma^2 + \rho) + 4\sigma^2} \tag{7.65}$$

Substituting the value of Lagrange multiplier in Eqs. (7.61), (7.62), (7.63),

$$C_1 = C_2 = C_{T-1} = C_T = \frac{\sigma^2}{(T-4)(\sigma^2 + \rho) + 4\sigma^2} \tag{7.66}$$

$$C_i = \frac{\sigma^2 + \rho}{(T-4)(\sigma^2 + \rho) + 4\sigma^2}; \quad i \neq 1, 2, T-1, T \tag{7.67}$$

7.4.4 REGRESSION FOREST WEIGHT OPTIMIZATION

In this section, we discuss two approaches to select the weights for the ensemble of trees based on MLE and incorporation of tree correlations.

7.4.4.1 MLE for mixture model

Consider N independent and identically distributed samples $(\mathbf{x_i}, y_i)$ for $i = \{1, \ldots, N\}$ used for the generation of the T trees. Let $\alpha_1, \ldots, \alpha_T$ denote the weight of the trees, then the likelihood (conditional) will be given by:

$$\mathcal{L}(x_1, y_1, \ldots, x_N, y_N) = \prod_{i=1}^{N} \left(\sum_{j=1}^{T} \alpha_j P(y_i | \mathbf{x_i}, \phi_j) \right) \tag{7.68}$$

To ensure that $\sum_{j=1}^{T} \alpha_j P(y_i | \mathbf{x_i}, \phi_j)$ represents a valid probability distribution function, the weights has to satisfy the following constraints $\alpha_i \geq 0$ for $i = 1, \ldots, T$ and $\sum_{j=1}^{T} \alpha_j = 1$.

If we denote $P(y_i | \mathbf{x_i}, \phi_j)$ by $\xi_{i,j}$ for $i = 1, 2, \ldots, N$ samples and $j = 1, \ldots, T$ trees, a compact form of representation of the likelihood of the samples as an N length vector $\mathbf{f} = [f_1, \ldots, f_N]^T$ is $\mathbf{f} = \boldsymbol{\xi}\boldsymbol{\alpha}$:

$$\begin{bmatrix} f_1 \\ \cdot \\ \cdot \\ \cdot \\ f_N \end{bmatrix} = \begin{vmatrix} \xi_{1,1} & \cdots & \xi_{1,K} \\ & \cdots & \\ & \cdots & \\ & \cdots & \\ \xi_{N,1} & \cdots & \xi_{N,K} \end{vmatrix} \begin{bmatrix} \alpha_1 \\ \cdot \\ \cdot \\ \alpha_K \end{bmatrix}$$

The goal is to maximize the product $\prod_{i=1}^{N} f_i$ with constraints $\boldsymbol{\alpha} \geq 0$ and $\sum_{j=1}^{K} \alpha_j - 1 = 0$.

We solve this optimization problem using Matlab *fmincon* function that utilizes an interior point approach to find the minimum of a constrained nonlinear function.

7.4.4.2 Weight distribution based on correlation of trees for weighted sum of random variables model

Among T trees of the forest, consider that some of them have high correlation between themselves while others have limited correlations which can be clustered as correlated and uncorrelated trees. The idea is to give higher weight to the uncorrelated trees as compared to the uncorrelated trees, which is shown algorithmically in Algorithm 7.1.

ALGORITHM 7.1 ALGORITHMIC REPRESENTATION OF WEIGHT SELECTION

STEP 1: Cluster Trees Based on Correlations
STEP 2: Let the k clusters be $[\alpha_1, \ldots, \alpha_{\rho_1}], \ldots, [\alpha_{T-\rho_k+1}, \ldots, \alpha_T]$
STEP 3: Assign equal weight $\frac{1}{k}$ to each cluster
that is, assign weight $\frac{1}{\rho_r k}$ to each tree in cluster r

---

7.4.5 APPLICATION TO SYNTHETIC DATASET

Maximum likelihood estimate for mixture model: We have generated a random dataset of 100 samples and 10 input features for an output that is a function of four input features. The output response is generated using Eq. (7.69).

$$\mathbf{Y}_1 = 2\mathbf{x}_1 + 5\mathbf{x}_2 - 1.5\mathbf{x}_3 + \mathbf{x}_4 \tag{7.69}$$

One of our objectives is to check if we are able to reduce the mean square error (MSE) in prediction along with lowering the width of the CI by using MLE of the tree weights. "Henceforth, the probabilistic random forest with tree weights generated by MLE will be termed PRF and the probabilistic random forest with equal tree weights will be denoted as RF."

To see the comparative effect of RF and PRF on MSE and width of the CI, we have used 75 samples (75%) as training and 25 samples (25%) for testing

Table 7.3 MSE, Correlation Between Actual and Predicted Responses and Change in CI Width for Different Confidence Level for 100 Samples for Different Number of Trees in the Forest Is Reported Here

	MSE		Correlation		Different Confidence Level				
Tree	RF	PRF	RF	PRF	50%	70%	80%	95%	99%
2	0.0346	0.0045	0.7412	0.9698	10.0239	8.3458	8.5816	5.7037	4.2077
3	0.0128	0.0045	0.9072	0.9655	5.1107	5.3763	5.3435	3.5975	2.5964
5	0.0268	0.0037	0.8445	0.9721	6.5495	6.9138	6.8993	4.7179	3.5474
10	0.0087	0.0052	0.9556	0.9538	4.2105	4.1349	3.8429	3.9399	3.4521
20	0.0079	0.0070	0.9734	0.9712	9.8182	9.2428	8.6281	7.2049	6.3845
100	0.0065	0.0060	0.9547	0.9724	3.1898	2.4582	1.9939	4.0987	5.9993

Notes: Minimum size of the leaf is 3, while for each split of the node 5 features have been selected [39].

(hold-out validation). We have considered different numbers of trees to build the two models (RF and PRF) and the change in MSE, correlation between actual and predicted values and width of CI is reported in Table 7.3. Here, a 10% change in the width of the CI denotes that on an average, PRF generated CI is 10% lower than the RF generated CI. Table 7.3 shows improvement in all the cases for PRF, as compared to RF in lowering the CI and MSE and increasing the correlation between actual and predicted responses.

Table 7.4 shows the change in MSE, correlation between actual and predicted responses, and width of CI for a simulation with 500 samples. Similar to the case with 100 samples, we observe improvement for PRF as compared to RF in terms of lowering the CI and MSE and increasing the correlation between actual and predicted responses.

Table 7.4 MSE, Correlation Between Actual and Predicted Output, and Change in CI Width for Different Confidence Levels for 500 Samples for Different Number of Trees in the Forest

	MSE		Correlation		Different Confidence Level				
Tree	RF	PRF	RF	PRF	50%	70%	80%	95%	99%
2	0.0051	0.0033	0.9576	0.9739	4.0496	3.9550	3.6326	3.9503	3.2406
3	0.0056	0.0031	0.9477	0.9682	2.7099	3.5152	3.3654	3.1243	2.5936
5	0.0025	0.0019	0.9723	0.9834	3.6237	3.1812	3.2234	2.8925	2.1825
10	0.0021	0.0017	0.9825	0.9848	4.5110	4.6718	4.9503	6.2598	6.3468
20	0.0020	0.0014	0.9819	0.9861	3.2640	2.8151	2.6526	2.7465	2.5435

Notes: Minimum size of the leaf is 3, while for each split of the node 5 features have been selected [39].

Weighted sum of random variables: In this section, we consider the effect of tree weights on the MSE and prediction Variance for the weighted sum of random variables scenario. We have generated a synthetic matrix of 500×1000 where $\frac{2}{3}$ of the samples (334) have been considered for model training and the rest are used for testing. The output response has been generated using Eq. (7.70).

$$Y_1 = 2x_1 + 10x_2 - 1.5x_3 + x_4 \tag{7.70}$$

We used the filter feature selection approach *RRelieff* [42] to reduce the initial set of 1000 features to 100 (10 among these 100 are randomly considered for each node splitting) for training the regression trees.

The random forest model has been generated with five trees and the covariance matrix for training samples of the model is given by Eq. (7.71).

$$V = \begin{bmatrix} 0.0664 & 0.0316 & 0.0297 & 0.0405 & 0.0344 \\ 0.0316 & 0.0548 & 0.0249 & 0.0304 & 0.0280 \\ 0.0298 & 0.0250 & 0.0582 & 0.0328 & 0.0257 \\ 0.0408 & 0.0299 & 0.0334 & 0.0606 & 0.0357 \\ 0.0342 & 0.0277 & 0.0258 & 0.0345 & 0.0595 \end{bmatrix} \tag{7.71}$$

In Eq. (7.71), the diagonal elements are the variance of each tree with itself, while the nondiagonal elements are covariance between different trees. We note that the covariance between trees 1 and 4 is high (more than $\frac{2}{3}$ of average variance of individual trees) compared to other covariances. Using the weighting method, the variance will be lowered when the weight of trees 1 and 4 is $\dfrac{\sigma^2}{T * \sigma^2 + \rho(T-2)} =$ 0.1378 and weights of trees 2, 3, and 5 is $\dfrac{\sigma^2 + \rho}{T * \sigma^2 + \rho(T-2)} = 0.2413$.

The covariance matrix for testing samples of the model is

$$\Sigma = \begin{bmatrix} 0.0462 & 0.0071 & 0.0060 & 0.0264 & 0.0078 \\ 0.0065 & 0.0513 & 0.0035 & 0.0194 & 0.0042 \\ 0.0065 & 0.0040 & 0.0580 & 0.0027 & 0.0015 \\ 0.0269 & 0.0194 & 0.0020 & 0.0664 & 0.0114 \\ 0.0085 & 0.0043 & 0.0007 & 0.0101 & 0.0636 \end{bmatrix} \tag{7.72}$$

Thus the variance of the forest with equal weight of the trees is

$$[1/5 \ 1/5 \ 1/5 \ 1/5 \ 1/5] * \Sigma * [1/5 \ 1/5 \ 1/5 \ 1/5 \ 1/5]' = 0.0186 \tag{7.73}$$

and variance of the forest with unequal weight of the trees.
$(C = [0.1378 \ 0.2413 \ 0.2413 \ 0.1378 \ 0.2413])$ is

$$C * \Sigma * C' = 0.0179 \tag{7.74}$$

The above results illustrate that for the weighted sum of random variables scenario, the variance of the forest prediction can be reduced by generating the weight of the trees based on tree clusters, as compared to using equal weights for all trees.

7.4.6 MAXIMUM LIKELIHOOD ESTIMATE OF MIXTURE MODEL APPLIED TO CCLE DATASET

CCLE dataset has been downloaded from http://www.broadinstitute.org/ccle/home. CCLE dataset has two types of genetic characterization information: (i) Gene Expression and (ii) Single Nucleotide Polymorphism (SNP6). Gene expression has been downloaded from *CCLE_Expression _Entrez_2012-09-29.gct*. In this dataset, there are 18,988 gene features with no missing values for 1037 cell lines. The SNP6 dataset has been extracted from *CCLE_copynumber _byGene_2013-12-03.txt*. For 1043 cell lines, there are 233,316 features. For our experiments, we have selected 1012 cell lines that are common to both gene expression and SNP6 dataset.

The drug sensitivity data have been downloaded from the addendum published by Barretina et al. [28]. These data provide 24 drug responses for 504 cell lines. Drug sensitivity data of *Area Under Curve* have been collected from *ActArea* and normalized to [0 1]. The SNP6 and gene expression data integrated model was constructed based on individual random forest models combined with a linear regression stacking approach [2].

CI and variance: For the calculation of the CI, we have considered 15 drugs in the CCLE database and considered samples with drug sensitivity higher than 0.1 so as to have noticeable variance among the output responses. The number of samples used for the experiments for the 15 drugs varies from 70 to 395. We have used fivefold cross-validation for all our computations where the data samples are randomly partitioned into five equal parts and four parts are used for training and the remaining part used for testing, and the process repeated five times corresponding to the five different testing partitions.

Based on the model inferred from the training samples, the mean and variance of the output of the leaf node for the testing set has been calculated. Thus, for a testing set of 20 samples and 10 trees, we have a matrix of mean and variance of size 20×10. Based on the calculated means and variances, a Gaussian mixture distribution has been derived. Cumulative distribution function has been eventually derived from this distribution to calculate the CIs for different confidence levels.

To analyze the estimated CIs, we have considered the ratio of the number of experimental testing responses contained in the predicted CI to the total number of testing samples. We will term the ratio as the *coverage probability* of the CI. Note that we are calculating the coverage probability from cross-validation data as compared to resubstitution data and thus there can be significant differences from the CI level for limited samples.

The coverage probability for different confidence levels for all the 15 drugs is shown in Table 7.5. We observe that the RF and PRF coverage probabilities are quite similar and PRF coverage probability is closer to the actual confidence level than the RF coverage probability. As expected, the coverage probability is increasing with the increase in confidence level for both RF and PRF models.

For the results shown in Table 7.5, we also calculated the *P*-values of paired *t*-test between PRF and RF predictions and actual responses. The *P*-values of paired

Table 7.5 Coverage Probabilities for Four Confidence Levels (CL) for PRF and RF Predictions for Different Drugs

| Drug | Coverage Probability | | | | | | | |
| | 50% CL | | 70% CL | | 80% CL | | 95% CL | |
	RF	PRF	RF	PRF	RF	PRF	RF	PRF
17-AAG	0.686	0.650	0.911	0.883	0.977	0.959	1	1
AZD0530	0.829	0.764	0.934	0.929	0.949	0.959	0.994	0.989
AZD6244	0.743	0.712	0.920	0.893	0.951	0.938	0.991	0.986
Erlotinib	0.844	0.836	0.931	0.939	0.982	0.991	0.991	1
Lapatinib	0.838	0.788	0.932	0.898	0.974	0.949	1	1
Nilotinib	0.795	0.742	0.913	0.881	0.956	0.913	0.986	0.989
Nutlin-3	0.872	0.825	0.941	0.953	0.953	0.965	1	1
Paclitaxel	0.707	0.671	0.909	0.886	0.969	0.959	1	0.997
PD-0325901	0.686	0.665	0.893	0.872	0.965	0.944	0.996	0.996
PD-0332991	0.849	0.831	0.973	0.929	0.991	0.964	1	0.991
PF2341066	0.842	0.808	0.931	0.938	0.972	0.965	0.993	1
PHA-665752	0.855	0.842	0.973	0.960	1	1	1	1
PLX4720	0.9	0.785	0.971	0.942	0.985	0.957	0.985	0.985
Sorafenib	0.901	0.862	0.950	0.950	0.980	0.970	0.99	0.99
TAE684	0.816	0.771	0.955	0.948	0.982	0.965	0.996	0.996

Notes: We have used $T = 10$ trees and the following constraints for the weights of the trees for PRF model $\frac{1}{3T} \leq \alpha_i \leq 1$ and $\sum_{j=1}^{T} \alpha_j = 1$ [39].

t-test between (a) PRF prediction and actual responses turned out to be 0.6172, and between (b) RF prediction and actual responses turned out to be 0.6052. A higher value for the PRF scenario represents that the PRF predictions are closer to the actual responses as compared to the RF predictions.

The change in coverage probability with number of trees (T) for drug 17-AAG is shown in Table 7.6. We observe that the coverage probabilities are closer to actual confidence levels with lower number of trees. However, the increase in number of trees in the forest produces lower variance and higher prediction accuracy.

From Tables 7.5 and 7.6, we observe that both RF and PRF provide similar coverage probabilities for the generated CIs.

We next analyzed the error in prediction using different error metrics (MSE, MAE, and Normalized Root Mean Square Error [NRMSE]) and the length of the CIs for PRF in comparison to RF and weighted random forest (wRF) [43].

We first explored whether PRF in comparison to RF and wRF can reduce prediction error (as measured by different metrics), while decreasing the CI in the majority of the cases. The ratio of number of testing samples where the PRF model generated CI is lower than the RF model generated CI to all samples is defined as

Table 7.6 Coverage Probabilities for Four Confidence Levels for Different Number of Trees (From 2 to 100) for Drug 17-AAG With 395 Samples

| No. of Trees (T) | Coverage Probability | | | | | | | |
| | 50% CL | | 70% CL | | 80% CL | | 95% CL | |
	RF	PRF	RF	PRF	RF	PRF	RF	PRF
2	0.6532	0.6228	0.8228	0.8101	0.8911	0.8886	0.9848	0.9873
5	0.6937	0.6658	0.9038	0.8861	0.9570	0.9418	0.9975	0.9949
10	0.686	0.650	0.911	0.883	0.977	0.959	1	1
20	0.7089	0.6886	0.9139	0.9089	0.9747	0.9722	1	1
100	0.7291	0.7241	0.9342	0.9266	0.9823	0.9772	1	1

Notes: Results for both RF and PRF models show similar type of behavior [39].

PRF CI ratio. For example, a PRF CI ratio of 0.60 will denote that for 60% of the testing samples, PRF model generated CI is lower than RF model generated CI.

The MSE, MAE, and NRMSE for different drugs are shown in Table 7.7 while Table 7.8 shows the PRF CI ratio in comparison to RF and wRF for different confidence levels. Tables 7.7 and 7.8 show that the average errors for PRF in

Table 7.7 Performance for All the Drugs in Terms of MSE, MAE, and NRMSE

| Drug | MSE | | | MAE | | | NRMSE | | |
	RF	wRF	PRF	RF	wRF	PRF	RF	wRF	PRF
17-AAG	0.0175	0.0167	0.0185	0.1075	0.1051	0.1108	1.0055	0.9828	1.0363
AZD0530	0.0071	0.0066	0.0078	0.0628	0.0605	0.0650	1.0023	0.9637	1.0502
AZD6244	0.0157	0.0160	0.0169	0.0983	0.1018	0.1011	0.9567	0.9642	0.9946
Erlotinib	0.0047	0.0049	0.0057	0.0513	0.0528	0.0573	0.9956	1.0021	1.0894
Lapatinib	0.0070	0.0073	0.0079	0.0629	0.0616	0.0649	0.9799	0.9992	1.0418
Nilotinib	0.0241	0.0226	0.0230	0.1015	0.0931	0.0974	1.0326	1.0021	1.0116
Nutlin-3	0.0034	0.0038	0.0037	0.0435	0.0449	0.0438	0.9762	1.0415	1.0296
Paclitaxel	0.0237	0.0236	0.0243	0.1226	0.1240	0.1257	0.9229	0.9205	0.9354
PD-0325901	0.0259	0.0254	0.0279	0.1312	0.1306	0.1364	0.9534	0.9446	0.9873
PD-0332991	0.0053	0.0045	0.0058	0.0573	0.0524	0.0609	0.9825	0.9139	1.0379
PF2341066	0.0075	0.0074	0.0060	0.0646	0.0603	0.0564	1.0578	1.0458	0.9519
PHA-665752	0.0039	0.0039	0.0039	0.0509	0.0497	0.0488	1.0667	1.0700	1.0616
PLX4720	0.0100	0.0107	0.0106	0.0730	0.0775	0.0720	1.0011	1.0270	1.0283
Sorafenib	0.0072	0.0069	0.0063	0.0568	0.0533	0.0505	1.0471	1.0309	0.9923
TAE684	0.0087	0.0075	0.0089	0.0696	0.0644	0.0707	0.9682	0.8957	0.9759
Average	0.0114	0.0112	0.0118	0.0769	0.0755	0.0774	0.9966	0.9869	1.0149

Notes: We have used $T = 10$ and the following constraints for the weight of the trees for the PRF model $\frac{1}{4T} \le \alpha_i \le 1$ and $\sum_{j=1}^{T} \alpha_j = 1$ [39].

Table 7.8 Performance for All the Drugs in Terms of CI

Drug	PRF CI Ratio Compared to RF					PRF CI Ratio Compared to wRF				
	50% CL	70% CL	80% CL	95% CL	99% CL	50% CL	70% CL	80% CL	95% CL	99% CL
17-AAG	0.537	0.572	0.560	0.570	0.575	0.549	0.560	0.557	0.565	0.572
AZD0530	0.508	0.533	0.583	0.598	0.583	0.503	0.533	0.578	0.603	0.578
AZD6244	0.628	0.637	0.602	0.642	0.655	0.628	0.637	0.606	0.642	0.659
Erlotinib	0.423	0.423	0.431	0.405	0.431	0.422	0.422	0.431	0.405	0.431
Lapatinib	0.534	0.525	0.568	0.576	0.517	0.534	0.525	0.559	0.576	0.525
Nilotinib	0.624	0.624	0.667	0.669	0.624	0.591	0.624	0.667	0.731	0.667
Nutlin-3	0.511	0.640	0.628	0.663	0.651	0.570	0.640	0.628	0.663	0.651
Paclitaxel	0.520	0.545	0.578	0.614	0.669	0.525	0.530	0.581	0.619	0.669
PD-0325901	0.510	0.541	0.559	0.600	0.621	0.528	0.538	0.572	0.607	0.624
PD-0332991	0.522	0.500	0.500	0.500	0.540	0.531	0.487	0.487	0.478	0.531
PF2341066	0.562	0.582	0.616	0.582	0.562	0.562	0.589	0.616	0.582	0.555
PHA-665752	0.566	0.645	0.618	0.645	0.724	0.553	0.632	0.618	0.645	0.724
PLX4720	0.700	0.700	0.686	0.729	0.729	0.700	0.686	0.686	0.729	0.743
Sorafenib	0.451	0.481	0.549	0.539	0.578	0.461	0.490	0.559	0.549	0.578
TAE684	0.554	0.592	0.595	0.599	0.602	0.557	0.599	0.595	0.599	0.602
Average	0.541	0.564	0.577	0.593	0.605	0.548	0.566	0.583	0.599	0.607

Notes: PRF CI ratio denotes the ratio of samples where PRF CI is lower than RF CI or wRF CI. We have used $T = 10$ and the following constraints for the weight of the trees for the PRF model $\frac{1}{4T} \leq \alpha_j \leq 1$ and $\sum_{j=1}^{T} \alpha_j = 1$ [39].

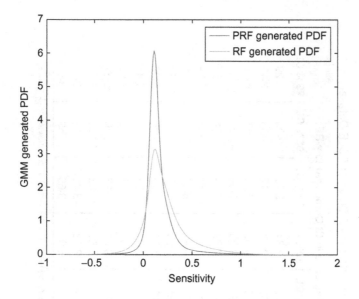

FIG. 7.7

RF generated PDF is more spread out than PRF generated pdf, which implies that the CI of RF generated pdf is higher than PRF generated pdf [39].

comparison to RF and wRF are similar based on multiple error metrics, whereas the PRF CI ratio is larger than 0.5 (between 0.54 and 0.6) for all confidence levels. Thus the results support the idea that as compared to using equal weights for all trees, weight optimization using MLE can potentially predict drug sensitivity with higher confidence while maintaining similar error. Fig. 7.7 represents two example pdfs generated by PRF and RF, which show that the PRF predicted distribution has lower variance as compared to RF while maintaining similar mean.

The % decreases in mean CI with PRF as compared to RF and wRF are shown in Table 7.9. We note that the average CI for PRF is lower than RF and wRF in overwhelming majority of the cases.

We also compared our approach with Quantile Regression Forests (QRF) [44] that uses nonparametric empirical distributions to model the distributions at the leaf nodes. We observed (results not included) that QRF can produce smaller CIs than RF and PRF, but the coverage probability of PRF is significantly lower. It appears that the empirical distributions based on a few samples can provide smaller variance, but has limited coverage which defeats the purpose of designing the CIs.

Prior feature selection: In this experiment, we have used filter feature selection algorithm *RRelieff* [42] to reduce the initial set of features used for training the RF, PRF, and wRF models. We have considered the CCLE cell lines that are common to all 15 drugs, resulting in 396 samples. Features election has been used to reduce the number of features to 50 for each dataset.

Table 7.9 % Decrease in Mean CI With PRF as Compared to RF and wRF for 15 Drugs of CCLE Dataset

Drug	% Decrease in CI Compared to RF					% Decrease in CI Compared to wRF				
	50% CL	70% CL	80% CL	95% CL	99% CL	50% CL	70% CL	80% CL	95% CL	99% CL
17-AAG	2.43	2.39	2.11	2.03	2.28	2.41	2.34	2.11	2.00	2.29
AZD0530	3.04	2.20	2.18	2.56	3.15	3.09	2.20	2.17	2.56	3.15
AZD6244	4.6	4.57	4.48	4.76	5.10	4.60	4.61	4.50	4.77	5.12
Erlotinib	-1.92	-2.13	-2.02	-1.34	1.20	-1.81	-2.26	-1.98	-1.27	1.24
Lapatinib	4.25	4.77	4.70	4.04	4.62	4.21	4.89	4.64	3.94	4.65
Nilotinib	7.87	9.07	11.76	13.12	9.60	7.39	9.20	11.87	13.29	9.84
Nutlin-3	8.76	6.76	5.80	5.66	7.73	8.94	6.87	5.92	5.81	7.76
Paclitaxel	3.02	2.97	3.07	3.26	3.35	2.96	2.93	3.12	3.25	3.37
PD-0325901	1.10	1.84	2.07	2.11	2.12	1.10	1.75	2.09	2.05	2.08
PD-0332991	2.23	1.22	0.36	0.74	1.52	2.24	0.93	0.35	0.64	1.46
PF2341066	3.69	4.63	4.26	3.45	3.75	3.56	4.55	4.21	3.43	3.69
PHA-665752	9.65	9.11	9.06	8.79	8.35	9.47	9.01	8.98	8.70	8.37
PLX4720	13.12	11.08	11.85	12.85	11.41	13.13	11.55	12.03	13.12	11.51
Sorafenib	-2.46	-3.15	-3.65	-1.50	2.34	-2.39	-3.23	-3.48	-1.47	2.31
TAE684	4.08	3.90	3.64	3.59	3.62	4.17	3.91	3.65	3.52	3.63

Notes: Number of features used for each split is 10, minimum number of sample in a leaf node = 5, $T = 10$, and PRF constraints of $\frac{1}{4T} \leq \alpha_i \leq 1$ and $\sum_{j=1}^{T} \alpha_j = 1$ [39].

Table 7.10 Performance for All the Drugs in Terms of MSE and MAE for PRF Compared With RF and wRF

Drug	MSE			MAE		
	RF	wRF	PRF	RF	wRF	PRF
17-AAG	0.0179	0.0186	0.0171	0.1088	0.1138	0.1044
AZD0530	0.0103	0.0102	0.0079	0.0839	0.0739	0.0737
AZD6244	0.0162	0.0128	0.0133	0.1037	0.0920	0.0927
Erlotinib	0.0042	0.0043	0.0045	0.0502	0.0524	0.0503
Lapatinib	0.0052	0.0052	0.0058	0.0513	0.0512	0.0565
Nilotinib	0.0056	0.0069	0.0042	0.0527	0.0520	0.0490
Nutlin-3	0.0042	0.0035	0.0030	0.0436	0.0441	0.0450
Paclitaxel	0.0192	0.0219	0.0182	0.1122	0.1180	0.1098
PD-0325901	0.0244	0.0231	0.0230	0.1313	0.1233	0.1218
PD-0332991	0.0050	0.0039	0.0046	0.0535	0.0532	0.0501
PF2341066	0.0046	0.0041	0.0068	0.0445	0.0440	0.0536
PHA-665752	0.0043	0.0046	0.0030	0.0453	0.0458	0.0427
PLX4720	0.0042	0.0030	0.0068	0.0445	0.0413	0.0512
Sorafenib	0.0055	0.0034	0.0031	0.0460	0.0439	0.0443
TAE684	0.0120	0.0093	0.0089	0.0850	0.0758	0.0764
Average	0.0095	0.0090	0.0087	0.0704	0.0683	0.0681

Notes: Number of features used for building the model is 50 and the number of trees considered is 40. $\frac{1}{4T} \leq \alpha_i \leq 1$ and $\sum_{j=1}^{T} \alpha_j = 1$ [39].

Table 7.10 shows the average errors in terms of MSE and MAE for the 15 drugs with 50 selected features for RF, wRF, and PRF. We observe that PRF performs better in comparison to RF and wRF in terms of both average MSE and MAE.

The prediction performance can also be measured in terms of the *bias* and *variance* of the error distributions produced by different predictive models. The bias will be an inverse measure of *accuracy* and variance will be an inverse measure of *Precision*. Table 7.11 shows the bias and variance for RF, wRF, and PRF for different drugs. We note that the average absolute bias (measure of inaccuracy) is lower for PRF (0.0014) as compared to RF (0.0020) and wRF (0.0025). Similarly, the variance (measure of imprecision) for PRF (0.0087) is smaller than variances for RF (0.0096) and wRF (0.0090).

The results shown in Tables 7.10 and 7.11 show that the PRF provides improvement in terms of average error, accuracy, and precision. We next consider the length of the CI with PRF as compared to RF and wRF. Table 7.12 shows that the percentage decrease in average CI for PRF when compared to RF and wRF is positive for majority of the drugs.

Table 7.11 Performance for All the Drugs in Terms of Bias and Variance for PRF Compared With RF and wRF

Drug	Bias			Variance		
	RF	wRF	PRF	RF	wRF	PRF
17-AAG	0.0023	0.0087	0.0020	0.0182	0.0187	0.0173
AZD0530	−0.0055	−0.0150	−0.0005	0.0104	0.0101	0.0080
AZD6244	−0.0133	0.0190	0.0037	0.0162	0.0126	0.0135
Erlotinib	0.0054	0.0110	−0.0054	0.0042	0.0042	0.0045
Lapatinib	−0.0116	−0.0080	−0.0142	0.0052	0.0052	0.0057
Nilotinib	−0.0062	0.0002	0.0012	0.0057	0.0070	0.0043
Nutlin-3	−0.0102	0.0020	−0.0057	0.0042	0.0036	0.0031
Paclitaxel	0.0040	0.0192	0.0351	0.0194	0.0217	0.0171
PD-0325901	0.0086	0.0025	−0.0048	0.0246	0.0234	0.0232
PD-0332991	0.0118	0.0064	0.0083	0.0050	0.0039	0.0047
PF2341066	0.0012	−0.0053	−0.0003	0.0047	0.0041	0.0069
PHA-665752	−0.0045	−0.0108	0.0058	0.0044	0.0046	0.0030
PLX4720	−0.0027	0.0030	−0.0038	0.0042	0.0030	0.0069
Sorafenib	−0.0081	0.0005	−0.0028	0.0055	0.0034	0.0032
TAE684	−0.0009	0.0048	0.0030	0.0122	0.0094	0.0090
Average	−0.0020	0.0025	0.0014	0.0096	0.0090	0.0087

Notes: Number of features used for building the model is 50 and the number of trees considered is 40. $\frac{1}{4T} \le \alpha_i \le 1$ and $\sum_{j=1}^{T} \alpha_j = 1$ [39].

DISCUSSION

This section presented a probabilistic analysis of Random Forests by representing a forest as an ensemble of probabilistic regression trees. The two perspectives presented in this section are based on how we would like to treat a probabilistic ensemble of regression trees. We can consider that we would like to select one tree from the available trees conditioned on the weights and predict the output response based on the tree distribution resulting in the mixture distribution scenario. The second scenario is where the output response is considered as a weighted average of all realizations of the trees, similar to the averaging of responses from different trees as considered in conventional Random Forest. Thus, if individual trees have large biases (measure of inaccuracy) that are both positive and negative, considering a Weighted Sum of Random Variables can provide a better representation. If we consider the mixture distribution approach for this case, selecting an individual tree for each prediction might not remove the bias. However, the mixture distribution approach is reasonable in selecting tree weights to reduce the CIs while maintaining coverage and MSE as shown in the reported results.

To summarize, the probabilistic representation allowed us to generate and analyze the CIs of individual predictions. We explored various structures of covariance

Table 7.12 Performance for All the Drugs in Terms of % Decrease in Mean CI With PRF as Compared to RF and wRF

Drug	% Decrease in CI Compared to RF					% Decrease in CI Compared to wRF				
	50% CL	70% CL	80% CL	95% CL	99% CL	50% CL	70% CL	80% CL	95% CL	99% CL
17-AAG	2·79	2·37	2·48	2·59	2·54	2·87	2·38	2·52	2·61	2·54
AZD0530	3·10	2·47	2·86	2·52	3·40	3·11	2·35	2·98	2·58	3·37
AZD6244	1·52	1·68	1·88	0·87	0·70	1·53	1·83	1·79	0·93	0·79
Erlotinib	3·02	2·13	1·75	1·37	1·63	3·03	2·17	1·76	1·39	1·64
Lapatinib	1·49	1·34	1·29	1·02	1·97	1·37	1·31	1·29	0·96	1·95
Nilotinib	1·62	0·64	0·49	2·12	4·50	1·44	0·57	0·44	2·13	4·49
Nutlin-3	0·01	−0·61	−1·01	−1·18	0·06	0·07	−0·65	−1·01	−1·14	0·08
Paclitaxel	3·92	2·91	2·38	1·92	2·03	3·90	2·79	2·48	1·91	2·01
PD-0325901	−0·09	−0·07	−0·03	−0·49	−0·40	0·00	−0·01	−0·06	−0·43	−0·41
PD-0332991	3·38	3·13	2·58	1·65	2·35	3·39	3·13	2·55	1·57	2·29
PF2341066	5·76	6·09	5·08	5·06	5·88	5·82	6·02	5·15	5·05	5·80
PHA-665752	1·83	0·97	0·37	−0·59	−0·23	1·76	0·93	0·37	−0·66	−0·23
PLX4720	0·80	0·59	0·97	−0·23	0·04	0·80	0·64	0·94	−0·20	0·01
Sorafenib	3·95	4·16	3·57	2·81	4·33	3·95	4·29	3·55	2·87	4·32
TAE684	0·61	0·40	−0·02	−0·68	−0·33	0·70	0·44	0·02	−0·62	−0·33

Notes: Number of features used for building the model is 50 and the number of trees considered is $T = 40$. $\frac{1}{4T} \le \alpha_i \le 1$ and $\sum_{j=1}^{T} \alpha_j = 1$ [39].

matrices representing the relationships between the generated probabilistic regression trees and the corresponding tree weights that will optimally reduce the variance of prediction. We studied the effect of tree weights generated, using MLE on different error measures and prediction CIs. The application of the maximum likelihood estimates of tree weights on the CCLE drug sensitivity prediction problem illustrated the average reduction in CI while maintaining or lowering MSE.

REFERENCES

[1] J.C. Costello, et al., A community effort to assess and improve drug sensitivity prediction algorithms, Nat. Biotechnol. (2014), http://dx.doi.org/10.1038/nbt.2877.

[2] Q. Wan, R. Pal, An ensemble based top performing approach for NCI-DREAM drug sensitivity prediction challenge, PLoS ONE 9 (6) (2014) e101183.

[3] G. Riddick, H. Song, S. Ahn, J. Walling, D. Borges-Rivera, W. Zhang, H.A. Fine, Predicting in vitro drug sensitivity using random forests, Bioinformatics 27 (2) (2011) 220–224.

[4] K.L. Lunetta, L.B. Hayward, J. Segal, P. Van Eerdewegh, Screening large-scale association study data: exploiting interactions using random forests, BMC Genet. 5 (2004) 32.

[5] A. Bureau, J. Dupuis, K. Falls, K.L. Lunetta, B. Hayward, T.P. Keith, P. Van Eerdewegh, Identifying SNPs predictive of phenotype using random forests, Genet. Epidemiol. 28 (2) (2005) 171–182.

[6] R. Diaz-Uriarte, S.A. de Andres, Gene selection and classification of microarray data using random forest, BMC Bioinform. 7 (2006) 3.

[7] D. Yao, J. Yang, X. Zhan, X. Zhan, Z. Xie, A novel random forests-based feature selection method for microarray expression data analysis, Int. J. Data Min. Bioinform. 13 (1) (2015) 84–101.

[8] H. Pang, S.L. George, K. Hui, T. Tong, Gene selection using iterative feature elimination random forests for survival outcomes, IEEE/ACM Trans. Comput. Biol. Bioinform. 9 (5) (2012) 1422–1431.

[9] K. Moorthy, M.S. Mohamad, Random forest for gene selection and microarray data classification, Bioinformation 7 (3) (2011) 142–146.

[10] X.Y. Wu, Z.Y. Wu, K. Li, Identification of differential gene expression for microarray data using recursive random forest, Chin. Med. J. 121 (24) (2008) 2492–2496.

[11] D.J. Yu, Y. Li, J. Hu, X. Yang, J.Y. Yang, H.B. Shen, Disulfide connectivity prediction based on modelled protein 3D structural information and random forest regression, IEEE/ACM Trans. Comput. Biol. Bioinform. 12 (3) (2015) 611–621.

[12] H. Li, K.S. Leung, M.H. Wong, P.J. Ballester, Substituting random forest for multiple linear regression improves binding affinity prediction of scoring functions: cyscore as a case study, BMC Bioinform. 15 (2014) 291.

[13] Y. Qi, Z. Bar-Joseph, J. Klein-Seetharaman, Evaluation of different biological data and computational classification methods for use in protein interaction prediction, Proteins 63 (3) (2006) 490–500.

[14] J.D. Ospina, J. Zhu, C. Chira, A. Bossi, J.B. Delobel, V. Beckendorf, B. Dubray, J.L. Lagrange, J.C. Correa, A. Simon, O. Acosta, R. de Crevoisier, Random forests to predict rectal toxicity following prostate cancer radiation therapy, Int. J. Radiat. Oncol. Biol. Phys. 89 (5) (2014) 1024–1031.

[15] D.S. Cao, Q.N. Hu, Q.S. Xu, Y.N. Yang, J.C. Zhao, H.M. Lu, L.X. Zhang, Y.Z. Liang, In silico classification of human maximum recommended daily dose based on modified random forest and substructure fingerprint, Anal. Chim. Acta 692 (1–2) (2011) 50–56.

[16] L. Breiman, Bagging predictors, Mach. Learn. 24 (2) (1996) 123–140.

[17] T.K. Ho, Random decision forests, in: Proceedings of the Third International Conference on Document Analysis and Recognition (ICDAR'95), vol. 1, 1995, p. 278.

[18] T.K. Ho, The random subspace method for constructing decision forests, IEEE Trans. Pattern Anal. Mach. Intell. 20 (8) (1998) 832–844.

[19] Y. Amit, D. Geman, Shape quantization and recognition with randomized trees, Neural Comput. 9 (7) (1997) 1545–1588.

[20] Y. Freund, R.E. Schapire, A decision-theoretic generalization of on-line learning and an application to boosting, J. Comput. Syst. Sci. 55 (1) (1997) 119–139.

[21] K. Monteith, J.L. Carroll, K.D. Seppi, T.R. Martinez, Turning Bayesian model averaging into Bayesian model combination, in: IJCNN, IEEE, 2011, pp. 2657–2663.

[22] B. Clarke, Comparing Bayes model averaging and stacking when model approximation error cannot be ignored, J. Mach. Learn. Res. 4 (2003) 683–712.

[23] L. Breiman, Random forests, Mach. Learn. 45 (1) (2001) 5–32.

[24] N. Meinshausen, Quantile regression forests, J. Mach. Learn. Res. 7 (2006) 983–999.

[25] G. Biau, Analysis of a random forests model, J. Mach. Learn. Res. 98888 (2012) 1063–1095.

[26] C. Strobl, A.L. Boulesteix, A. Zeileis, T. Hothorn, Bias in random forest variable importance measures: illustrations, sources and a solution, BMC Bioinform. 8 (2007) 25.

[27] T. Hothorn, K. Hornik, A. Zeileis, Unbiased recursive partitioning: a conditional inference framework, J. Comput. Graph. Stat. 15 (3) (2006) 651–674.

[28] J. Barretina, et al., The cancer cell line encyclopedia enables predictive modelling of anticancer drug sensitivity, Nature 483 (7391) (2012) 603–607.

[29] D.H. Wolpert, Stacked generalization, Neural Netw. 5 (1992) 241–259.

[30] B. Efron, Bootstrap methods: another look at jackknife, Ann. Statist. 7 (1979) 1–26.

[31] W.J. Krzanowski, Data-based interval estimation of classification error rates, J. Appl. Stat. 28 (5) (2001) 585–595.

[32] K. Kira, L.A. Rendell, A practical approach to feature selection, in: Proceedings of the Ninth International Workshop on Machine Learning (ML92), Morgan Kaufmann Publishers Inc., San Francisco, CA, 1992, pp. 249–256.

[33] I. Kononenko, Estimating attributes: analysis and extensions of relief, in: Proceedings of the European Conference on Machine Learning on Machine Learning (ECML-94), Springer-Verlag New York, Inc., Secaucus, NJ, 1994, pp. 171–182.

[34] Broad-Novartis Cancer Cell Line Encyclopedia, Genetic and pharmacologic characterization of a large panel of human cancer cell lines, http://www.broadinstitute.org/ccle/home.

[35] M.W. Karaman, S. Herrgard, D.K. Treiber, P. Gallant, C.E. Atteridge, B.T. Campbell, K.W. Chan, P. Ciceri, M.I. Davis, P.T. Edeen, R. Faraoni, M. Floyd, J.P. Hunt, D.J. Lockhart, Z.V. Milanov, M.J. Morrison, G. Pallares, H.K. Patel, S. Pritchard, L.M. Wodicka, P.P. Zarrinkar, A quantitative analysis of kinase inhibitor selectivity, Nat. Biotechnol. 26 (1) (2008) 127–132.

[36] P.P. Zarrinkar, R.N. Gunawardane, M.D. Cramer, M.F. Gardner, D. Brigham, B. Belli, M.W. Karaman, K.W. Pratz, G. Pallares, Q. Chao, K.G. Sprankle, H.K. Patel, M. Levis, R.C. Armstrong, J. James, S.S. Bhagwat, AC220 is a uniquely potent and selective

inhibitor of FLT3 for the treatment of acute myeloid leukemia (AML), Blood 114 (14) (2009) 2984–2992.

[37] M. Jordan, A statistical approach to decision tree modeling, in: M. Warmuth (Ed.), Proceedings of the Seventh Annual ACM Conference on Computational Learning Theory, ACM Press, 1994, pp. 13–20.

[38] M. Jordan, Z. Ghahramani, L. Saul, Hidden Markov decision trees, in: Neural Information Processing Systems, 1996.

[39] R. Rahman, S. Haider, S. Ghosh, R. Pal, Design of probabilistic random forests with applications to anticancer drug sensitivity prediction, Cancer Inform. 14 (Suppl. 5) (2015) 57–73.

[40] W.K. Newey, K.D. West, A Simple, Positive Semi-Definite, Heteroskedasticity and Autocorrelation Consistent Covariance Matrix, National Bureau of Economic Research, Cambridge, MA, 1986.

[41] N.J. Higham, Analysis of the Cholesky decomposition of a semi-definite matrix, Reliab. Numer. Comput. (1990) 161–185.

[42] M. Robnik-Sikonja, I. Kononenko, An adaptation of Relief for attribute estimation in regression, in: Proceedings of the Fourteenth International Conference on Machine Learning (ICML'97), Morgan Kaufmann Publishers Inc., San Francisco, CA, 1997, pp. 296–304.

[43] S.J. Winham, R.R. Freimuth, J.M. Biernacka, A weighted random forests approach to improve predictive performance, Stat. Anal. Data Min. 6 (6) (2013) 496–505.

[44] N. Meinshausen, Quantile regression forests, J. Mach. Learn. Res. 7 (2006) 983–999.

Predictive modeling based on multivariate random forests

8.1 INTRODUCTION

An important goal of systems medicine is to utilize the genetic characterization of individual tumors to select a drug or drug combination with high sensitivity. The ability of designed models to accurately predict sensitivity of an individual tumor to a drug or drug combination can lead us to personalized cancer therapy treatments with expected efficacy significantly higher than current standard of care approaches. A recent community-based effort organized by the Dialogue on Reverse Engineering Assessment and Methods (DREAM) project [1] and the National Cancer Institute (NCI) explored multiple drug sensitivity prediction algorithms when applied to a single dataset. As discussed in the previous chapter, one of the top performing algorithms was based on Random Forests [2] where individual models

Predictive Modeling of Drug Sensitivity. http://dx.doi.org/10.1016/B978-0-12-805274-7.00008-7

were designed for each drug. In this chapter, we show that correlations in the responses across multiple drugs can be utilized to increase the prediction accuracy of the random forest-based ensemble framework. We expect that drugs with common targets will have relationships between their responses that can be exploited in the design of multivariate random forests (MRFs). The framework presented in this article extends ensemble-based drug sensitivity prediction to incorporate Multitask Learning [3].

The first approach to design MRFs will be based on incorporating the covariance matrix of the output responses in the design of regression trees. In the second part of the chapter, we will consider a novel cost criterion that captures the dissimilarity in the output response structure between the training data and node samples, as the difference in two empirical copulas. We will illustrate that copulas are suitable for capturing the multivariate structure of output responses independent of the marginal distributions, and the copula-based MRF (CMRF) framework can provide higher accuracy prediction and improved variable selection.

8.2 MRF BASED ON COVARIANCE APPROACH

Let us consider the multiple response scenario with output $y(i,k)(i = 1,\ldots,n; k = 1,\ldots,r)$ for n samples and r drugs. The primary difference between MRF and RF is in generation of the trees with different node costs denoted by $D_m(\eta)$ and $D(\eta)$, respectively [4].

The node cost $D(\eta_P) = \sum_{i \in \eta_P} (y(i) - \mu(\psi_P))^2$ for the univariate case is the sum of squares of the differences between the output response and the mean output response for the node. For the multivariate case, we would like to use a multivariate node cost that calculates the difference between a sample point and the multivariate mean distribution. One possible measure is the sum of the squares of Mahalanobis Distances [5] as shown next:

$$D_m(\eta_P) = \sum_{i \in \eta_P} (\mathbf{y}(i) - \boldsymbol{\mu}(\eta_P))\Lambda^{-1}(\eta_P)(\mathbf{y}(i) - \boldsymbol{\mu}(\eta_P))^T \tag{8.1}$$

where Λ is the covariance matrix between the output responses, $\mathbf{y}(i)$ is the row vector $(y(i,1),\ldots,y(i,r))$, and $\boldsymbol{\mu}(\eta_P)$ is the row vector denoting the mean of $\mathbf{y}(i)$ in node η_P.

The inverse covariance matrix (Λ^{-1}) is a precision matrix [6] which is helpful to test conditional dependence between multiple random variables. The Mahalanobis distance square normalizes the output responses by their standard deviations and in the case of Λ being diagonal, it represents the normalized Euclidean distance. For the bivariate case with covariance Λ, the node cost is increased when the deviations of the two output responses from the mean responses are in opposite directions.

8.2.1 SIMPLE EXAMPLE TO ILLUSTRATE THE GENERATION OF UNIVARIATE AND MULTIVARIATE REGRESSION TREE

Let us consider a node of a tree that contains five samples with four input features shown as matrix X and two output responses shown as matrix Y. The 1st and 2nd column of matrix Y represents drug response for drug-1 and drug-2 for the five samples.

$$X = \begin{bmatrix} 1 & 3 & 1 & 9 \\ 2 & 6 & 5 & 4 \\ 5 & 8 & 9 & 5 \\ 4 & 9 & 3 & 3 \\ 6 & 5 & 8 & 2 \end{bmatrix} \quad Y = \begin{bmatrix} 2 & 1 \\ 3 & 3 \\ 7 & 5 \\ 8 & 6 \\ 6 & 2 \end{bmatrix}$$

If $m = 2$, there are six possible combinations of two features from set X. For each combination, we will consider the maximum reduction in node cost using the two different features and select the feature with minimum cost. For instance, if we consider the first and second features for splitting a node of a regression tree where we are trying to predict the sensitivity of drug 2 (i.e., a univariate RF model predicting $Y(:, 2)$), then the original cost $D(\eta)$ is $(1-3.4)^2+(2-3.4)^2+(3-3.4)^2+(5-3.4)^2+(6-3.4)^2 = 17.2$ where 3.4 denotes the mean of $Y(:, 2)$. For the split using feature 2 with threshold 7 (i.e., $X(i, 2) < 7$ or ≥ 7), we have $D(\eta_L) = (1-2)^2 + (2-2)^2 + (3-2)^2 = 2$ and $D(\eta_R) = (5-5.5)^2+(6-5.5)^2 = 0.5$ resulting in total reduction in node cost of $17.2-2-0.5 = 14.7$. Similarly, if we use feature 1 with threshold 3, we have $D(\eta_L) = (1-2)^2 + (3-2)^2 = 2$ and $D(\eta_R) = (5-4.33)^2 + (6-4.33)^2 + (2-4.33)^2 = 8.67$ resulting in total reduction in node cost of $17.2-2-8.67 = 6.53$. Thus, for splitting at this node, feature 2 with threshold 7 produces a higher reduction in cost as compared to feature 1 with threshold 3. For each node, we check the reduction in cost for the randomly selected features and all possible thresholds for each feature. The one with the highest reduction in cost is selected.

If we consider multivariate tree generation, then the reduction in cost is based on Eq. (8.1). If we use the sample covariance,

$$\Lambda = \begin{bmatrix} 6.7 & 4.4 \\ 4.4 & 4.3 \end{bmatrix}$$

the multivariate cost at this node is calculated, as shown next:

$$D_m(\eta) = \sum_{i=1}^{5} (Y(i, 1) - 5.2 \, Y(i, 2) - 3.4) \begin{vmatrix} 0.455 & -0.466 \\ -0.466 & 0.71 \end{vmatrix} \begin{pmatrix} Y(i, 1) - 5.2 \\ Y(i, 2) - 3.4 \end{pmatrix} = 8 \quad (8.2)$$

If we consider feature 1 for splitting with threshold 3, the left and right node costs are given by

$$D_m(\eta_L) = \sum_{i=1}^{2}(Y(i,1) - 2.5\,Y(i,2) - 2) \begin{vmatrix} 0.455 & -0.466 \\ -0.466 & 0.71 \end{vmatrix} \begin{pmatrix} Y(i,1) - 2.5 \\ Y(i,2) - 2 \end{pmatrix} = 0.714$$

(8.3)

$$D_m(\eta_R) = \sum_{i=3}^{5}(Y(i,1) - 7\,Y(i,2) - 3.33) \begin{vmatrix} 0.455 & -0.466 \\ -0.466 & 0.71 \end{vmatrix} \begin{pmatrix} Y(i,1) - 7 \\ Y(i,2) - 3.33 \end{pmatrix} = 3.33$$

(8.4)

Thus the reduction in cost using feature 1 with threshold 3 is $8 - 0.714 - 3.33 = 3.96$. On the other hand, if we consider splitting using feature 2 with threshold 7, the reduction in cost is $8 - 4.43 - 0.116 = 3.45$. Thus, for the specific node η, feature 1 with threshold 3 is a better choice than feature 2 with threshold 7 for the multivariate tree. Note that, for predicting the 2nd column of Y individually, feature 2 was a better choice, as shown earlier. However, $Y(:, 1)$ can be best divided by $X(:, 1)$ because they have a direct functional relationship: $X < 3$, $Y(i, 1) = X(i, 1) + 1$ and for $X \geq 3$, $Y(i, 1) = 12 - X(i, 1)$. And $Y(:, 2)$ being correlated with $Y(:, 1)$ with a correlation coefficient of 0.82, the MRF approach selects feature 1 for splitting the node of the multivariate tree.

The above node splitting procedure is implemented for each node to construct a nonpruned multivariate regression tree. A collection of multivariate trees is then combined to form a multivariate regression forest.

We will present the performance of multivariate regression trees based on co-variance matrices in later sections, while comparing with copula-based multivariate regression trees. In the next section, we present an approach for optimizing the weights of covariance-based multivariate regression tree ensembles using multiobjective optimization.

8.2.2 MULTIOBJECTIVE OPTIMIZATION-BASED GENERATION OF MRF

For generation of covariance-based MRF (VMRF), each multivariate regression tree attempts to form child nodes that are homogeneous with respect to output responses. However, the generation of the forest from the multivariate regression trees follows the regular averaging of all constructed trees without incorporating any weighting of trees that favors an ensemble with better multivariate response. In this section, we approach the problem from a multiobjective optimization framework where we plan to optimize the error across each output. Each forest consists of a subset of trees from the set of initially generated multivariate trees and optimization is approached using genetic algorithms (GAs). We start with a random set of L forests and after each iteration consisting of crossovers and mutations, we form the next generation of L forests based on selecting forests from the Pareto frontier. After multiple iterations, the forests in the Pareto front provide choices for selecting predictive models with different optimization objectives.

An ensemble Λ of n_{tree} multivariate regression trees is initially generated. Afterward, a population of N forests consisting of $L < n_{tree}$ trees is generated by random sampling without replacement from Λ. GAs are applied to optimize the forests, with the multiple objectives being optimized in terms of Pareto efficiency.

8.2.2.1 Pareto optimality

Let k denote the number of objectives to be optimized and the errors across the k objectives for forest f be denoted by $e_1^f, \ldots, e_k^f$. If forests f_1 and f_2 satisfy the following relation: $e_i^{f_1} \leq e_i^{f_2}$ for $i = 1, \ldots, k$ and $e_i^{f_1} < e_i^{f_2}$ for at least one $i \in \{1, \ldots, k\}$, then f_1 is considered to dominate f_2 from the multiobjective Pareto sense. The forests that are not dominated by any other forest will form the Pareto efficient front.

We next consider the prediction problem with multiple output responses based on an ensemble of multivariate regression trees. The selection of the optimal ensemble is formulated as a multiobjective optimization problem and solved using GAs. We illustrate the application of the approach on the drug sensitivity prediction problem, where the proposed methodology outperforms regular MRFs in terms of correlation coefficients between predicted and experimental sensitivities. It is also demonstrated that generating the Pareto-optimal front provides a choice of ensembles for different optimization objectives.

8.2.2.2 Genetic algorithm

GA is inspired by evolutionary theory where strong species have a higher opportunity to pass their genes to offspring via reproduction [7] and weaker chromosomes are eliminated by natural selection. Each generation or population consists of diverse individuals or chromosomes and in our Genetic Algorithm-Based Multivariate Random Forests (GAMRF), each forest is regarded as a chromosome comprised of different trees. During each reproduction process, crossover and mutation operators are applied for the purpose of generating new solutions from existing ones. In order to select the best solutions (forests) for the next generation, the fitness of each solution is computed. In the GAMRF case, we consider the prediction error as our fitness function. We first divide each dataset into three parts: training, validation, and testing. Specifically, the validating datasets are used to calculate errors between actual and predicted output responses. Note that we use the validation dataset for fitness generation in GAMRF, but the training and validation datasets are combined to form a training set for other predictive models. The goal in the multiobjective GAMRF is to minimize the prediction errors across multiple output variables and to find nondominated forests in each generation. From the original ensemble of multivariate regression trees, we form the initial population P_0 of N forests, where each forest consists of L trees. After calculating the fitness functions for the existing population, we calculate different Pareto front layers according to their dominance relationship. A specific number (here it is set to 20) of top Pareto optimal points are selected to pairwise conduct crossover and mutations to form offsprings. After merging these offsprings with their parent population P_{t-1}, we extract top N forests to generate population P_t. We iterate our algorithm for *totalG* number of generations. Note that evolutionary algorithms like GA will not guarantee convergence of the Pareto front, but the performance of our forests will improve if the Pareto front moves toward our desired direction with subsequent GA iterations. The detailed procedure for multiobjective GAMRF is shown in Algorithm 8.1 and the approach to select the top forests is shown in Algorithm 8.2.

ALGORITHM 8.1 MULTIOBJECTIVE GAMRF

Require: n_{tree}, m_{tree}, N, L, *totalG*
Ensure: Population P_t
  Initializing population:
  MRF = Train MRF(n_{tree}, m_{tree})
  for $i = 1$ to N **do**
    mrf_i = **Random**(MRF, L)
    Add *mr* f_i to P_0
  end for
  f_0 = **EvaluateFitness**(P_0)
  for $t = 1$ to *totalG* **do**
    Compute layers of Pareto front:
    $front_t$ = **TopFront**(f_{t-1}, 20)
    crossover(*front_t*, *rate_c*)
    mutation(*front_t*, *rate_m*)
    f_t = **EvaluateFitness**($P_{t-1} \cup front_t$)
    P_t = **TopFront**(f_t, N)
  end for
  return P_t

ALGORITHM 8.2 PICK TOP *F* PARETO FRONT POINTS FROM *F*

function TopFront(f, F)
repeat
  $layer$ = **ParetoFront**(f)
  Add *layer* to *front*
  $F = F - |front|$
  $f = f \setminus layer$
until $F \leq 0$
return *front*

8.2.2.3 Application of GAMRF

For analyzing the prediction performance of GAMRF, we consider the Cancer Cell Line Encyclopedia (CCLE) [8] database that includes genomic characterization and drug responses of 24 drugs for over 490 cell lines. Since MRFs are suitable for predicting responses of cell lines to drugs with common targets, we considered two *EGFR* inhibitors: *Erlotinib* and *Lapatinib* [9,10]. For this analysis, we utilize a smaller set of features containing around 400 kinase producing genes selected from drug target inhibition profile studies [11,12].

We set the number of trees in the MRF to $n_{tree} = 500$ and the splitting in each node as $m_{tree} = 10$. Consequently, $N = 100$ forests each containing $L = 300$ trees are formed. In multiobjective GAMRF, 80%, 10%, and 10% of the samples are randomly selected as training, validation, and testing data. Once we have the initial population from the training samples, the mean absolute errors (MAEs) between predicted and experimental drug sensitivities in the validation cell lines are computed. Fig. 8.1 shows the Pareto front for the first generation. Fig. 8.2 shows the application of Algorithm 8.2 to select top 20 forests for producing offsprings through crossover.

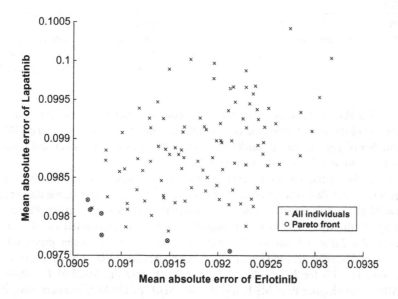

FIG. 8.1

Initial population with Pareto front.

(©2014 IEEE. Reprinted, with permission, from Q. Wan, R. Pal, Multi-objective optimization of ensemble of regression trees using genetic algorithms, in: 2014 IEEE Global Conference on Signal and Information Processing (GlobalSIP), 2014, pp. 1356–1359.)

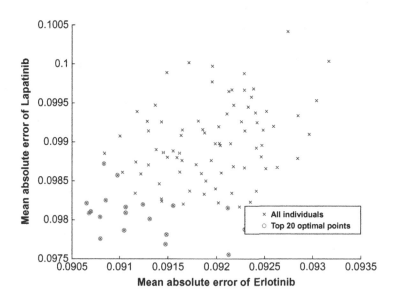

FIG. 8.2

Plot of top 20 (Pareto) optimal individuals of the initial population.

(©2014 IEEE. Reprinted, with permission, from Q. Wan, R. Pal, Multi-objective optimization of ensemble of regression trees using genetic algorithms, in: 2014 IEEE Global Conference on Signal and Information Processing (GlobalSIP), 2014, pp. 1356–1359.)

Fig. 8.3 shows the Pareto front for subsequent generations and illustrates the movement of the front toward the origin with each iteration and convergence. We set the number of generations at *totalG* = 150 and the rates for crossover and mutation are selected to be 0.2 and 0.01, respectively.

For validating the performance of multiobjective GAMRF, we randomly selected a forest from the Pareto front of the last generation. For comparison, we also created a new MRF of 500 trees by using the earlier training and validation datasets for training. Both MRF and GAMRF models are then implemented on the testing datasets. As Table 8.1 shows, the correlation coefficient between predicted and experimental sensitivities of GAMRF noticeably exceeds MRF, with an average gain of around 11.43% for *Erlotinib* and *Lapatinib*. Despite the MAE of *Lapatinib* for GAMRF being higher than MRF, the average MAE of GAMRF outperforms MRF. Another drug pair (*Crizotinib* and *PHA-665752*) with common targeted gene *MET* [13,14] is also utilized to verify the effectiveness of the GAMRF framework. As elucidated in Table 8.1, GAMRF shows superior performance compared to regular MRF. Specifically, we observe a reduction in MAE using GAMRF as compared to MRF for both the drugs and an average increase of 13.49% in correlation coefficients between predicted and experimental sensitivities for GAMRF as compared to RF.

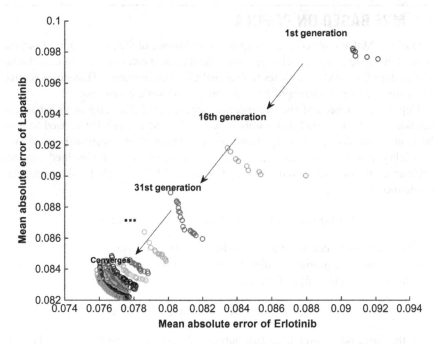

FIG. 8.3

Optimization of Pareto front through genetic algorithm. Only the Pareto front layers of every 15 generations are plotted, a total of 11 Pareto front layers are illustrated.

(©2014 IEEE. Reprinted, with permission, from Q. Wan, R. Pal, Multi-objective optimization of ensemble of regression trees using genetic algorithms, in: 2014 IEEE Global Conference on Signal and Information Processing (GlobalSIP), 2014, pp. 1356–1359.)

Table 8.1 Results of CCLE Drug Sensitivity Prediction for Two Drug Pairs in the Form of Correlation Coefficients and MAE

Drug Name	Correlation Coefficients		MAE	
	MRF	*GAMRF*	*MRF*	*GAMRF*
Erlotinib	0.289	0.343	0.0935	0.0901
Lapatinib	0.409	0.426	0.0926	0.0939
Crizotinib	0.247	0.296	0.0830	0.0812
PHA-665752	0.140	0.150	0.0847	0.0842

Notes: GAMRF, Genetic Algorithm-based MRF; MRF, Multivariate Random Forest.
©2014 IEEE. Reprinted, with permission, from Q. Wan, R. Pal, Multi-objective optimization of ensemble of regression trees using genetic algorithms, in: 2014 IEEE Global Conference on Signal and Information Processing (GlobalSIP), 2014, pp. 1356–1359.

8.3 MRF BASED ON COPULA

Because the Mahalanobis distance captures the distance of the sample point from the mean of the node along the principal component axes, it can fail to capture nonlinear relationships that produce a closer to diagonal Covariance matrix. Thus our objective is to introduce Copulas to capture the nonlinear multivariate structure.

Copulas can represent the dependence between multiple random variables independent of the marginal distributions. A copula function [15,16] is used to map the joint cumulative probability distribution in terms of the marginal cumulative probability distributions. Let Ψ_1, Ψ_2, ..., Ψ_N represent N real-valued random variables uniformly distributed on $[0, 1]$. Copula $C: [0, 1]^N \rightarrow [0, 1]$ with parameter θ is defined as:

$$C_\theta(u_1, u_2, \ldots, u_N) = P(\Psi_1 \leq u_1, \Psi_2 \leq u_2, \ldots, \Psi_N \leq u_N) \tag{8.5}$$

The multivariate cumulative probability distribution $F_X(x_1, x_2, \ldots, x_N)$ and the marginal cumulative probability distributions $F_i(x_i)$ for ($i \in \{1, 2, \ldots, N\}$) are related by Sklar's theorem [15,16] as follows:

$$F_X(x_1, x_2, \ldots, x_N) = C(F_1(x_1), F_2(x_2), \ldots, F_N(x_N)) \tag{8.6}$$

If the marginal cumulative distributions ($F_i(x)$) are continuous, copula C is unique [16].

Some copulas can be parameterized using a few parameters, for instance, the clayton copula [17] for bivariate distribution is defined as follows, using parameter ξ:

$$C(u_1, u_2; \xi) = (u_1^{-\xi} + u_2^{-\xi} - 1)^{-1/\xi} \quad \xi \in (0, \infty) \tag{8.7}$$

Similarly, the copula characterizing two independent variables will have the form $C(u_1, u_2) = u_1 u_2$. Some other common forms of parameterized copulas include Gaussian Copula [18], Frank Copula [19], Student's t-copula [20], and Gumbel copula [21]. However, the standard forms of parameterized copulas may not capture all forms of relationships. We can consider the use of empirical copulas that are estimated directly from the cumulative multivariate distribution. Note that the calculation of empirical copulas will have higher computational complexity than parameterized copulas, but they can capture a broad range of relationships. We utilize empirical copulas to represent our multivariate structures.

8.3.1 NODE SPLIT CRITERIA USING COPULA

As described earlier, the regression tree generation process involves partition of a node into two branches based on optimizing a cost criterion. The node cost for multivariate regression trees utilizing Mahalanobis distance is shown in Eq. (8.1). The feature and threshold that results in maximum cost reduction for that node is selected for splitting.

We next discuss the design of node cost function based on copulas to capture the output dependencies. We expect that the dependency structure among the samples in a node should be similar to the dependency structure observed in the original training data. Consider the node η_P with N_P samples and let Ψ denote the integral of the difference in the empirical copulas observed at node η_P and the root node (this is same as the empirical copula for the training data). We design the node cost $D_C(\eta_P)$ for a copula-based multivariate regression tree as follows:

$$D_C(\eta_P) = D_1 + \alpha D_2 \quad \text{where} \tag{8.8}$$
$$D_1 = 6 N_P \Psi \quad \text{and}$$
$$D_2 = \sum_{j=1}^{r} \frac{\sum_{i \in \eta_{P_j}} [y(i,j) - \mu(\eta_{P_j})]^2}{\sigma_j^2} \tag{8.9}$$

where α denotes a scaling factor determining the relative weight of the two components of the node cost, η_{P_j} for $j \in \{1, \ldots, r\}$ denotes the set of jth output responses at node η_P and σ_j^2 for $j \in \{1, \ldots, r\}$ denotes the variance of the jth output response at root node.

We next present the motivation for selecting the weight 6 for the integral of copula distance along with approaches to select the scaling factor α. For maintaining D_1 and D_2 in the same range, we analyzed the range of Ψ as compared to D_2.

Hereafter, the MRF approach that uses copula-based node splitting criteria (based on Eq. 8.8) will be termed *CMRF* and the MRF approach using covariance-based node splitting criteria (based on Eq. 8.1) will be termed *VMRF*.

8.3.2 ANALYZING INTEGRAL OF DIFFERENCES IN COPULAS

We first analyze the upperbound on the integral of the difference between two bivariate copulas and subsequently explore further multivariate copulas. Based on Frechet-Hoeffding bounds [22], any bivariate copula $C(u, v)$ is bounded by the following:

$$C_L(u, v) = \max[u + v - 1, 0] \le C(u, v) \le \min[u, v] = C_U(u, v)$$

Thus for any two copulas $C_1(u, v)$ and $C_2(u, v)$, we have

$$|C_1(u, v) - C_2(u, v)| \le C_U(u, v) - C_L(u, v) \tag{8.10}$$
$$\forall u, v \in [0 \ 1] \tag{8.11}$$

Consequently,

$$\int_{v=0}^{1} \int_{u=0}^{1} |C_1(u, v) - C_2(u, v)| du \, dv \le \int_{v=0}^{1} \int_{u=0}^{1} [C_U(u, v) \tag{8.12}$$
$$- C_L(u, v)] du \, dv \tag{8.13}$$

Using the two diagonals in the unit square ($u = v$ and $u+v = 1$), we can divide the unit square into four triangles where the values of $C_U(u, v)$ and $C_L(u, v)$ are simple functions of u and v. For region 1, we have $u > v$ and $u + v > 1$ and

$$C_U(u, v) - C_L(u, v) = v - (u + v - 1) = 1 - u \qquad (8.14)$$

For region 2, we have $u > v$ and $u + v \leq 1$ and

$$C_U(u, v) - C_L(u, v) = v - 0 = v$$

Similarly for region 3, we have $u \leq v$ and $u + v \leq 1$ and

$$C_U(u, v) - C_L(u, v) = u - 0 = u$$

And for region 4, we have $u \leq v$ and $u + v > 1$ and

$$C_U(u, v) - C_L(u, v) = u - (u + v - 1) = 1 - v$$

The integral over region 1 is as follows:

$$\iint_{\text{Region 1}} (C_U(u, v) - C_L(u, v)) du\, dv \qquad (8.15)$$

$$= \int_{v=0.5}^{1} \int_{u=v}^{1} (1 - u) du\, dv + \int_{v=0}^{0.5} \int_{u=1-v}^{1} (1 - u) du\, dv = 1/24 \qquad (8.16)$$

We can likewise show that the value of the integral for each of the other regions is $\frac{1}{24}$. Thus $\int_{v=0}^{1} \int_{u=0}^{1} (C_U(u, v) - C_L(u, v)) du\, dv = 1/6$ and the upper bound on the surface integral of the difference between any two bivariate copulas is $1/6$. Similarly, if we consider the independent copula $C_I(u, v) = uv$, then we can show that $\int_{v=0}^{1} \int_{u=0}^{1} (C_U(u, v) - C_I(u, v)) du\, dv = 1/12$. For $n > 2$, we conducted simulations to estimate the value of the integrals which is shown in Table 8.2.

Thus, because the upper bound of $\int (C_U - C_L)$ lies in the range of $1/6$ to 0.21, $D_1 = 6N_P\Psi$ will be upper bounded by N_P for a bivariate copula. If the regression tree is unable to reduce the initial variance in each output response, the value of D_2 will be in the range of rN_P as the jth numerator term will be close to $N_P\sigma_j^2$. However, because the regression tree will likely reduce the variance in the output response at nodes further away from the root, the value of D_2 will be much lower than rN_P.

Table 8.2 Integral of Copula Differences for Different Dimensions

Dimensions	$\int C_U - C_L$	$\int C_U - C_I$	$\int C_U$
2	$\frac{1}{6}$	$\frac{1}{12}$	$\frac{1}{3}$
3	0.2093	0.1255	0.252
4	0.197	0.1425	0.2072

8.3.3 SELECTION OF α

Our previous analysis of Ψ provided a range of the integral difference between two copulas, but was unable to provide a weight factor for combining D_1 and D_2 that is optimal in terms of predictive performance. We expect that the behavior of D_1 and D_2 will change significantly for different training datasets and thus we select α based on each training dataset. We next describe two techniques to select the weight factor α to achieve higher prediction accuracy.

8.3.3.1 Method 1: Evaluating and selecting from a set of α's

This is a straightforward approach where different values of α (we considered 10 values of α spaced between 0.1 and 10) are evaluated and the one with best predictive performance selected. The original training data are subdivided into secondary training and secondary testing samples. The secondary training samples are used to create MRF models that are used to find prediction of secondary testing samples. This process is repeated for a set of possible values of α. The correlation coefficient between predicted secondary testing samples and original secondary testing samples are recorded, and the α corresponding to highest correlation coefficient is selected (α_S). This α_S is then used to create an MRF model using the original training samples and tested on original testing samples. For our examples, we have applied 10-fold CV on the original data and thus for each fold of training data, we may select different α. However, for each specific fold, α will be fixed for all the trees generated. The above method increases the computational complexity due to the evaluation of multiple values of α. We next present another approach that attempts to reduce the evaluation of numerous values of α.

8.3.3.2 Method 2: Pareto frontier approach to select α

In this approach, we consider the node cost function minimization from a multiobjective optimization problem perspective where we aim to jointly minimize both D_1 and D_2. From the multiobjective perspective, if we plot the D_2 versus D_1 for all possible feature and threshold combinations for a specific node (if we have n samples at a specific node and m features, the number of partitions to be evaluated is mn), we should select a feature and threshold combination that lies in the Pareto frontier. In other words, we look for solutions that are not dominated by any other solution: for instance, if the D_1 and D_2 values for w different feature and threshold combinations are denoted by $\{\epsilon_1(i), \epsilon_2(i)\}$ for $i \in \{1, \ldots, w\}$, a combination i is considered dominated by j if either (a) $\epsilon_1(i) > \epsilon_1(j)$ and $\epsilon_2(i) \geq \epsilon_2(j)$ or (b) $\epsilon_1(i) \geq \epsilon_1(j)$ and $\epsilon_2(i) > \epsilon_2(j)$ is valid. The feature and threshold combinations that are not dominated by any of the other $w-1$ combinations form the Pareto frontier. For instance, Fig. 8.4A shows an example Pareto frontier (outermost circles) for the left child node for the first split of a specific tree (the D_1 and D_2 values are denoted by D_{1L} and D_{2L}, respectively) generated from a synthetic example described in next section. Similarly, Fig. 8.5A shows the Pareto frontier (outermost circles) for the right child node for the first split of a specific tree (the D_1 and D_2 values are denoted by D_{1R} and D_{2R}, respectively).

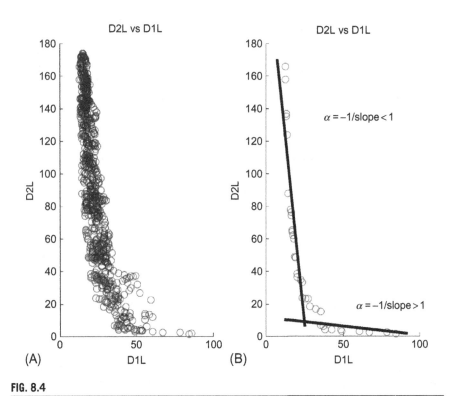

FIG. 8.4

D_2 vs D_1 for left node. (A) An example Pareto frontier (*outermost circles*) for the left child node for the first split of a specific tree (the D_1 and D_2 values are denoted by D_{1L} and D_{2L}, respectively). (B) The Pareto frontier can be approximated by *two straight lines*: one with slope greater than 1 and another with slope less than 1 [23].

Our idea is to approximate the Pareto frontier using straight lines and utilize the slope of the lines to design α. Figs. 8.4B and 8.5B show that the Pareto frontier can be approximated by two straight lines: one with slope greater than 1 and another with slope less than 1. Consequently, the value of α can be approximated by the following equation.

$$\alpha = -1/\rho$$

where ρ denotes the slope of the straight line fitted to the Pareto frontier. Thus we have two possible values of α from the very first split of a specific tree. If we prepare scatter-plots for the first split of all the trees for $\alpha > 1$ and $\alpha < 1$, we arrive at plots similar to Fig. 8.6A and B, respectively. In this method, we calculate the predictive performance of the two average α's and select the one with the best performance to design the overall forest.

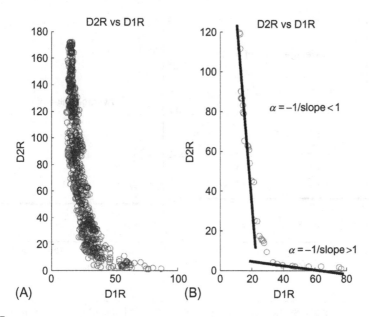

FIG. 8.5

D_2 vs D_1 for right node. (A) An example Pareto frontier (*outermost circles*) for the right child node for the first split of a specific tree (the D_1 and D_2 values are denoted by D_{1R} and D_{2R}, respectively). (B) The Pareto frontier can be approximated by *two straight lines*: one with slope greater than 1 and another with slope less than 1 [23].

AN EXAMPLE TO ILLUSTRATE THE APPROPRIATENESS OF COPULA-BASED NODE COST FOR DESIGN OF MULTIVARIATE REGRESSION TREES

We have observed that MRFs incorporating covariance (Mahalanobis distance square) between output responses is more suitable for predicting output responses with linear relationships, as compared to output responses with nonlinear relationships. Copula presents a methodology to capture the nonlinear dependence relationships between multiple variables; and we anticipate that copula will be suitable for predicting output drug responses with nonlinear relationships between them. We next present a synthetic example with nonlinear relationship to investigate the performance of the proposed approach as compared to VMRF design.

We consider a 50×10 input data matrix (50 samples and 10 features) denoted by $\mathbf{X}$ that was created randomly from a normal distribution $\mathcal{N}(0, 1)$. We next generated two output responses $\mathbf{Y}_1$ and $\mathbf{Y}_2$ based on functions of the input features. Let column vectors $\mathbf{x}_i (i = 1, 2, \ldots, 10)$ denote the 10 features and the output responses $\mathbf{Y}_1$ and $\mathbf{Y}_2$ are defined as follows:

$$\mathbf{Y}_1 = 2\mathbf{x}_1 + 5\mathbf{x}_2 - 1.5\mathbf{x}_3 + \mathbf{x}_4 \tag{8.17}$$

$$\mathbf{Y}_2 = (\mathbf{Y}_1 - \mathbb{E}(\mathbf{Y}_1))^2 \tag{8.18}$$

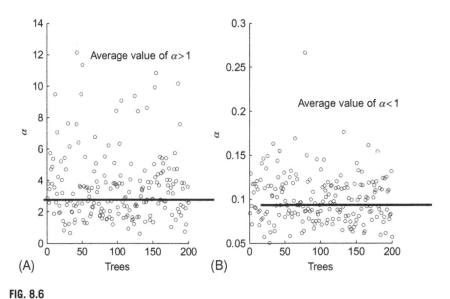

FIG. 8.6

Scatter plot of α's across the trees. Scatter-plots for the first split of all the trees for $\alpha > 1$ (A) and $\alpha < 1$ (B) [23].

Note that the output responses are dependent on only 4 features out of the 10 possible input features. Based on the relative weights, x_2 is the most weighted feature and should play a critical role while growing the trees at the beginning. Note that Y_1 and Y_2 have a quadratic relationship.

We consider two multivariate regression trees trained on the same input $\mathbf{X}$ and same output responses $[\mathbf{Y}_1, \mathbf{Y}_2]$, but different node splitting criterion. The regression trees denoted by $Tree\{\mathbf{V}, [\mathbf{Y}_1, \mathbf{Y}_2]\}$ and $Tree\{\mathbf{C}, [\mathbf{Y}_1, \mathbf{Y}_2]\}$ are inferred using the covariance (Eq. 8.1) and copula-based (Eq. 8.8) node cost functions, respectively. For this example, all the features were considered at each node, that is, $m = M$ and randomly chosen 80% of 50 samples with bootstrapping were used for generating the regression trees.

Fig. 8.7 shows two multivariate regression trees generated using copula-based (Eq. 8.8) and covariance-based (Eq. 8.1) node cost functions. Fig. 8.7 illustrates that the splitting process for each tree is dependent on different features at each node, which eventually leads to two totally dissimilar trees. The empty circles denote leaf nodes; the circles enclosing a number signify a split node and the number inside the circle indicates the featured selected on that node for splitting.

We expect that the regression tree generation based on copula as compared to covariance will be better able to capture the nonlinear relationship between Y_1 and Y_2. Fig. 8.7 demonstrates that the copula-based $Tree\{\mathbf{C}, [\mathbf{Y}_1, \mathbf{Y}_2]\}$ has selected the most significant features (features 2 and 1 that have the highest weights during generation of Y_1 and Y_2) while generating the multivariate regression tree. On the other hand,

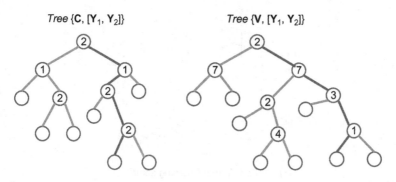

FIG. 8.7

Two multivariate regression trees trained on the same input **X** and same output responses $[\mathbf{Y}_1, \mathbf{Y}_2]$ but the node cost criteria being copula-based ($Tree\{\mathbf{C}, [\mathbf{Y}_1, \mathbf{Y}_2]\}$) and covariance-based ($Tree\{\mathbf{V}, [\mathbf{Y}_1, \mathbf{Y}_2]\}$), respectively. The *empty circles* represent leaf nodes and the *circles enclosing a number* signifies a split node; the *number inside the circle* indicates the featured selected on that node for splitting [23].

the covariance-based $Tree\{\mathbf{V}, [\mathbf{Y}_1, \mathbf{Y}_2]\}$ trained on the same data selected a spurious feature 7 which was not involved in the generation of either Y_1 or Y_2.

To visually compare the multivariate structure during regression tree splits, we plotted the cumulative distribution functions (CDFs) for the original data and after splitting using copula- and covariance-based node cost functions. Fig. 8.8 shows the original CDF and the CDFs at the left and right child nodes when the node split is based on Eq. (8.8) (CMRF). Likewise, Fig. 8.9 shows the original CDF and the CDFs at the left and right child nodes when the node split is based on Eq. (8.1) (VMRF). We observe that the node split using copula-based node cost better maintains the CDF observed in the original data (Fig. 8.8B and C are similar to A) as compared to the split using covariance-based node cost (Fig. 8.9B is significantly different from A).

8.3.3.3 Variable importance measure

In this section, we consider the issue of feature selection for MRFs. We would like to generate and compare the Variable Importance Measure (VIM) for CMRF and VMRF. We expect that CMRF will have higher feature scores for the significant features as compared to VMRF. Typical VIMs for random forest consider the frequency of feature selection, out of bag error, or permutation measures [24]. We consider the basic approach of calculating the number of times each feature gets selected and the VIM for each forest will be the sum of these frequencies across all trees normalized to the range between 0 and 1. Based on the synthetic data, we generated 100 Multivariate Regression Trees using CMRF (with fixed α) and VMRF with output responses Y_1 and Y_2 and generated the variable importance of the 10 input features. The normalized variable importance scores reported in Table 8.3 illustrate that the top four features selected by CMRF (X_2, X_1, X_3, X_4) are the same as the four

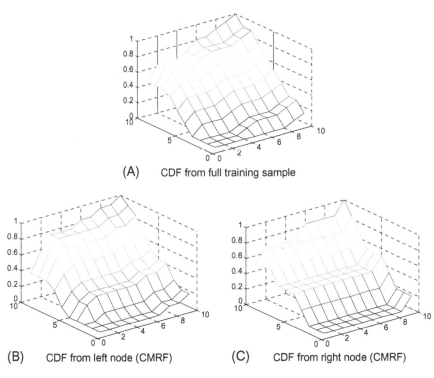

(A)   CDF from full training sample

(B)   CDF from left node (CMRF)   (C)   CDF from right node (CMRF)

FIG. 8.8

CDF created from left and right child node for a single split using CMRF. It is compared visually with the original CDF created from the training samples [23]. (A) CDF from full training sample. (B) CDF from left node (CMRF). (B) CDF from right node (CMRF).

features that were used to generate Y_1, Y_2 using Eqs. (8.17), (8.18), respectively. Furthermore, the ordering of the scores $VIM(X_2) > VIM(X_1) > VIM(X_3) > VIM(X_4)$ is the same as the ordering of the absolute weights of the four features in generation of Y_1 where X_2 has the largest weight followed by X_1, X_3, and X_4. On the other hand, the top four features selected by $VMRF$ X_2, X_1, X_3, X_6 fail to pick X_4 and includes a spurious feature X_6 that was not involved in the generation of output response Y_1. Thus the example supports that CMRF might be better suitable to select top features as compared to VMRF.

8.3.4 PREDICTIVE PERFORMANCE ON BIOLOGICAL DATASETS

For analyzing the prediction capabilities of our framework, we considered two different datasets: GDSC and CCLE. Both include genomic characterization of numerous cell lines and different drug responses for each cell line. For the current analysis, we consider the gene expression data as the genomic characterization

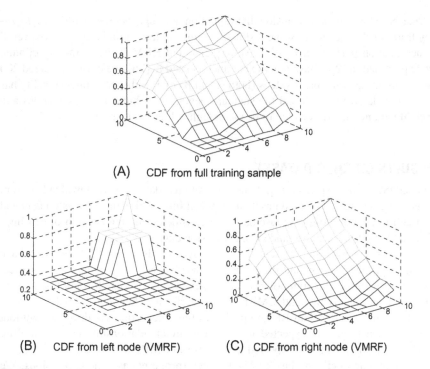

(A)   CDF from full training sample

(B)   CDF from left node (VMRF)   (C)   CDF from right node (VMRF)

FIG. 8.9

CDF created from left and right child node for a single split using VMRF. It is compared visually with the original CDF created from the training samples [23]. (A) CDF from full training sample. (B) CDF from left node (VMRF). (B) CDF from right node (VMRF).

Table 8.3 Variable Importance Measure Calculated Using $CMRF_{Y_1, Y_2}$ and $VMRF_{Y_1, Y_2}$ [23]

Feature	X_1	X_2	X_3	X_4	X_5	X_6	X_7	X_8	X_9	X_{10}
$CMRF_{Y_1, Y_2}$	0.2440	0.3720	0.0952	0.0744	0.0179	0.0625	0.0238	0.0417	0.0327	0.0357
$VMRF_{Y_1, Y_2}$	0.2340	0.3700	0.0836	0.0529	0.0056	0.0729	0.0418	0.0418	0.0474	0.0501

information for both datasets. Area under the curve (AUC) is used as representation of drug responses for both GDSC and CCLE. Both datasets are high dimensional in the number of features (gene expressions). For all performance comparison results presented in this article, a prior feature selection method (RELIEFF [25]) is applied to reduce the number of features to be used for training.

For performance comparison purposes, we report results of CMRF along with univariate RF (denoted by RF), VMRF, and Kernelized Bayesian Multitask Learning (KBMTL) [26] approaches. KBMTL is a Bayesian formulation that combines kernel-based nonlinear dimensionality reduction and regression in a *multitask learning* framework that tries to solve distinct, but related, tasks jointly to improve overall generalization performance. We have implemented KBMTL using the algorithmic code provided in [26]. Based on the parameters used in [26], we have considered 200 iterations and gamma prior values (both α and β) of 1. Subspace dimensionality has been considered to be 20 and the standard deviation of hidden representations and weight parameters are selected to be the default 0.1 and 1, respectively.

RESULTS ON GDSC DATASET

The GDSC gene expression and drug sensitivity dataset was downloaded from Cancerrxgene.org [27]. The dataset has 789 cell lines with gene expression data and 714 cell lines with drug response data. We considered the intersection of cell lines that had both drug response and gene expression data.

For our experiments, we consider four sets of drug pairs where three have common primary targets and the remaining pair has no common target. We expect that the drug pairs with common primary targets will have some form of relationship among their sensitivities and CMRF should perform better than VMRF, and both should perform better than RF approach. On the other hand, the drug pair without any common targets is expected to have a minimal relationship among the drug sensitivities and thus RF is expected to outperform CMRF and VMRF.

Initially each cell line has 22,277 features (probeset) as gene expressions. We have reduced it to 500 for each drug response, using RELIEFF [25] and used a union of the 500 features in each of the four sets of drugs.

The first selected set S_{C2} consisting of {*Erlotinib, Lapatinib*} has common target *EGFR* [9,10]. The second set S_{C3} consisting of {*AZD-0530, TAE-684*} has common target *ABL1* [28]. The third set S_{C1} was {*AZD6244, PD-0325901*} with common target *MEK* [28–30]. The fourth set S_U consisting of {*17-AAG, Erlotinib*} has no common target.

As mentioned earlier, each drug has some missing responses across the 714 cell lines. The drug sets S_{C1}, S_{C2}, S_{C3}, and S_U have drug responses in both drugs for 316, 349, 645, and 300 cell lines, respectively. To report our results, we compared fivefold cross-validated Pearson correlation coefficients, MAE, and Normalized Root Mean Square Error (NRMSE) between predicted and experimental responses for RF, VMRF, CMRF, and KBMTL. NRMSE of drug m can be calculated as [26]:

$$NRMSE_m = \sqrt{\frac{(y_m - \hat{y}_m)^T (y_m - \hat{y}_m)}{(y_m - 1 \cdot \mathbf{E}(y_m))^T (y_m - 1 \cdot \mathbf{E}(y_m))}} \tag{8.19}$$

where y_m and $\hat{y}_m$ denote the vector of actual and predicted drug sensitivities, respectively, and $\mathbf{E}(y_m)$ denote mean of vector y_m. For both VMRF and CMRF, we set the minimum size of samples in each leaf to $n_{size} = 5$, the number of trees in the forest to $T = 150$ and the splitting in each node considers $m = 10$ random features.

The correlation coefficients using fivefold cross-validation error estimation are illustrated for each drug set in Table 8.4. The corresponding MAE and NRMSE behaviors are illustrated in Table 8.5.

For CMRF, results with scaling factor α selected using *Method-1* discussed earlier has been used. Table 8.4 shows that CMRF outperformed (in terms of correlation coefficients) VMRF, RF, and KBMTL for the related drug pairs S_{C1}, S_{C2}, S_{C3}, whereas CMRF is outperformed by the other approaches for the unrelated drug pair S_U. Table 8.5 shows that *CMRF outperforms VMRF, KBMTL, and RF in terms of average NRMSE for the related pairs of drugs S_{c_1}, S_{c_2}, and S_{c_3}. For the unrelated pair S_U, univariate RF outperforms the multivariate approaches for both average correlation coefficients and NRMSE. The scatter plots of predicted response versus original response for drug set S_{C1} using RF, VMRF, and CMRF are shown in Fig. 8.10.

RESULTS ON CCLE DATASET

The CCLE [8] database includes genomic characterization for 1037 cell lines and drug responses over 24 drugs for over 480 cell lines. For the purpose of predicting responses, four sets of drugs were selected. The first set $S_{C1} = \{Erlotinib, Lapatinib\}$ has *EGFR* as a common target [9,10], the second set $S_{C2} = \{PF-02341066$ *(Crizotinib), PHA-665752\}* has *MET* as a common target [13,14], the third set $S_{C3} = \{ZD6474$ *(Vandetanib), AZD0530 (Saracatinib)\}* has *EGFR* as a common target [31,32], and the fourth set $S_4 = \{17\text{-}AAG, Erlotinib\}$ has no common target.

Initially, each cell line had 18,988 features (probeset) as gene expressions. We reduced it to 500 for each drug response, using RELIEFF [25] feature selection and considered a union of the 500 features in each of the four sets of drugs. We have

Table 8.4 Fivefold CV Results for GDSC Dataset Drug Sensitivity Prediction for Four Drug Sets in the Form of Correlation Coefficients

Drug Set	Common Target	Drug Name	Correlation Coefficients			
			RF	VMRF	CMRF	KBMTL
S_{C1}	EGFR	Erlotinib	0.5156	0.5193	0.5301	0.2500
		Lapatinib	0.5544	0.5742	0.5699	0.1132
S_{C2}	ABL1	AZD-0530	0.3553	0.3810	0.3990	0.3181
		TAE-684	0.4060	0.4100	0.4338	0.2420
S_{C3}	MEK	AZD6244	0.4625	0.4508	0.4590	0.0950
		PD-0325901	0.5890	0.6022	0.6016	0.3236
S_U	None	17-AAG	0.6304	0.6244	0.6167	0.4375
		Erlotinib	0.5859	0.5906	0.5708	0.4081

Notes: VMRF and CMRF represent MRF using Covariance and Copula, respectively. KBMTL represents Kernelized Bayesian multitask learning (parameters considered are 200 iterations, $\alpha = \beta = 1$ and subspace dimensionality = 20) [23].

Table 8.5 Fivefold CV Results for GDSC Dataset Drug Sensitivity Prediction for Four Drug Sets in the Form of MAE and NRMSE for RF, VMRF, CMRF, and KBMTL Approaches [23]

Drug Set	Common Target	Drug Name	MAE				NRMSE			
			RF	VMRF	CMRF	KBMTL	RF	VMRF	CMRF	KBMTL
S_{C1}	EGFR	Erlotinib	0.0319	0.0322	0.0314	0.0503	0.8733	0.8749	0.8719	1.3365
		Lapatinib	0.0292	0.0294	0.0286	0.0488	0.8516	0.8459	0.8486	1.3538
S_{C2}	ABL1	AZD-0530	0.0446	0.0448	0.0442	0.0613	0.9407	0.9378	0.9291	1.2344
		TAE-684	0.0829	0.0829	0.0821	0.1159	0.9285	0.9299	0.9195	1.3698
S_{C3}	MEK	AZD6244	0.0584	0.0590	0.0584	0.1138	0.8949	0.9034	0.8962	1.8016
		PD-0325901	0.0723	0.0727	0.0717	0.1199	0.8263	0.8230	0.8193	1.4028
S_U	None	17-AAG	0.0584	0.0590	0.0584	0.1198	0.7840	0.7894	0.7955	1.1624
		Erlotinib	0.0723	0.0727	0.0717	0.0410	0.8335	0.8441	0.8505	1.1013

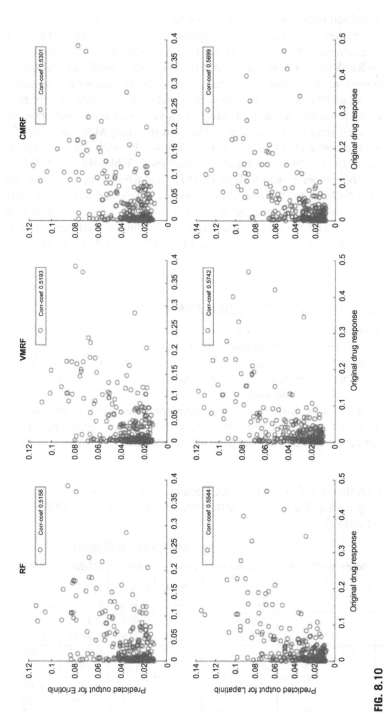

FIG. 8.10

Scatter plots of predicted response vs original response for Erlotinib and Lapatinib (GDSC). Here *corr-coef* stands for correlation coefficient between predicted response and output response [23].

used the first 300 cell lines that have gene expression and drug responses for specific pairs of drugs. To report our results, we compared fivefold cross-validated Pearson correlation coefficients, MAE and NRMSE between predicted and experimental responses for RF, VMRF, CMRF, and KBMTL. For both VMRF and CMRF, we set the minimum size of samples in each leaf to $n_{size} = 5$, the number of trees in the forest to $T = 150$ and the splitting in each node considers $m = 10$ random features.

The correlation coefficients using fivefold cross-validation error estimation are illustrated for each drug set in Table 8.6. The corresponding MAE and NRMSE behaviors are illustrated in Table 8.7. For CMRF, results with scaling factor α selected using *Method-1* discussed earlier has been used. Tables 8.6 and 8.7 show that CMRF performed better than VMRF, KBMTL, and RF in terms of correlation coefficients and NRMSE for the related drug pairs S_{C1}, S_{C2}, and S_{C3}. When there is no relationship in the drug pair as in S_U, univariate RF performs better than the multivariate approaches on an average. The scatter plots of predicted response versus original response for drug set S_{C2} using RF, VMRF, and CMRF are shown in Fig. 8.11.

RESULTS OF VARIABLE IMPORTANCE ANALYSIS

We have examined the VIM for GDSC data using VMRF and CMRF in terms of protein interaction network enrichment analysis. In this section, we will primarily provide the detailed results for S_{C1} in GDSC. To avoid any bias due to feature selection in variable importance, we consider the full set of probe set IDs without application of RELIEFF for this analysis.

In both VMRF and CMRF, the 50 top-ranked probesets were generated separately. It should be noted that multiple probeset IDs can map to a single Gene Symbol of a protein. This mapping was done in Genome Medicine Database of Japan

Table 8.6 Fivefold CV Results for CCLE Dataset Drug Sensitivity Prediction for Four Drug Sets in the Form of Correlation Coefficients for *RF*, *VMRF*, *CMRF*, and *KBMTL*

Drug Set	Common Target	Drug Name	Correlation Coefficients			
			RF	*VMRF*	*CMRF*	*KBMTL*
S_{C1}	EGFR	Erlotinib	0.3916	0.3980	0.3927	0.3457
		Lapatinib	0.4460	0.4468	0.4673	0.2609
S_{C2}	MET	Crizotinib	0.4813	0.4719	0.4882	0.4519
		PHA-665752	0.3547	0.3587	0.3746	0.2250
S_{C3}	EGFR	ZD-6474	0.2355	0.2535	0.2627	0.1304
		AZD-0530	0.1990	0.1844	0.1957	0.1973
S_U	None	17-AAG	0.3620	0.3337	0.3255	0.4100
		Erlotinib	0.3818	0.3852	0.3718	0.2828

Table 8.7 Fivefold CV Results for CCLE Dataset Drug Sensitivity Prediction for Four Drug Sets in the Form of MAE and NRMSE for *RF*, *VMRF*, *CMRF*, and *KBMTL* [23]

Drug Set	Common Target	Drug Name	MAE				NRMSE			
			RF	VMRF	CMRF	KBMTL	RF	VMRF	CMRF	KBMTL
S_{C1}	EGFR	Erlotinib	0.0522	0.0520	0.0515	0.0612	0.9223	0.9210	0.9218	1.0593
		Lapatinib	0.0513	0.0520	0.0509	0.0654	0.8976	0.8977	0.8895	1.1398
S_{C2}	MET	Crizotinib	0.0484	0.0483	0.0477	0.0546	0.8836	0.8921	0.8828	0.9674
		PHA-665752	0.0492	0.0496	0.0489	0.0614	0.9367	0.9367	0.9307	1.1573
S_{C3}	EGFR	ZD-6474	0.0660	0.0659	0.0656	0.0876	0.9721	0.9674	0.9650	1.3037
		AZD-0530	0.0728	0.0728	0.0727	0.0866	0.9801	0.9834	0.9810	1.2188
S_U	None	17-AAG	0.1003	0.1005	0.1008	0.0997	0.9553	0.9614	0.9644	0.9740
		Erlotinib	0.0517	0.0519	0.0520	0.0612	0.9258	0.9260	0.9311	1.0957

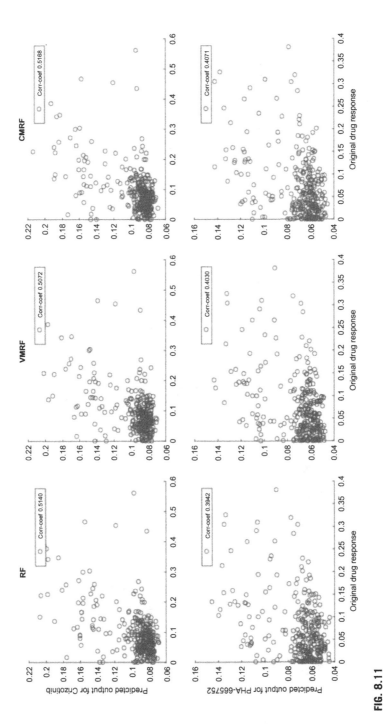

FIG. 8.11

Scatter plots of predicted response vs original response for Crizotinib and PHA-665752 (CCLE). Here *corr-coef* stands for correlation coefficient between predicted response and output response [23].

(GeMDBJ) ID conversion tool (https://gemdbj.nibio.go.jp/dgdb/ConvertOperation. do). Based on this mapping, we arrived at 58 top-ranked proteins for VMRF and 70 top-ranked proteins for CMRF. These proteins were provided as inputs to the string-db database (http://string-db.org/) for known protein-protein interactions (PPIs). The PPI networks for top proteins using CMRF and VMRF are shown in Figs. 8.12 and 8.13, respectively. The enrichment analysis for both the networks are shown alongside each network. We observe that the network generated using CMRF is more enriched in connectivity than the network generated using VMRF. Eighteen interactions with a P-value of 0.132 were observed for the VMRF PPI network, whereas a total of 35 interactions with a P-value of 0.00775 were observed for the CMRF network. Moreover, the common target EGFR is picked in the top 50 targets and is well connected to other targets of CMRF, whereas EGFR is not selected even in top 150 targets of VMRF.

Similarly, in drug set S_{C2} of GDSC (network not shown), there are 42 interactions with 51 proteins in CMRF and 25 interactions with 54 proteins in VMRF.

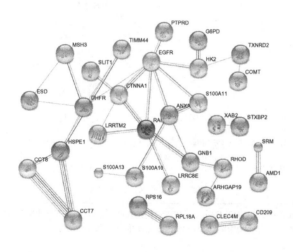

P-value:	7.75e − 3
Interactions observed:	35
Interactions expected:	2.20e + 1
Proteins:	70

FIG. 8.12

Protein-protein interaction network observed between top regulators found from CMRF in GDSC dataset S_{C1}. Disconnected nodes are hidden [23].

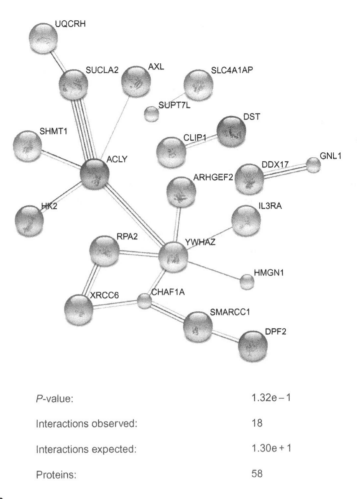

P-value:	1.32e−1
Interactions observed:	18
Interactions expected:	1.30e+1
Proteins:	58

FIG. 8.13

Protein-protein interaction network observed between top regulators found from VMRF in GDSC dataset S_{C1}. Disconnected nodes are hidden [23].

REFERENCES

[1] J.C. Costello, et al., A community effort to assess and improve drug sensitivity prediction algorithms, Nat. Biotechnol. (2014), http://dx.doi.org/10.1038/nbt.2877.

[2] Q. Wan, R. Pal, An ensemble based top performing approach for NCI-DREAM drug sensitivity prediction challenge, PLoS ONE 9 (6) (2014) e101183.

[3] R. Caruana, Multitask learning, Mach. Learn. 28 (1) (1997) 41–75.

[4] M. Segal, Y. Xiao, Multivariate random forests, Wiley Interdiscip. Rev. Data Min. Knowl. Discov. 1 (1) (2011) 80–87.

[5] P.C. Mahalanobis, On the generalised distance in statistics, in: Proceedings National Institute of Science, India, vol. 2, 1936, pp. 49–55.

[6] K.C. Sim, M.J.F. Gales, Precision Matrix Modelling for Large Vocabulary Continuous Speech Recognition, Appendix B1, Cambridge University Engineering Department, 2004.

[7] D. Goldberg, Genetic Algorithms in Search, Optimization, and Machine Learning, Addison-Wesley, Reading, MA, 1989.

[8] Broad-Novartis Cancer Cell Line Encyclopedia, Genetic and pharmacologic characterization of a large panel of human cancer cell lines, http://www.broadinstitute.org/ccle/home.

[9] Y.-H. Ling, T. Li, Z. Yuan, M. Haigentz, T.K. Weber, R. Perez-Soler, Erlotinib, an effective epidermal growth factor receptor tyrosine kinase inhibitor, induces p27KIP1 up-regulation and nuclear translocation in association with cell growth inhibition and G1/S phase arrest in human non-small-cell lung cancer cell lines, Mol. Pharmacol. 72 (2) (2007) 248–258.

[10] S.R. Johnston, A. Leary, Lapatinib: a novel EGFR/HER2 tyrosine kinase inhibitor for cancer, Drugs Today (Barc) 42 (7) (2006) 441–453.

[11] M.W. Karaman, S. Herrgard, D.K. Treiber, P. Gallant, C.E. Atteridge, B.T. Campbell, K.W. Chan, P. Ciceri, M.I. Davis, P.T. Edeen, R. Faraoni, M. Floyd, J.P. Hunt, D.J. Lockhart, Z.V. Milanov, M.J. Morrison, G. Pallares, H.K. Patel, S. Pritchard, L.M. Wodicka, P.P. Zarrinkar, A quantitative analysis of kinase inhibitor selectivity, Nat. Biotechnol. 26 (1) (2008) 127–132.

[12] P.P. Zarrinkar, R.N. Gunawardane, M.D. Cramer, M.F. Gardner, D. Brigham, B. Belli, M.W. Karaman, K.W. Pratz, G. Pallares, Q. Chao, K.G. Sprankle, H.K. Patel, M. Levis, R.C. Armstrong, J. James, S.S. Bhagwat, AC220 is a uniquely potent and selective inhibitor of FLT3 for the treatment of acute myeloid leukemia (AML), Blood 114 (14) (2009) 2984–2992.

[13] J. Tanizaki, I. Okamoto, K. Okamoto, K. Takezawa, K. Kuwata, H. Yamaguchi, K. Nakagawa, MET tyrosine kinase inhibitor crizotinib (PF-02341066) shows differential antitumor effects in non-small cell lung cancer according to MET alterations, J. Thorac. Oncol. 6 (10) (2011) 1624–1631.

[14] P.C. Ma, E. Schaefer, J.G. Christensen, R. Salgia, A selective small molecule c-MET Inhibitor, PHA665752, cooperates with rapamycin, Clin. Cancer Res. 11 (6) (2005) 2312–2319.

[15] A. Sklar, Fonctions de repartition an dimensions et leurs marges, Publ. Inst. Statist. Univ. Paris 8 (1959) 229–231.

[16] A. Sklar, Random variables, joint distribution functions, and copulas, Kybernetika 9 (1973) 449–460.

[17] D.G. Clayton, A model for association in bivariate life tables and its application in epidemiological studies of familial tendency in chronic disease incidence, Int. Stat. Rev. 65 (1978) 141–151.

[18] L. Lee, Generalized econometric models with selectivity, Econometrica 51 (1983) 507–512.

[19] M.J. Frank, On the simultaneous associativity of $F(x, y)$ and $x + y - F(x, y)$, Aequationes Math. 19 (1979) 194–226.

[20] S. Demarta, A.J. McNeil, The T copula and related copulas, Int. Stat. Rev. 73 (2005) 111–129.

[21] E.J. Gumbel, Distributions des Valeurs Extremes en Plusieurs Dimensions, Publ. Inst. Stat. Univ. Paris 9 (1960) 171–173.

[22] P.K.S.K. Owzar, Copulas: concepts and novel applications, Int. J. Stat. LXI (3) (2003) 323–353.

[23] S. Haider, R. Rahman, S. Ghosh, R. Pal, A copula based approach for design of multivariate random forests for drug sensitivity prediction, PLoS ONE 10 (12) (2015) e0144490.

[24] K.J. Archer, R.V. Kimes, Empirical characterization of random forest variable importance measures, Comput. Stat. Data Anal. 52 (4) (2008) 2249–2260.

[25] M. Robnik-Sikonja, I. Kononenko, Theoretical and empirical analysis of ReliefF and RReliefF, Mach. Learn. 53 (2003) 23–69.

[26] M. Gonen, A.A. Margolin, Drug susceptibility prediction against a panel of drugs using kernelized Bayesian multitask learning, Bioinformatics 30 (17) (2014) i556–i563.

[27] W. Yang, et al., Genomics of Drug Sensitivity in Cancer (GDSC): a resource for therapeutic biomarker discovery in cancer cells, Nucleic Acids Res. 41 (D1) (2013) D955–D961.

[28] Super Target, http://bioinf-apache.charite.de/supertarget_v2/index.php?site=home.

[29] G.S. Falchook, K.D. Lewis, J.R. Infante, M.S. Gordon, N.J. Vogelzang, D.J. DeMarini, P. Sun, C. Moy, S.A. Szabo, L.T. Roadcap, et al., Activity of the oral MEK inhibitor trametinib in patients with advanced melanoma: a phase 1 dose-escalation trial, Lancet Oncol. 13 (8) (2012) 782–789.

[30] L. Ciuffreda, D. Del Bufalo, M. Desideri, C. Di Sanza, A. Stoppacciaro, M.R. Ricciardi, S. Chiaretti, S. Tavolaro, B. Benassi, A. Bellacosa, et al., Growth-inhibitory and antiangiogenic activity of the MEK inhibitor PD0325901 in malignant melanoma with or without BRAF mutations, Neoplasia (New York, NY) 11 (8) (2009) 720.

[31] A. Morabito, M.C. Piccirillo, F. Falasconi, G. De Feo, A. Del Giudice, J. Bryce, M. Di Maio, E. De Maio, N. Normanno, F. Perrone, Vandetanib (ZD6474), a dual inhibitor of vascular endothelial growth factor receptor (VEGFR) and epidermal growth factor receptor (EGFR) tyrosine kinases: current status and future directions, Oncologist 14 (4) (2009) 378–390.

[32] A.B. Larsen, M.-T. Stockhausen, H.S. Poulsen, Cell adhesion and EGFR activation regulate EphA2 expression in cancer, Cell. Signal. 22 (4) (2010) 636–644.

Predictive modeling based on functional and genomic characterizations

9

CHAPTER OUTLINE

9.1 INTRODUCTION

A primary objective of precision medicine is to generate high accuracy predictive models for diseases that can be utilized to generate an optimal therapeutic strategy. The model generation is highly challenging for diseases that exhibit high variability among patients, such as cancer. Personalized medicine for cancer involves the design of drug sensitivity prediction models that can predict patient response to various drugs. Current approaches to drug sensitivity modeling are primarily considered as a model inference problem based on the genomic characterizations and drug responses to training cell lines. However, the genetic characterizations observed under normal growth conditions can only provide a single snapshot of the biological system.

Predictive Modeling of Drug Sensitivity. http://dx.doi.org/10.1016/B978-0-12-805274-7.00009-9

The model inferred from analyzing multiple samples is primarily an aggregate model, with the expectation that new tumor samples will activate distinct parts of the model, which will be sufficient to distinguish their diverse responses to drugs. This methodology will provide reliable results when the tumor samples have limited variations in their pathways and the tumor type is well characterized. However, for less studied tumors such as sarcoma and tumors exhibiting numerous aberrations in the molecular pathways, it is desirable to investigate further personalized inference of pathway structure.

Here we present an approach to generate predictive models termed *Target Inhibition Maps* (TIMs) based on responses to a set of targeted drugs with known target inhibition profiles [1–3]. The framework considers each tumor sample as a distinct system and the output response to multiple drugs with known targets integrated with any available limited state measurements is utilized to infer the model of the system. The appropriateness of the framework in considerably increasing the accuracy of drug sensitivity prediction and design of synergistic combination therapies have been shown using human patient derived DIPG primary cell cultures [4], canine UPS primary cell cultures [1,2], and human cancer cell line databases [3].

9.2 MATHEMATICAL FORMULATION

Existing drug sensitivity prediction approaches as described in Chapters 6–8 are primarily based on genetic and epigenetic characterizations alone without incorporation of drug target profiles. The common methodology is to consider a training set of cell lines with experimentally measured genomic characterizations (such as RNA expression, Protein Expression, Methylation, SNPs, Copy Number Variations, etc.) and responses to different drugs, and design supervised predictive models for each individual drug based on one or more genomic characterizations. Examples of such approaches include using genetic mutations [5]; based on gene expression profiles [6]; based on phosphor-proteomic signals and prior biological knowledge of generic pathway [7]; utilizing RNA expression, mutational status of specific genes, and SNPs [8]; and multiple approaches tried using Gene Expression, Methylation, RNASeq, SNP6, Exome Sequencing, and RPPA [9,10].

The technique described in this chapter tackles the problem from an alternative perspective. Rather than designing a model based on the response of a single drug to multiple different genetic samples, we consider the response of one genetic sample to multiple different drugs. The multiple drugs are applied through a drug screen with known target inhibition profiles and their steady-state response allows us to create the TIM which can predict the sensitivity to any combination of target inhibitions. The drugs included in the drug screen are often kinase inhibitors and thus the starting number of targets are limited to around 500 kinases of the human kinome.

The difference in the datasets used in this approach as compared to earlier approaches can be illustrated through Fig. 9.1, which shows the various datasets

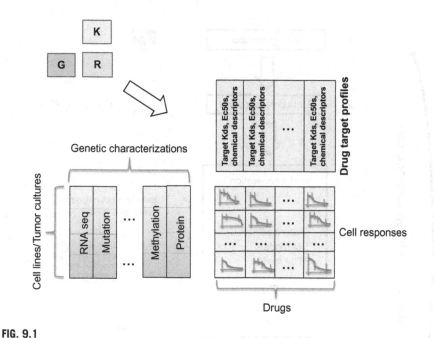

FIG. 9.1

Representation of data types used for drug sensitivity prediction.

used for generating drug sensitivity models. G denotes the matrix of genetic characterizations of size $m \times N$, where $m =$ number of cell lines or tumor cultures, $N =$ total number of genomic features. R denotes the drug response matrix of size $m \times D$ in some form of characteristics of the drug response curves (such as IC_{50}, area under the curve, etc.), where $D =$ number of drugs. K denotes the drug target profile matrix of size $L \times D$, where L denotes the total number of drug features. The currently available prediction approaches based on genetic characterizations primarily consider the matrix G and a column of R to generate a model that can predict the response to that drug for a new genomic characterization. Our approach considers K and a row of R, along with the corresponding row of G to generate a predictive model that can predict the response to a new drug or drug combination with known target inhibition profiles.

The steps involved in generating the integrated model are as follows:

(1) Data generation: Establish tumor culture or cell line and screen for cell viability after exposure to targeted drugs.

(2) Data preprocessing: Process the data such that each drug is represented by a binary vector of inhibited or unchanged targets.

(3) Model generation: Generate TIM and Probabilistic Target Inhibition Map (PTIM) with smallest error for the tumor culture. This step also includes feature selection to select the set of targets relevant for that specific tumor culture.

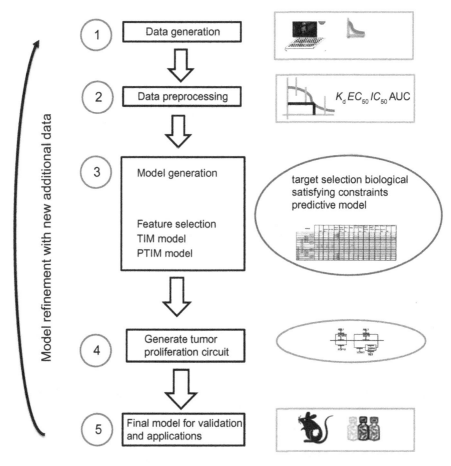

FIG. 9.2

Flowchart for model generation.

(4) **Generate tumor proliferation circuit**: Utilize the PTIM to create a visual tumor proliferation circuit.

(5) **Model refinement**: Utilize additional information for model refinement.

The flowchart for the steps to generate the model is shown in Fig. 9.2.

9.3 DATA PREPROCESSING: DRUG TARGET AND OUTPUT SENSITIVITY NORMALIZATION

This section presents algorithms for generation of binarized drug targets that are used in the next steps of the model design. Following binarization, **1** denotes that the target is inhibited by the drug and **0** denotes that the target is not modified by the drug. This section also considers approaches to map the IC_{50}s of the drugs to a continuous sensitivity score for each drug (a number between 0 and 1 with a higher value denoting that the drug is effective on the tumor).

In order to perform the binarization of the drug target profile, we need to consider the nature of the data that include the EC_{50}s or K_ds of the drug targets, the IC_{50}s of the drugs when applied to a tumor culture, and the expected drug concentrations to be used without significant adverse effects. We can binarize the data in multiple ways. One simple approach is to consider the target to be inhibited (i.e., 1) if its EC_{50} or K_d is less than β multiplied by the drug concentration to be applied where β is a constant.

Another approach of binarization is based on considering the link between the drug response curve and the target inhibition curves. If the primary mechanism of tumor suppression is assumed to be the inhibition of protein kinases realized by these targeted drugs, a natural consequence would be the existence of a relationship between the IC_{50} and EC_{50} values.

This relationship is explained as such: suppose for a drug S_i, the IC_{50} value of S_i and the EC_{50} of protein target k_j (denoted by $e_{i,j}$) are of similar value, then it can be reasonably assumed that protein target k_j is possibly a primary mechanism in the effectiveness of the drug. In other words, if 50% inhibition of a protein target directly correlates with 50% of the tumor cells losing viability, then inhibition of the protein target is most likely one of the causes of cell death. Hence, the target that matches the drug IC_{50} is binarized as a *target hit* for the drug. However, this assumption is too restrictive due to (a) potential noises in EC_{50} and IC_{50} measurements, (b) assumption that drugs operate on single points of failure, and (c) the link between IC_{50} and EC_{50} may not be direct and latent mechanisms might be more important. To address these issues, we consider the binarization range of targets for a drug as $\alpha \cdot \log(IC_{50}) \leq \log(EC_{50}) \leq \beta \cdot \log(IC_{50})$ where $0 \leq \alpha \ll \beta$. We require that β is a smaller constant, such as 3 or 4. For the situation where the above bounds do not result in at least one binarized target, the immediate option is to eliminate the drug from the dataset before target selection. This prevents incomplete information from affecting the desired target set. α is usually considered to be 0 or small in the range of 0.1.

Sensitivity mapping of IC_{50}: The IC_{50}s are usually converted to sensitivities between 0 and 1 using a logarithmic mapping function such as $y = 1 - \frac{\log(IC_{50})}{\log(MaxDose)}$ [2,3]. When maximum achievable clinical dose of a drug is known, we also use the following binarization to incorporate a measure of drug toxicity in the sensitivity:

$$y_i = \begin{cases} 1, & \text{if } IC_{50,i} < Cmax_i \\ c * (1 - \log(IC_{50,i})/\log(MaxDose_i)), & \text{if } Cmax \leq x \leq MaxDose_i \end{cases}$$

where $MaxDose_i$ is the maximum given dose of drug S_i, $Cmax_i$ is the maximum achievable clinical dose of drug S_i, and $c = 1 - \log(Cmax_i)/\log(MaxDose_i)$ to ensure that the scoring function is continuous.

After the preprocessing steps, we have binarized drug target profiles (K) matrix and normalized cell responses (R) matrix.

9.4 MODEL GENERATION

9.4.1 TARGET INHIBITION MAP

The initial model building process will be based on binarized sensitivity, that is, the normalized cell response matrix R will be binarized using a threshold. For L different binarized drug targets, the maximum possible number of distinct multiple target inhibitor activities exhibited by drugs is 2^L because each target can be either inhibited or not and there are L targets, thus the number of possibilities is $(2 \times 2 \times \cdots \times 2)_{L \text{ times}} = 2^L$. Utilizing the information of experimental drug responses on D drugs ($D \ll 2^L$), we attempt to design a model that can predict the response for all possible 2^L target inhibitor combinations. To solve this inference problem, we will incorporate biologically motivated constraints to facilitate selection of a relevant target set with minimal prediction error.

To explain the reasoning for TIMs, let us consider the abstract representation of a biological tumor proliferation pathway shown in Fig. 9.3.

We will consider that a drug works (i.e., able to stop tumor proliferation) if it inhibits all the paths between X and Y. The circuit shown in Fig. 9.3 has three paths from X to Y through targets $\{k_1, k_2\}$, $\{k_1, k_3\}$, and $\{k_4\}$. To block all the paths from X to Y, we have to inhibit all these parallel pathways. The first two pathways can be blocked by either inhibiting k_1 or inhibiting both k_2 and k_3. The third pathway has to be blocked by inhibiting k_4. If we represent this information as a map where a 1 indicates all pathways being blocked for that set of target inhibitions, then the

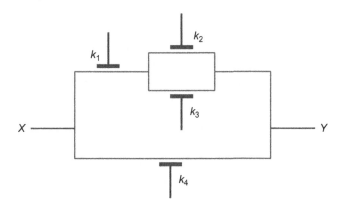

FIG. 9.3

Circuit 1 [1].

$k_1 k_2$ \ $k_3 k_4$	0 0	0 1	1 1	1 0
0 0	0	0	0	0
0 1	0	0	1	0
1 1	0	1	1	0
1 0	0	1	1	0

FIG. 9.4

TIM corresponding to circuit 1 [1].

resultant mapping will be as shown in Fig. 9.4. For example, a 1 in the 2nd row and 3rd column of the 4 × 4 matrix denotes that all the pathways for circuit 1 can be blocked by the combination $k_1 = 0, k_2 = 1, k_3 = 1, k_4 = 1$, that is, any drug inhibiting k_2, k_3, and k_4 will have high sensitivity for circuit 1. Similarly, a 0 in the 2nd row and 2nd column denotes that inhibiting k_2 (i.e., $k_2 = 1$) and inhibiting k_4 (i.e., $k_4 = 1$) cannot produce high sensitivity for circuit 1.

The map shown in Fig. 9.4 will be termed TIM and can be constructed from circuit 1 by checking the output for all the 2^4 combinations of target activities. The Boolean logic for the circuit can be constructed based on the fact that each series connection is like Boolean **OR** function and each parallel connection is like Boolean **AND** function, thus circuit 1 refers to $(k_1$ **OR** $(k_2$ **AND** $k_3))$ **AND** (k_4).

9.4.2 CONSTRUCTING TIMs FROM DRUG TARGETS AND SENSITIVITY DATA

For the construction of the TIMs, the following two sets of rules relevant to our problem will assist in filling the entries of the map.

Rule 1: If S_i is the inhibiting set of targets for drug i and the drug is successful in inhibiting the circuit, then any set B containing the set S_i (i.e., B is a superset of S_i, $B \supset S_i$) will also be successful in inhibiting the circuit.

Rule 2: If S_i is the inhibiting set of targets for drug i and the drug is unsuccessful in inhibiting the circuit, then any set B that is the subset of set S_i (i.e., $B \subset S_i$) will also be unsuccessful in inhibiting the circuit.

Rule 1 essentially says that if inhibiting a number of targets has blocked all the paths, then inhibiting more targets will not open any path. For instance, if we consider circuit 1 and our experiments denote that the set $\{k_1, k_4\}$ is able to inhibit the circuit, then any superset of $\{k_1, k_4\}$ such as $\{k_1, k_4, k_2\}$ will also inhibit the circuit. The number of possible supersets of a set S_i containing a_i elements among possible a elements is 2^{a-a_i}. For circuit 1, if $S_i = \{k_1, k_4\}$, the number of possible supersets of S_i is $2^{4-2} = 4$ and they are $\{k_1, k_4\}, \{k_1, k_4, k_2\}, \{k_1, k_4, k_3\}$, and $\{k_1, k_4, k_2, k_3\}$. If we consider the TIM in Fig. 9.4, then knowing the information that $\{k_1, k_4\}$ inhibits the circuit, we can fill the entries of its superset as 1. The entries corresponding to its superset are $[k_1\ k_2\ k_3\ k_4] = \{1001, 1101, 1011, 1111\}$ which fills the TIM(3, 2), TIM(3, 3), TIM(4, 2), and TIM(4, 3) entries of the inhibition map. Here TIM(i, j) denotes the ith row and jth column entry of the Inhibition Map. The TIM is a matrix of size $2^{p_1} \times 2^{p_2}$ where $p_1 + p_2 = a$ where a is the number of targets and $p_1 = \lfloor a/2 \rfloor$ and $p_2 = a - p_1$.

Rule 2 captures the fact that if a set of target inhibitors is unsuccessful in blocking the paths of a circuit, then any reduced number of target inhibitors among the inhibiting targets cannot block all the paths. For instance, in circuit 1, if our experiments denote that the set $\{k_1, k_2, k_3\}$ of kinase inhibition is not successful in blocking all the paths of the circuit, then any subset of $\{k_1, k_2, k_3\}$, such as $\{k_1, k_2\}$ will also be unsuccessful in blocking the paths of the circuit. The number of possible subsets of a set S_i containing a_i elements is 2^{a_i}. For circuit 1, if $S_i = \{k_1, k_2, k_3\}$, the number of possible subsets of S_i is $2^3 = 8$ and they are $\{k_1, k_2, k_3\}, \{k_1, k_2\}, \{k_1, k_3\}, \{k_2, k_3\}, \{k_1\}, \{k_2\}, \{k_3\}$, and $\{\}$. In the Boolean logic, the subsets will be as follows $[k_1\ k_2\ k_3\ k_4] = \{1110, 1100, 1010, 0110, 1000, 0100, 0010, 0000\}$. Thus, if a drug with set S_i of inhibitors is unsuccessful, then we can mark the states in the TIM that has zeros for the $S - S_i$ kinases as zeros. For our example, $S - S_i = \{4\}$ and any state having zero for k_4 will be marked zero in the kinase inhibition map. So the 1st and 4th column of the TIM will be marked 0 based on the experimental piece of information that the set $\{k_1, k_2, k_3\}$ is unsuccessful. This approach fills up a large number of entries of the TIM based on experimental knowledge on few drugs. For our example, the two experimental results that $\{k_1, k_4\}$ is successful and $\{k_1, k_2, k_3\}$ is not successful fills in 12 out of 16 entries in the TIM. Higher number of entries are filled when a drug with a large number of target inhibitors is unsuccessful or a drug with smaller set of target inhibitors is successful.

Note that the above rules assume that the targets in focus are oncogenes, genes that promote cancer growth and whose inhibition can prevent tumor development. The majority of kinases in the Drug Screen panel behave as oncogenes, and as such, our approach utilizes the above rules. In this chapter, we will develop the framework based on targets being oncogenes. If we consider tumor suppressor genes, complementary rules similar to these rules can be generated to apply to the problem. The tumor suppressor or oncogenic behavior of individual targets can be tested using an siRNA panel.

The algorithm for constructing the TIM is shown in Algorithm 9.1.

ALGORITHM 9.1 ALGORITHM FOR CONSTRUCTING TIM [1]

for $i = 1$ to m **do**
   for $j = 1$ to n **do**
      if Drug is successful **then**
         {Use Rule 1 to fill up entries in the TIM for cell line j (TIM$_j$)}
         for all R such that $R \supset S_i$ **do**
            TIM$_j$ (R) = 1;
         end for
      else
         {Use Rule 2 to fill up entries in the TIM for cell line j (TIM$_j$)}
         for all R such that $R \subset S_i$ **do**
            TIM$_j$ (R) = 0;
         end for
      end if
   end for
end for

9.4.2.1 Feature selection

Algorithm 9.1 will be intractable when all the drug targets are considered for modeling. For instance, $L = 400$ kinases can produce 2^{400} inhibition combinations. Furthermore, searching for subset and superset combinations using large number of targets will produce few matches among the experimental datasets. From a biological context, not all targets will be important with respect to tumor proliferation for a specific tumor, and each tumor is expected to have a selected set of targets that are relevant for inhibiting tumor progression.

Thus we consider the use of feature selection approaches for selecting the relevant set of targets for a specific tumor culture. We can use gene or protein expression data, if available, for an initial feature selection. Because the drugs considered are target inhibitors, any nonexpressed proteins are unlikely to be play a significant role in reducing tumor proliferation. Thus, if we have gene or protein expression data for the tumor culture matched to normal sample, we utilize the information to remove all nonexpressed proteins from our set of potential targets for subsequent feature selection analysis. If we have no genomic characterization data available, all the targets are considered as potential targets for the next part of our analysis.

Feature selection refers to selecting a subset of features from the full set of input features based on some design criteria. The details on feature selection approaches have been described earlier in Chapter 3. For our purposes, we will consider the wrapper-based feature selection approach of *sequential floating forward search* (SFFS) to select our relevant targets. The objective function $J(S_m)$ used to evaluate the goodness of a particular feature subset S_m was the Normalized Mean Absolute Error between predicted and experimental responses using *Leave-one-out* or *10-fold cross-validation* error estimation approaches.

The predicted response for a new drug based on the inferred TIM model is described in the next section.

9.4.2.2 Predicting sensitivity of a new drug based on its set of target inhibitions

If the set of target inhibitions is known for a new drug, then we can check the TIM entry for that set of inhibition for each cell line and predict the outcome of the drug when applied to that cell line.

Example 9.1. Let us consider that we have two cell lines whose abstract circuit representation of the pathways are shown in Figs. 9.5 and 9.6. Based on experimental data on these cell lines, we can construct the target inhibition maps TIM1 and TIM2 as shown in Figs. 9.7 and 9.8. Here, the set of targets $S = \{k_1, k_2, k_3, k_4\}$ and target k_4 is not directly involved in pathway 1 whereas target k_1 is not directly involved in pathway 2. If a new drug D_{D+1} has the kinase inhibition set $S_{m+1} = \{k_2, k_3, k_4\} = \{0111\}$, then based on the 2nd row, 3rd column entries of the inhibition maps TIM1 and TIM2, the drug will be ineffective for CP1 and effective for CP2.

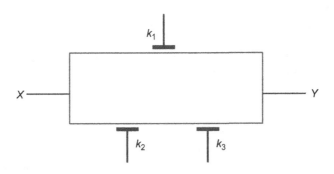

FIG. 9.5

CP1: Cellular pathway representation 1 [1].

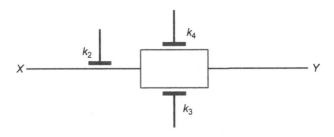

FIG. 9.6

CP2: Cellular pathway representation 2 [1].

$k_1 k_2$ \ $k_3 k_4$	0 0	0 1	1 1	1 0
0 0	0	0	0	0
0 1	0	0	0	0
1 1	1	1	1	1
1 0	0	0	1	1

FIG. 9.7

TIM1: Target inhibition map corresponding to Cellular pathway representation 1 shown in Fig. 9.5 [1].

$k_1 k_2$ \ $k_3 k_4$	0 0	0 1	1 1	1 0
0 0	0	0	1	0
0 1	1	1	1	1
1 1	1	1	1	1
1 0	0	0	1	0

FIG. 9.8

TIM2: Target inhibition map corresponding to Cellular pathway representation 2 shown in Fig. 9.6 [1].

9.4.2.3 Predicting target inhibitions of a new drug based on its sensitivities over cell lines

This framework can also be applied to estimate the potential target inhibitions of a new drug based on the responses of the drug over different cell lines. Utilizing prior experiments and biological knowledge, we can construct the inhibition maps corresponding to each of the n cell lines $CL_1, CL_2, \ldots, CL_n$. Let A_i denote the set of indices for which the entry in TIM i is 1. Similarly, A_i^c denotes the set of indices for which the entry in TIM i is 0. Let V denote the set of all indices of a inhibition map, that is, $V = A_i \cup A_i^c$ and the set V has 2^a entries. Let us consider that a new drug is applied to the n cell lines and the sensitivities are given as $Z_1, Z_2, \ldots, Z_n$ where Z_is are binary with 1 denoting that the drug was successful for cell line i and 0 denoting that the drug was unsuccessful for cell line i. The possible set of target inhibitions of the new drug is constructed using Algorithm 9.2.

ALGORITHM 9.2 ALGORITHM FOR PREDICTING TARGET INHIBITIONS OF A NEW DRUG

> **for** $i = 1$ to n **do**
>   Initialize $G = V$
>   **if** Drug is successful, that is, $Z_i = 1$ **then**
> $G = G \cap A_i$
>   **else**
> $G = G \cap A_i^c$
>   **end if**
> **end for**
>
> **return** G

Example 9.2. Let us consider that we have two cell lines with pathways as shown in Figs. 9.5 and 9.6. Experimental data on application of drugs on these pathways resulted in the inhibition maps shown in Figs. 9.7 and 9.8. Suppose a new drug is applied to these cell lines and the drug is effective for cell line 1 and ineffective for cell line 2. Because $A_1 = \{1100, 1101, 1111, 1110, 1011, 1010\}$ and $A_2 = \{0000, 0001, 0010, 1000, 1001, 1010\}$, using Algorithm 9.2, we have our possible set of target inhibitions of the new drug as $A_1 \cap A_2^c = \{1010\}$. So the new drug inhibits targets k_1 and k_3.

9.4.3 PROBABILISTIC TARGET INHIBITION MAP

In this section, we will extend the TIM model to encompass continuous sensitivity responses. Because the continuous normalized sensitivities can be considered as probability of the sensitivity being 1, we will term this approach as PTIM modeling. To generate a PTIM from the training data, we require the capability to predict the continuous sensitivity for a new drug with target inhibition C_i. For this purpose, we utilize the following third rule that expresses sensitivity of a target set as a function of its most similar target combinations:

Rule 3: If $(C_i|T)$ is the inhibiting set of a target combination with unknown sensitivity, then the sensitivity of $(C_i|T)$ will be at least that of maximum of $(C_i|T)$'s subsets, and at most, that of minimum of $(C_i|T)$'s supersets.

Rule 3 follows from the first two rules; rule 1 provides that any superset will have greater sensitivity, and rule 2 provides that any subset will have lower sensitivity.

To apply rule 3 in practical situations, we must guarantee that every combination $(C_i|T)$ will have a subset and superset with an experimental value. We will assume that the target combination that inhibits all targets in T will be highly effective with sensitivity 1 and the target combination that consists of no inhibition of any target will have no effectiveness, and as such will have a sensitivity of 0. Either of these can be substituted with experimental sensitivity values that have the corresponding target combination.

With the lower bound (maximum of subset sensitivities) and upper bound (minimum of superset sensitivities) of the target combination sensitivity fixed, we interpolate the unknown target combination sensitivity.

Let $(C_{\max}|T)$ denote the target combinations of the subset of $(C_i|T)$ with the maximum sensitivity and $(C_{\min}|T)$ denote the superset target combination with the minimum sensitivity. Let the sensitivities of $(C_{\max}|T)$ and $(C_{\min}|T)$ be y_l and y_u, respectively. Let the hamming distance between $C_{\max}$ and $C_{\min}$ be $h = (C_{\max}|T) \oplus (C_{\min}|T)$, and the hamming distance between $(C_i|T)$ and $(C_{\max}|T)$ be $d = (C_i|T) \oplus (C_{\max}|T)$. The basic interpolation of sensitivity y_i for inhibition C_i is as follows:

$$y_i = y_l + (y_u - y_l) \cdot \left(\frac{d}{h}\right)^{\gamma} \tag{9.1}$$

where γ is a tunable inference discount parameter, where decreasing γ increases y_i and presents an optimistic estimate of sensitivity. We usually consider $\gamma = 1$.

Other different interpolation approaches can be used based on the importance of each target when available as numeric weights or by fitting linear and nonlinear regression models.

Fig. 9.9 provides a pictorial example for the sensitivity prediction. The set of targets $S_{T_0} = \{T_1, \ldots, T_9\}$. The oval shapes represent a drug and targets of those drugs are mentioned inside the oval shape. For instance, the oval shape with target T_3 represents drug b_1 with $K(S_{T_0}, b_1) = \{0, 0, 1, 0, 0, 0, 0, 0, 0\}^T$. Similarly, $K(S_{T_0}, b_4) = \{1, 1, 1, 1, 1, 0, 1, 0, 0\}$ representing the oval shape with targets $T_1, T_2, T_3, T_4, T_5, T_7$. For this example, $y_1 = R(i, b_1)$, $y_2 = R(i, b_2)$, and $y_3 = R(i, b_3)$ are sensitivities of the drugs that inhibit a subset of targets of the drug to be predicted (which inhibits T_1, T_2, T_3, T_4, i.e., $K(S_{T_0}, i) = \{1, 1, 1, 1, 0, 0, 0, 0, 0\}$). Similarly, $y_4 = R(i, b_4)$, $y_5 = R(i, b_5)$, and $y_6 = R(i, b_6)$ are sensitivities of drugs that inhibit supersets of drug j. The desired sensitivity of the new drug y_p will be between $\max(y_1, y_2, y_3)$ and $\min(y_4, y_5, y_6)$. The exact prediction is determined using the interpolation approach described earlier.

The described approach can be used to generate the sensitivities for all possible combinations of the selected target set. Fig. 9.10 shows an example PTIM with 8 targets and $2^8 = 64$ different inhibition combinations. The PTIM was generated based on application of 60 kinase inhibitors on a canine osteosarcoma tumor culture (details available at [1]). For a large target set, it is advisable to avoid predicting the sensitivities of all possible target combinations, but only the ones that are required for validation or error calculation purposes.

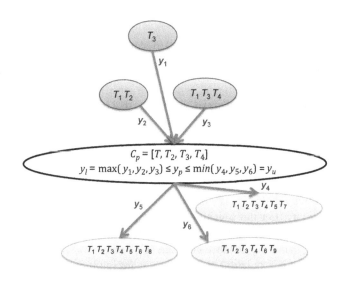

FIG. 9.9

Sensitivity prediction y_p for inhibitions $C_p = [T_1, T_2, T_3, T_4]$.

KINASES	MET	MET	MAP4K5	MAP4K5	MAP4K5 / MET	MAP4K5 / MET	EPHA3	EPHA3	EPHA3 / MET	EPHA3 / MET	PIK3CA / EPHA3	PIK3CA / EPHA3	PIK3CA / EPHA3 / MAP4K5 / MET	PIK3CA / MAP4K5	PIK3CA / MAP4K5 / MET	PIK3CA / MET
ACVR1B	0	0	0	0	0.01	0.3	0	0	0	0	1	1	0	0	0	0
MRCKA ACVR1B	0	0	0	0	0.49	1	0	0	0	0	1	1	0	0	0	0
MRCKA	0	0	0	0	0.07	1	0	0	1	1	1	1	1	1	1	0.91
PIM1 MRCKA	1	1	1	1	1	1	1	1	1	1	1	1	1	1	1	1
PIM1 MRCKA ACVR1B	1	1	1	1	1	1	1	1	1	1	1	1	1	1	1	1
PIM1 ACVR1B	1	1	1	1	1	1	1	1	1	1	1	1	1	1	1	1
PIM1	0.26	0.91	0.91	0.91	1	1	1	0.96	1	1	1	1	1	1	1	1
EGFR PIM1	0.91	0.91	0.96	0.91	1	1	1	1	1	1	1	1	1	1	1	1
EGFR PIM1 ACVR1B	1	1	1	1	1	1	1	1	1	1	1	1	1	1	1	1
EGFR PIM1 MRCKA ACVR1B	1	1	1	1	1	1	1	1	1	1	1	1	1	1	1	1
EGFR PIM1 MRCKA	1	1	1	1	1	1	1	1	1	1	1	1	1	1	1	1
EGFR MRCKA	0	0	0	0	0.64	1	0	0	1	1	1	1	1	1	1	1
EGFR MRCKA ACVR1B	0	0	0	0	0.96	1	0	0	1	1	1	1	1	1	1	1
EGFR ACVR1B	0	0	0	0	0.94	1	0	0	0	0	1	1	0	0	0	0
EGFR	0	0	0	0	0.3	1	0	0	0	0	1	1	0	0	0	0

FIG. 9.10

Example PTIM generated from canine osteosarcoma tumor culture and a drug screen of 60 kinase inhibitors [1].

9.5 GENERATE TUMOR PROLIFERATION CIRCUIT

In this section, we present algorithms for inference of blocks of targets whose inhibition can reduce tumor survival. The resulting combination of blocks can be represented as an abstract tumor survival pathway which will be termed the TIM circuit.

The following two types of Boolean relationships are usually expected for oncogenes in a tumor proliferation circuit: logical AND relationships where an effective treatment consists of inhibiting two or more targets simultaneously, and logical OR relationships where inhibiting one of two or more sets of targets will result in an effective treatment. Here, effectiveness is determined by the desired level of sensitivity before which a treatment will not be considered satisfactory. The two Boolean relationships are reflected in the two rules presented previously. By extension, a NOT relationship would capture the behavior of tumor suppressor targets which is not directly considered in this chapter. Another possibility is XOR (exclusive or) which is not included in the current formulation due to the absence of sufficient evidence for existence of such behavior at the protein target inhibition level.

Thus our underlying network consists of a Boolean equation with numerous terms. To construct the minimal Boolean equation that describes the underlying network, we utilize the concept of PTIM presented in the previous section. We fix a maximum size for the number of targets in each target combination to limit the number of required inference steps. Let this maximum number of targets considered be M. We usually select M to be 10 or less.

We consider all nonexperimental sensitivity combinations with fewer or equal to M targets. We binarize the resulting inferred sensitivities using the binarization threshold for inferred sensitivity values $\theta_i \in [0, 1]$. As $\theta_i \rightarrow 1$, an effective combination becomes more restrictive, and the resulting Boolean equations will have fewer effective terms. There is an equivalent threshold for target combinations with experimental sensitivity, denoted by θ_e.

We start with the target combinations with experimental sensitivities and binarize them using threshold θ_e. The terms that represent a successful treatment are added to the Boolean equation. To find the remaining terms of the Boolean equation, we begin with all possible target combinations of size 1. If the sensitivity of these single targets are sufficient relative to θ_i and θ_e, the target is binarized; any further addition of targets will only improve the sensitivity as per rule 3. If the target is not binarized at that level, we expand it by including all possible combinations of two targets, including the target in focus. We continue expanding this method, pruning search threads once the binarization threshold has been reached. The method essentially resembles a breadth or depth-first search routine over n branches to a maximum depth of M. The minimal Boolean equation generation algorithm is shown in Algorithm 9.3 where the function $binary(x|T)$ returns the binary equivalent of x given the number of targets in T, and $sensitivity(x|T)$ returns the sensitivity of the inhibition combination x for the target set T.

ALGORITHM 9.3 ALGORITHM FOR GENERATION OF MINIMAL BOOLEAN EQUATION [2]

Inputs: T, the set of kinase targets

M, the maximum number of targets in an inferred Boolean term

$\theta_e, \theta_i \in [0, 1]$, the threshold for experimental and inferred sensitivity binarization

$bins$, the target bins to separate training drugs

Output: *Terms*, the set of Boolean terms in the minimal equation

Terms $= \emptyset$

```
Queue = {0}
for b ∈ bins do
    if sensitivity(b|T) ≥ θ_e then
        Terms = Terms ∪ b
    end if
end for
while Queue ≠ ∅ do
    n = Pop(Queue)
    if sensitivity(n|T) ≥ θ_i then
        Terms = Terms ∪ n
    else
        if ∑ binary(n|T) < M then
            for x in binary(n|T) s.t. x is 0 do
                Push(n + 2^x, Queue)
            end for
        end if
    end if
end while
for n ∈ Terms do
    for m ∈ Terms do
        if binary(n|T) ⊂ binary(m|T)  then
            Terms = Terms \ m
        end if
    end for
end for

return Terms
```

To convey the minimal Boolean equation to clinicians and researchers unfamiliar with Boolean equations, we utilize a convenient circuit representation as shown in Fig. 9.11. The circuit was generated from the PTIM shown in Fig. 9.10 with $\theta_e = \theta_i = 0.25$. The circuit diagrams are organized by grouped terms, which we denote as *blocks*. Blocks in the TIM circuit act as possible treatment combinations. The blocks are organized in a linear OR structure; treatment of any one block should result in high sensitivity. As such, inhibition of each target results in its line being broken. When there are no available paths between the beginning and end of the circuit, the treatment is considered effective. As such, each block is essentially a modified AND/OR structure. Within the blocks, parallel lines denote an AND relationship, and adjacent lines represent an OR relationship. The goal of an effective treatment then, from the perspective of the network circuit diagram, is to prevent the tumor from having a pathway by which it can continue to grow.

9.6 MODEL REFINEMENT

The model can be further refined with the availability of new information in terms of genomic characterizations or responses to a new set of drugs. As discussed earlier,

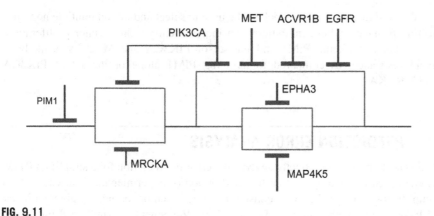

FIG. 9.11

Example TIM circuit generated from the PTIM shown in Fig. 9.10 [1].

gene or protein expression data can be used to remove false-positive targets from the model. To incorporate the responses to a new set of drugs, if the drugs target the same set of kinases, repeating of steps 2–4 is desirable with new appended dataset to generate the refined model. If the target set of the new drugs is significantly different from the original drugs, it is better to design a separate model with a different set of relevant targets from each dataset.

For validation, experiments can be designed to check the accuracy of the model predictions. For instance, kinases that appear in one block representing a AND relationship such as PIK3CA and MRCKA in Fig. 9.11 should have a synergistic effect: that is, the effect of targeting both PIK3CA and MRCKA together will be more than the sum of effects of targeting PIK3CA and MRCKA individually. Examples of experimental validation based on TIM circuits are available in Supplementary Figs. S3 and S4 of Grasso et al. [4].

The model can be utilized to select a drug or combination of drugs that will have high sensitivity, while maintaining toxicity below a threshold. One of the rudimentary ways to estimate the toxicity is by considering it to be proportional to the number of target inhibitions [11–13] and the targets of a drug combination considered as the union of targets of the individual drugs [14]. Other approaches to toxicity estimation can also be considered based on existing side effects data for individual drugs as available from publicly available databases, such as Side Effect Resource (SIDER) [15] or information on the absorption, distribution, metabolism, excretion, and toxicity (ADMET) properties of different drugs, such as ChEMBL [16]. A toxicity model can be potentially designed based on the individual side effects, chemical descriptors, and drug-protein interactions.

Another direction that can be explored is the design of a dynamical model from the estimated static PTIM [17]. Because a large number of dynamical models can potentially satisfy a given PTIM, additional state measurements need to be conducted to narrow down the feasible dynamic models [17].

To avoid resistance, the model can be used to select and inhibit multiple noninter-secting pathways whose inhibitions can independently reduce tumor proliferation. For instance, targeting PIM1 and the set of PIK3CA and MRCKA is likely to avoid resistance, as compared to targeting PIM1 alone or the set of PIK3CA and MRCKA.

9.7 PREDICTION ERROR ANALYSIS

In this section, we theoretically analyze the error in prediction for a simplified PTIM model based on assumptions on the distribution of experimental sensitivities. As a simplification, the prediction is considered to be based on an individual target block of PTIM model with β targets: $T_1, T_2, \ldots, T_\beta$. We consider a single cell line and η samples of drug responses for that cell line. Let X_i for $i \in \{1, 2, \ldots, \beta\}$ represent the random variable denoting the target inhibition of target T_i for $i \in \{1, 2, \ldots, \beta\}$. The range of the X_i for $i \in \{1, 2, \ldots, \beta\}$ is $[0, 1]$. Let us consider that the training set consists of η samples that are independently selected based on the probability distribution $f_{X_1, X_2, \ldots, X_\beta}$. We are interested in predicting the sensitivity of a new sample with the following target inhibition profile $V = y_1, y_2, \ldots, y_\beta$. Let A denote the event that a sample $x_1, x_2, \ldots, x_\beta$ selected based on the probability distribution $f_{X_1, X_2, \ldots, X_\beta}$ has the relation $(x_1 \leq y_1) \wedge (x_2 \leq y_2) \wedge \cdots (x_\beta \leq y_\beta)$. Then,

$$P(A) = \int_0^{y_\beta} \cdots \int_0^{y_2} \int_0^{y_1} f_{X_1, X_2, \ldots, X_\beta}(x_1, x_2, \ldots, x_\beta) \\ \ldots dX_1 dX_2 \ldots dX_\beta \tag{9.2}$$

Similarly, let B denote the event that a sample $x_1, x_2, \ldots, x_\beta$ selected based on the probability distribution $f_{X_1, X_2, \ldots, X_\beta}$ has the relation $(x_1 \geq y_1) \wedge (x_2 \geq y_2) \wedge \cdots (x_\beta \geq y_\beta)$. Thus

$$P(B) = \int_{y_\beta}^1 \cdots \int_{y_2}^1 \int_{y_1}^1 f_{X_1, X_2, \ldots, X_\beta}(x_1, x_2, \ldots, x_\beta) \\ \cdots dX_1 dX_2 \ldots dX_\beta \tag{9.3}$$

If the random variables $X_1, X_2, \ldots, X_\beta$ are independent and identically distributed with PDF f_X and CDF F_X, then $f_{X_1, X_2, \ldots, X_\beta}(x_1, x_2, \ldots, x_\beta) = f_X(x_1) f_X(x_2) \ldots f_X(x_\beta)$.

We next consider the validity of the assumption that random variables $X_1, X_2, \ldots, X_\beta$ are independent and identically distributed. For ease of analysis, we consider $\beta = 2$ and generate two-dimensional distributions for f_{X_1, X_2} from real biological target inhibition data. We have utilized three sets of target inhibition data: (a) a 339-compound kinome screen consisting of 339 drugs and 203 targets (denoted by KS), (b) a 60-drug panel consisting of 404 targets, and (c) another 60-drug panel consisting of 404 targets.

We considered the following discrete points in the probability distribution $\{0, 0.1, 0.2, \ldots, 0.9, 1\}$ and thus the joint probability distribution can be represented by 11×11 size matrix termed M_N. We selected a random pair of targets and noted the frequency in the matrix M_N. We did this for 10,000 different tries and continued appending the frequencies in the matrix M_N. Finally, the matrix M_N is normalized to generate the approximate two-dimensional probability distribution. To check the independency validity, we generated the marginal distributions of X_1 and X_2 and created a two-dimensional probability matrix M_I from the marginal distributions of X_1 and X_2 based on independence assumption. We next computed the norm of $M_N - M_I$ to check the validity of the independence assumption. For comparison purposes, a random two-dimensional probability matrix M_R was also generated and norm of $M_N - M_R$ calculated. The results are summarized in Table 9.1.

The small value of $|M_N - M_I|_1$ for three different drug panel files shown in Table 9.1 supports the validity of the assumption that X_1 and X_2 are independent.

Thus, returning to the probabilities of events A and B, we have

$$P(A) = \left(\int_0^{y_\beta} f_X(x_\beta) dX_\beta \right) \cdots \left(\int_0^{y_2} f_X(x_2) dX_2 \right) \cdots$$
$$\cdots \left(\int_0^{y_1} f_X(x_1) dX_1 \right) \tag{9.4}$$
$$= F_X(y_\beta) \cdots F_X(y_2) F_X(y_1)$$

and

$$P(B) = \left(\int_{y_\beta}^1 f_X(x_\beta) dX_\beta \right) \cdots \left(\int_{y_2}^1 f_X(x_2) dX_2 \right) \cdots$$
$$\cdots \left(\int_{y_1}^1 f_X(x_1) dX_1 \right) \tag{9.5}$$
$$= (1 - F_X(y_\beta)) \cdots (1 - F_X(y_2))(1 - F_X(y_1))$$

Table 9.1 Norm 1 Differences Between Actual Two-Dimensional Probability Distribution of Target Pairs (M_N)

	$|M_N - M_I|_1$	$|M_N - M_R|_1$
KS panel	0.0262	0.8626
60-Drug panel 1	0.0181	0.9731
60-Drug panel 2	0.0075	0.9936

Notes: Two-dimensional probability distribution generated using independence assumption (M_I). Two-dimensional probability distribution generated randomly (M_R). ©2014 IEEE. Reprinted, with permission, from N. Berlow, S. Haider, Q. Wan, M. Geltzeiler, L.E. Davis, C. Keller, R. Pal, An integrated approach to anti-cancer drugs sensitivity prediction, IEEE/ACM Trans. Comput. Biol. Bioinform. (2014), http://dx.doi.org/10.1155/2014/873436.

We next consider the scenario when f_x in uniform and generate limits on $P(A) + P(B)$. We have also utilized biological drug panel files to get an estimate of expected $P(A) + P(B)$ which is discussed later.

$\mathbb{E}(P(A) + P(B))$ **when** f_X **follows uniform distribution**: If f_X is uniform, then

$$P(A) = y_\beta \cdots * y_2 * y_1 \tag{9.6}$$
$$P(B) = (1 - y_\beta) \cdots * (1 - y_2) * (1 - y_1) \tag{9.7}$$

The number of samples out of the training set that satisfies events A and B follows binomial distributions $Binomial(\eta, P(A))$ and $Binomial(\eta, P(B))$, respectively. Thus the expected number of samples for events A and B are $\eta * P(A)$ and $\eta * P(B)$, respectively. For the case of uniform RVs, the expected number of samples in A is $\eta * y_1 * y_2 * \cdots y_\beta$. Note that if $y_1 = 0.5, y_2 = 0.5, \ldots, y_\beta = 0.5$, then the expected number of points in A is $\eta * 2^{-\beta}$. Thus a general guideline to expect α points in event A is to have $\eta > \alpha * 2^\beta$. Note that the number of samples η to guarantee points from the training set satisfies A grows exponentially with β. Let us consider a general case of $\epsilon \geq y_i \leq 1 - \epsilon$ for $i = 1, \ldots, \beta$, then

$$\gamma * \left(\sqrt{\epsilon * (1 - \epsilon)}\right)^\beta \leq P(A) + P(B) \leq (1 - \epsilon)^\beta + \epsilon^\beta \tag{9.8}$$

where $\gamma = 2$ when β is even and $\gamma = 1/(\sqrt{\epsilon * (1 - \epsilon)})$ when β is odd. However, the expectation of $(P(A) + P(B))$ will be $2^{1-\beta}$. Here expectation is based on the distribution of $y_1, y_2, \ldots, y_\beta$ which is considered to be i.i.d. with uniform distribution between $(0, 1)$ for each y_i.

$\mathbb{E}(P(A) + P(B))$ **when distribution of** f_X **is generated based on biological drug panel inhibition matrices**: We consider the expected value of $P(A) + P(B)$ when f_X is generated from different biological drug panels rather than using uniform distribution assumption. Similar to the analysis for inspecting the validity of independence assumption, we considered the following three sets of target inhibition data for generating f_X: (a) a 339-compound kinome screen consisting of 339 drugs and 203 targets (denoted by KS), (b) a 60-drug panel consisting of 404 targets (denoted by panel 1), and (c) another 60-drug panel consisting of 404 targets (denoted by panel 2). Subsequently, we considered random values for $0 \leq y_1 \leq 1, 0 \leq y_2 \leq 1, \ldots, 0 \leq y_\beta \leq 1$ (selected from uniform distribution) and generated $P(A)$ and $P(B)$. We repeated this 10,000 times and the expected $P(A) + P(B)$ for the three panels are shown in Table 9.2.

As Table 9.2 shows, $\mathbb{E}(P(A) + P(B))$ is significantly higher than the value generated (i.e., $2^{1-\beta}$) using uniform distribution assumption and $y_1 = y_2 = \cdots = y_\beta = 0.5$. This shows that for a new samples with inhibition profile V and a range of β, we expect to have a number of points satisfying either event A or B from among η training points.

Next, let us estimate the error in prediction when we estimate based on maximum sensitivity among points in A and minimum sensitivity among points in B. We will

Table 9.2 $\mathbb{E}(P(A) + P(B))$ for Different βs for Three Drug Panels

β	339-Drug KS Panel	60-Drug Panel 1	60-Drug Panel 2
2	0.8789	0.932	0.9512
3	0.8158	0.8983	0.9274
4	0.7614	0.8667	0.9042
5	0.7105	0.8353	0.8815
6	0.6653	0.8059	0.8593
7	0.6208	0.7785	0.838
8	0.5797	0.7493	0.8168
9	0.5459	0.7229	0.796
10	0.5082	0.6978	0.7766

consider that A and B contain n_1 and n_2 points, respectively, with $n_1 + n_2 = \lambda$. Let the sensitivities of the λ points be distributed uniformly in [0 1]. As end points, we will consider two additional drugs D_0 and $D_{\eta+1}$ in the training panels. The drug D_0 with inhibition profile $0, 0, \ldots, 0$ denotes a placebo with no inhibition and thus will have a sensitivity of 0. The drug $D_{\eta+1}$ with inhibition profile $1, 1, \ldots, 1$ will denote the drug that maximally inhibits all possible targets and thus will have a sensitivity of 1. The drug D_0 will satisfy event A and drug $D_{\eta+1}$ will satisfy event B for any new drug. Thus we will consider $n_1 + 1 + n_2 + 1 = \lambda + 2$ points satisfying event A or B. Let the sorted $\lambda + 2$ sensitivities be $0 \le s_1 \le s_2 \cdots \le s_\lambda \le 1$. Let us denote the maximum sensitivity among the n_1 points in A by Y_l and the minimum sensitivity among the n_2 points in B by Y_h. Based on biological constraints, the actual sensitivity for $y_1, y_2, \ldots, y_\beta$ lies between Y_l and Y_h. Without any other information, we will consider that the actual sensitivity Y_{ac} follows a uniform distribution $f_{Y_{ac}}$ between Y_l and Y_h. Thus, if we consider a basic prediction of $Y_p = (Y_l + Y_h)/2$ for our unknown sensitivity, the expected error in prediction for given Y_l and Y_h is $\mathbb{E}_{Y_{ac}}(|Y_{ac} - Y_p| \,|Y_l, Y_h, \lambda) = \int_{Y_l}^{Y_h} |x - Y_p| f_{Y_{ac}}(x) dx$. For uniform sensitivity in the range $[Y_l \; Y_h]$, the expected error is $(Y_h - Y_l)/4$. Thus

$$\mathbb{E}_{Y_l, Y_h, Y_{ac}}(|Y_{ac} - Y_p| \,|\lambda) = \mathbb{E}_{Y_l, Y_h}(\mathbb{E}_{Y_{ac}}(|Y_{ac} - Y_p| \,|Y_l, Y_h, \lambda))$$
$$= \frac{\mathbb{E}(Y_h) - \mathbb{E}(Y_l)}{4}$$

For the sorted $\lambda + 2$ sensitivities $0 \le s_1 \le s_2 \cdots \le s_\lambda \le 1$, $\mathbb{E}(s_i) - \mathbb{E}(s_{i-1}) = 1/(\lambda + 1)$. Thus the expected error for a given λ is $\mathbb{E}(|Y_{ac} - Y_p| \,|\lambda) = \frac{1}{4*(\lambda+1)}$.

To calculate the expected error based on the possibilities of λ, we note that λ follows a bionomial distribution $Binomial(\eta, P(A)+P(B))$. Thus the expected error is

$$\mathbb{E}(|Y_{ac} - Y_p|) = \sum_{\lambda=0}^{\eta} \frac{1}{4(1+\lambda)} \binom{\eta}{\lambda}(P(A) + P(B))^{\lambda}(1 - P(A) - P(B))^{\eta-\lambda}$$

$$= \frac{1 - (1 - P(A) - P(B))^{\eta+1}}{4(\eta + 1) * (P(A) + P(B))}$$

As a numerical example, when $P(A) + P(B) = 2^{1-\beta}$, $\eta = 10 * 2^{\beta}$, then $\mathbb{E}(|Y_{ac} - Y_p|) = 0.0122, 0.0125, 0.0125$ for $\beta = 2, 5, 7$, respectively.

Thus we note that when we are able to extract blocks of β targets that have monotonic relationships with the sensitivities and we utilize $10 * 2^{\beta}$ drugs with random target inhibition profiles in the drug screen, the errors in prediction will remain quite low in the range of 1.25%. If $\beta = 3$, then $10 * 2^{\beta} = 80$. The β targets denote one minimal set of targets that might be effective in altering the tumor proliferation. There can be multiple such blocks of target and we will predict the final sensitivity based on weighted averaging of the individual predictions.

9.8 APPLICATION RESULTS

In this section, we consider the application of the PTIM framework on synthetic and experimental datasets for performance evaluation.

9.8.1 SYNTHETIC EXPERIMENTS

Synthetic experiment 1: We tested the TIM framework using synthetic data generated from a section of a human cancer pathway taken from the KEGG database [18]. Here, the objective is to show that the proposed TIM method generates models that highly represent the underlying biological network which was sampled via synthetic drug perturbation data. This experiment replicates in synthesis the actual biological experiments that are considered for validating the TIM framework. To utilize the TIM algorithm, a panel of 60 targeted drugs pulled from a library of 1000 is used as a training panel to sample the randomly generated network. Additionally, a panel of 40 drugs is drawn from the library to serve as a test panel. The training panel and the testing panel have no drugs in common. Each of the 60 training drugs is applied to the network, and the sensitivity for each drug is recorded. The generated TIM is then sampled using the test panel which determines the predicted sensitivities of the test panel. The synthetic experiments were performed for 40 randomly generated cancer subnetworks for each of $n = 6, \ldots, 10$ active targets in the network. The active targets are those which, when inhibited, may have some effect on the cancer downstream. To more accurately mimic the Boolean nature of the biological networks, a drug which does not satisfy any of the Boolean network equations will have sensitivity 0, a drug which satisfies at least one network equation will have sensitivity 1. The inhibition profile of the test drugs is used to predict the sensitivity (0 or 1) of the new drug. The average number of correctly predicted drugs for each n is reported in Table 9.3. This synthetic modeling approach generally produces respectable levels of accuracy, with accuracies ranging from 89% to 99%. The 60 drugs for training mimics the drug screen setup used by our collaborators, and testing 20 drugs for predicted sensitivity approximates a secondary drug screen to pinpoint optimal therapies. The performance of the synthetic data shows fairly high reliability of the predictions made by the TIM approach.

Synthetic experiment 2: For the second set of synthetic experiments, we consider that the biological tumor proliferation pathways consist of multiple blocks as shown in Fig. 9.11. The tumor proliferation can be reduced by inhibiting all the targets in a block. For the simulation, cases like the third block in Fig. 9.11 consisting of EPHA3, MAP4K5, PIK3CA, MET, ACVR1B, and EGFR are considered as four blocks consisting of {EPHA3, MAP4K5, PIK3CA}, {EPHA3, MAP4K5, MET}, {EPHA3, MAP4K5, ACVR1B}, and {EPHA3, MAP4K5, EGFR}. The number of blocks for each random synthetic pathway is selected to be a random number between NB_{min} and NB_{max}. The number of targets in a block is a random number between 1 and *MaxBlockSize*. The sensitivity of each block is selected to be a random number between 0 and 1 that satisfies the biological constraints of Rule 1 and Rule 2. The final sensitivity when multiple blocks are inhibited is calculated similar to the manner described in the preliminary example. For T number of targets, there can be 2^T different drugs and once a synthetic pathway is created, a set of N random dugs from the possible 2^T drugs is selected to create the training samples. The prediction results are then tested on separate M drugs selected randomly from the 2^T drugs. As an example, if $T = 12$, there can be $2^{12} = 4096$ possible drugs, and we tested with 100 training drugs which constitute less than 2.5% of possible drugs and the sensitivities of the remaining 97.5% drugs are unknown.

The simulation results are shown in Table 9.4. In Table 9.4, MB refers to the maximum number of targets in a single block; the number of blocks for each synthetic pathway is between N1 and N2; NT provides the number of targets; NTR and NTS provide the number of training and testing samples for each synthetic pathway,

Table 9.3 Results of Synthetic Experiments Based on KEGG Pathways

Targets (n)	Correct Prediction[a]	Accuracy Percentage
6	39.83	99.56
7	38.68	96.69
8	38.18	95.44
9	36.80	92.00
10	35.63	89.06

[a]*Average number of correct prediction for 40 testing drugs for 40 different synthetic pathways [2].*

Table 9.4 Simulation Results for Synthetic Pathway Experiment 1

MB	[N1 N2]	NT	NTR	NTS	NP	Proposed Algorithm			Random Predictions		
						MAE	MC	ESD	MAE	MC	ESD
4	[5 10]	12	100	100	25	0.093	0.91	0.17	0.44	0.02	0.51
5	[5 10]	12	200	400	25	0.07	0.93	0.15	0.44	−0.01	0.51
4	[5 10]	10	200	400	25	0.046	0.96	0.12	0.44	0.01	0.51
4	[5 10]	10	100	100	25	0.071	0.93	0.14	0.45	−0.03	0.52
5	[5 15]	15	250	750	25	0.09	0.92	0.17	0.44	0	0.52
5	[5 15]	15	100	200	25	0.11	0.88	0.19	0.45	0	0.51
4	[5 15]	20	200	300	10	0.115	0.88	0.19	0.44	0.02	0.51
4	[5 15]	20	500	500	10	0.091	0.91	0.17	0.43	0.02	0.5

Notes: ESD, mean error standard deviation; MAE, mean absolute error; MB, MaxBlockSize; MC, mean correlation coefficient; N1, NB_{min}; N2, NB_{max}; NP, number of different pathways; NT, number of targets; NTR, number of training samples; NTS, number of test samples [2].

respectively; NP provides the number of pathways generated. For these simulation parameters, the prediction results of our proposed algorithm is provided as the mean absolute error (MAE), which denotes the expectation over the pathways of the average absolute error of each pathway for the testing samples. The MC denotes the average correlation coefficient between the actual synthetic pathway sensitivities and predicted sensitivities for the testing samples. The ESD denotes the mean standard deviation of the error in prediction for the testing samples. The next three columns denotes the MAE, MC, and ESD if we predict the sensitivities using uniform random numbers between 0 and 1. The 1st row shows that for 12 targets and 100 training samples, we are able to achieve an MAE of 0.093 and an MC of 0.91, while random predictions have an MAE of 0.44 and an MC of 0.02. For our synthetic pathway simulation experiments, the correlation coefficient between predicted and actual sensitivity is close to 0.9 and the MAE is around 0.1, showing the high prediction accuracy of our proposed algorithm.

Fig. 9.12 shows the histogram of the errors in prediction for the set of synthetic pathway results shown in the 2nd row of Table 9.4. Note that with NTS = 400 and NP = 25, there were a total of $400 \times 25 = 10,000$ prediction errors. The histogram in Fig. 9.12 shows that a large number of sample errors are close to 0 with 10 percentile being −0.154 and 90 percentile being 0.051.

We have also tested our algorithm on another set of randomly generated synthetic pathways (detailed results not included). A large number of testing samples were used for each pathway prediction and the results indicate an average error of less than 10% for multiple scenarios. In comparison, the average error with random predictions was 44%. The average correlation coefficient of the prediction to actual sensitivity for the eight sets of experiments (each including 10 or 25 different pathways) was 0.91. The average correlation coefficient with random predictions was 0.

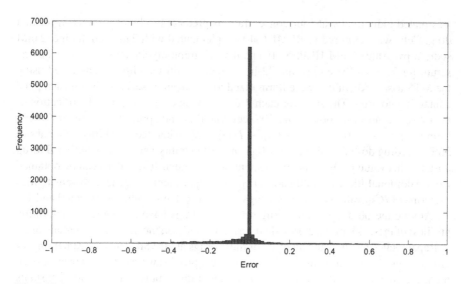

FIG. 9.12

Histogram of errors for $400 \times 25 = 10{,}000$ testing sample predictions for case 2 in Table 9.4 [2].

The results of the synthetic experiments on different randomly generated pathways show that the approach presented in this chapter is able to utilize a small set of training drugs from all possible drugs to generate a high accuracy predictive model.

9.8.2 EVALUATION ON CANINE OSTEOSARCOMA TUMOR CULTURES

In this section, we present results of application of the PTIM framework on *canine osteosarcoma tumor cultures*. The experimental data on 4 tumor cultures and 60 targeted drug screen panels were generated in the Keller Laboratory at Oregon Health and Sciences University. The tumor cultures applied to the drug screens were cultured from four distinct canines with spontaneously developed osteosarcoma, denoted by Bailey, Charley, Sy, and Cora. The tumor cultures were collected by Dr. Bernard Seguin of Oregon State University from canines that are part of an ongoing clinical trial for osteosarcoma (OSU IACUC approval numbers for this study are 4217 and 4273). The tumor samples were collected from client-owned animals that have developed the disease naturally. All procedures performed on these animals with regard to tumor collection were strictly for treatment purposes and nothing was done differently because of the drug perturbation study. All procedures were performed according to standard of care regardless of whether an animal had its tumor sampled.

For the generation of the experimental data, the canine osteosarcoma primary cell cultures were plated in 384 well plates at a seeding density of 2000 cells per well over graded concentrations of 60 small-molecule kinase inhibitors. Each inhibitor was plated individually at four concentrations predicted to bracket the IC_{50} for that drug. Cells were cultured in RPMI 1640 supplemented with 2 mM glutamine, 2 mM sodium pyruvate, 2 mM HEPES, 1% penicillin streptomycin, and 10% fetal bovine serum for 72 h. At the end of the 72-h incubation, cell viability was assessed using the MTS assay. All values were normalized to the mean of seven wells on each plate containing no drug. The IC_{50} for each drug was then determined by identification of the two concentrations bracketing 50% cell viability and application of the following formula: $[((A - 50)/(A - B)) * (D_B - D_A)] + D_A$ where cell viability value above 50% $= A$ (drug dose for this value is D_A) and cell viability value below 50% $= B$ (drug dose for this value is D_B). The experimentally generated IC_{50} values can be obtained from Additional file 1 of publication [2]. The experimentally generated sensitivities (in terms of IC_{50} values) of the 60 drugs are then scaled to values between 0 and 1.

Among the 60 drugs on the drug screen, 46 drugs have known target inhibition profiles; of these 46 drugs, 2 provide information only on the target mammalian target of Rapamycin (mTOR), and analysis of these drugs are trivial. Thus the remaining 44 drugs are used to generate the TIMs. These target profiles were extracted from several literature sources [11,19] based on experimental quantitative dissociation constants (k_d) which are treated as EC_{50} values (explained in the next section) for each drug across kinase target assays with more than 300 targets. The target profiles of the drugs can be accessed from Additional file 2 of publication [2]. Figs. 9.13 and 9.14 represent the equivalent TIM circuits generated from experimental data for Bailey and Sy, respectively.

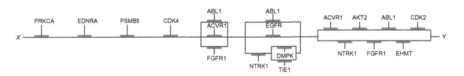

FIG. 9.13

TIM circuit for osteosarcoma primary culture Bailey [2].

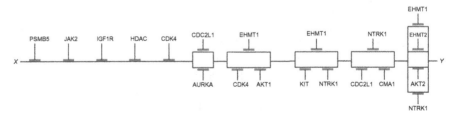

FIG. 9.14

TIM circuit for osteosarcoma primary culture Sy [2].

The inferred circuits for primary cultures Charley and Cora are shown in Figs. 9.15 and 9.16, respectively.

To emphasize the biological relevance provided by the TIM framework employed in the analysis of the biological data, we present a more in-depth analysis of the TIM circuit devised for the canine patient Bailey (shown in Fig. 9.13). The vast majority of human osteosarcomas contain genetic or posttranslational abnormalities in one or both of the tumor suppressors p53 [20–22] and pRb [23]. The first target identified

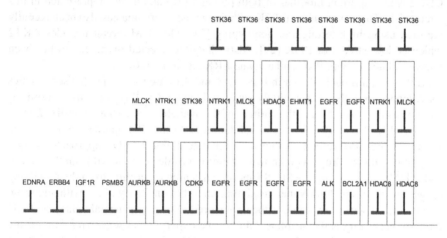

FIG. 9.15

TIM circuit for osteosarcoma primary culture Charley [2].

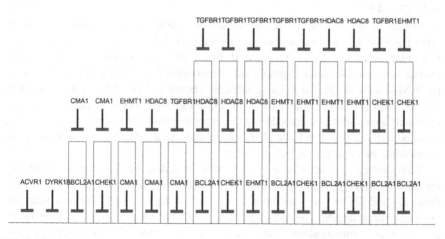

FIG. 9.16

TIM circuit for osteosarcoma primary culture Cora [2].

in this circuit is PKC alpha (PRKCA). PKC alpha modifies CDKN1A (p21), which is the primary mediator of p53 tumor suppressor activity [24]. PSMB5 represents the proteasome (specifically the beta 5 subunit). Previous studies [25] and early preclinical data from the Keller Laboratory confirms in vitro sensitivity of many osteosarcomas to proteasome inhibitors and this sensitivity is hypothesized to be due to the integral role of the proteasome in p53 regulation [26]. Interestingly, CDK4 is also prominent in this circuit, which is a primary inhibitor of the tumor suppressor pRb, which is also frequently abnormal in spontaneous human osteosarcoma [23]. CDK2 is an important modifier of both p53 and pRb and is also represented in this circuit [27]. The importance of PI3K pathway in osteosarcoma has also been recently reported using high-throughput genotyping [28]. Our TIM circuit includes AKT2 which is downstream of PI3K [29]. Also, EDNRA selected in the circuit has been known to interact with PKC and activate ERK signaling [30].

If the circuit models shown in Figs. 9.13 and 9.14 are used to predict sensitivities for comparison with experimentally generated data, we will get optimistic results as the models are trained using the entirety of the available data. Thus we utilize *Leave-one-out* (LOO) and 10-fold Cross-Validation (10-fold CV) approaches to test the validity of the TIM framework that we present next. For the LOO approach, a single drug among the 44 drugs with known inhibition profiles is removed from the dataset and a TIM is built, using the SFFS suboptimal search algorithm, from the remaining drugs. The resulting TIM is then used to predict the sensitivity of the withheld drug. The predicted sensitivity value is then compared to its experimental value; the LOO error for each drug is the absolute value of the experimental sensitivity y minus the predicted sensitivity $(y'|TIM)$; that is, $|y-(y'|TIM)|$. The closer the predicted value is to the experimentally generated sensitivity, the lower the error for the withheld drug. Tables 9.5–9.8 provide the complete LOO error tables and the average LOO error (MAE) for each primary culture. The average LOO error over the four cell cultures is 0.045 or 4.5%.

For the 10-fold CV error estimate, we divided the available drugs into 10 random sets of similar size and the testing is done on each fold while being trained on the remaining ninefolds. This is repeated 10 times and average error calculated on the testing samples. We again repeated this experiment five times and the average of those MAEs for the primary cell cultures are shown in Table 9.9. We note that both 10-fold CV and LOO estimates for all the cultures have errors less than 9%, which is extremely low, especially considering the experimental nature of the drug screening process performed in the Keller Laboratory and the available response of only 44 drugs with known target inhibition profile.

To provide a measure of the overlap between drugs, we considered a similarity measure $\Lambda(D_1, D_2)$ based on the EC_{50} of the drugs D_1 and D_2. Let the EC_{50}s of the drugs D_1 and D_2 be given by the n-length vectors E_1 and E_2 where n denotes the number of drug targets. The entries for the targets that are not inhibited by the drugs (i.e., no EC_{50} value available) are set to 0. Let the vectors V_1 and V_2 represent the binarized targets of the drugs, that is, it has a value of 1 if the target is inhibited by the drug and a value of 0 if the target is not inhibited by the drug. Then we define the

Table 9.5 Leave-One-Out Error Table for Osteosarcoma Primary Culture Bailey

Avg. Err.	Drug	Error	Pred. Sens.	Exp. Sens.
0.047	Veliparib (ABT-888)	0.00	0.00	0.00
	Selumetinib (AZD6244)	0.00	0.00	0.00
	Bortezomib	0.00	0.00	0.00
	Bosutinib (SKI-606)	0.02	0.27	0.29
	Dasatinib	0.00	0.91	0.91
	Erlotinib	0.00	0.00	0.00
	Panobinostat (LBH-589)	0.07	0.93	1.00
	Pazopanib (GW-786034)	0.00	0.00	0.00
	PI-103	0.00	0.00	0.00
	Sorafenib	0.00	0.00	0.00
	Vorinostat (SAHA)	0.08	0.93	0.85
	Obatoclax (GX15-070)	0.01	0.28	0.27
	Crizotinib (PF-2341066)	0.00	0.48	0.48
	MK-2206	0.00	0.65	0.65
	Vismodegib (GDC-0449)	0.00	0.00	0.00
	Alisertib (MLN8237)	0.00	0.00	0.00
	SNS-032 (BMS-387032)	0.03	0.66	0.69
	Carfilzomib	0.36	0.64	1.00
	Imatinib	0.02	0.21	0.19
	BIX 01294	0.18	0.65	0.83
	BMS-754807	0.00	1.00	1.00
	SJ-172550	0.00	0.00	0.00
	Barasertib (AZD1152-HQPA)	0.00	0.00	0.00
	Ruxolitinib (INCB018424)	0.00	0.00	0.00
	Cediranib (AZD2171)	0.03	0.45	0.48
	Lapatinib	0.00	0.00	0.00
	Sunitinib	0.01	0.19	0.20
	Trichostatin A	0.06	0.93	0.86
	Tozasertib (VX-680)	0.02	0.58	0.60
	Enzastaurin	0.36	0.64	1.00
	PD0332991	0.36	0.64	1.00
	Valproate	0.00	0.00	0.00
	Resveratrol	0.00	0.00	0.00
	Zibotentan (ZD4054)	0.36	0.64	1.00
	SP600125	0.00	0.00	0.00
	Ponatinib (AP24534)	0.00	0.00	0.00
	BIX 02188	0.00	0.00	0.00
	RO4929097	0.00	0.00	0.00
	Curcumin	0.00	1.00	1.00
	Sodium butyrate	0.00	0.00	0.00
	GANT61	0.11	0.58	0.47
	Aurothiomalate	0.00	0.00	0.00
	(OSI-906)	0.00	0.00	0.00
	Pelitinib (EKB-569)	0.03	0.91	0.88

Notes: Avg. Err., average of the 44 leave-one-out errors; Error = |Pred. Sens. − Exp. Sens.|;
Exp. Sens., experimental sensitivity; Pred. Sens., predicted sensitivity [2].

Table 9.6 Leave-One-Out Error Table for Osteosarcoma Primary Culture Charley

Avg. Err.	Drug	Error	Pred. Sens.	Exp. Sens.
0.040	Veliparib (ABT-888)	0.00	0.00	0.00
	Selumetinib (AZD6244)	0.00	0.00	0.00
	Bortezomib	0.00	0.00	0.00
	Bosutinib (SKI-606)	0.00	0.00	0.00
	Dasatinib	0.02	0.94	0.96
	Erlotinib	0.00	0.00	0.00
	Panobinostat (LBH-589)	0.00	1.00	1.00
	Pazopanib (GW-786034)	0.00	0.00	0.00
	PI-103	0.00	0.00	0.00
	Sorafenib	0.00	0.00	0.00
	Vorinostat (SAHA)	0.21	1.00	0.79
	Obatoclax (GX15-070)	0.00	0.30	0.30
	Crizotinib (PF-2341066)	0.01	0.64	0.63
	MK-2206	0.03	0.61	0.65
	Vismodegib (GDC-0449)	0.00	0.00	0.00
	Alisertib (MLN8237)	0.00	1.00	1.00
	SNS-032 (BMS-387032)	0.00	1.00	1.00
	Carfilzomib	0.33	0.67	1.00
	Imatinib	0.00	0.00	0.00
	BIX 01294	0.18	0.68	0.86
	BMS-754807	0.00	1.00	1.00
	SJ-172550	0.00	0.00	0.00
	Barasertib (AZD1152-HQPA)	0.00	0.00	0.00
	Ruxolitinib (INCB018424)	0.00	0.00	0.00
	Cediranib (AZD2171)	0.02	0.43	0.44
	Lapatinib	0.00	0.00	0.00
	Sunitinib	0.05	0.88	0.82
	Trichostatin A	0.00	1.00	1.00
	Tozasertib (VX-680)	0.00	1.00	1.00
	Enzastaurin	0.00	0.00	0.00
	PD0332991	0.00	0.00	0.00
	Valproate	0.00	0.00	0.00
	Resveratrol	0.00	0.00	0.00
	Zibotentan (ZD4054)	0.33	0.67	1.00
	SP600125	0.00	0.00	0.00
	Ponatinib (AP24534)	0.00	0.00	0.00
	BIX 02188	0.00	1.00	1.00
	RO4929097	0.00	0.00	0.00
	Curcumin	0.00	1.00	1.00
	Sodium butyrate	0.00	0.00	0.00
	GANT61	0.57	0.43	1.00
	Aurothiomalate	0.00	0.00	0.00
	(OSI-906)	0.00	1.00	1.00
	Pelitinib (EKB-569)	0.00	1.00	1.00

Notes: Avg. Err., average of the 44 leave-one-out errors; Error = |Pred. Sens. − Exp. Sens.|; Exp. Sens., experimental sensitivity; Pred. Sens., predicted sensitivity [2].

Table 9.7 Leave-One-Out Error Table for Osteosarcoma Primary Culture Cora

Avg. Err.	Drug	Error	Pred. Sens.	Exp. Sens.
0.036	Veliparib (ABT-888)	0.00	0.00	0.00
	Selumetinib (AZD6244)	0.00	0.00	0.00
	Bortezomib	0.01	1.00	0.99
	Bosutinib (SKI-606)	0.00	0.00	0.00
	Dasatinib	0.03	0.83	0.85
	Erlotinib	0.00	0.00	0.00
	Panobinostat (LBH-589)	0.05	0.96	1.00
	Pazopanib (GW-786034)	0.00	0.00	0.00
	PI-103	0.00	0.00	0.00
	Sorafenib	0.00	0.00	0.00
	Vorinostat (SAHA)	0.13	0.91	0.78
	Obatoclax (GX15-070)	0.01	0.42	0.44
	Crizotinib (PF-2341066)	0.04	0.66	0.69
	MK-2206	0.28	0.66	0.93
	Vismodegib (GDC-0449)	0.00	0.00	0.00
	Alisertib (MLN8237)	0.00	0.00	0.00
	SNS-032 (BMS-387032)	0.00	1.00	1.00
	Carfilzomib	0.01	0.99	1.00
	Imatinib	0.00	0.00	0.00
	BIX 01294	0.23	1.00	0.89
	BMS-754807	0.00	1.00	1.00
	SJ-172550	0.00	0.00	0.00
	Barasertib (AZD1152-HQPA)	0.00	0.00	0.00
	Ruxolitinib (INCB018424)	0.00	0.00	0.00
	Cediranib (AZD2171)	0.31	0.44	0.75
	Lapatinib	0.00	0.00	0.00
	Sunitinib	0.02	0.78	0.76
	Trichostatin A	0.05	0.89	0.96
	Tozasertib (VX-680)	0.00	1.00	1.00
	Enzastaurin	0.00	0.00	0.00
	PD0332991	0.24	0.76	1.00
	Valproate	0.00	0.00	0.00
	Resveratrol	0.00	0.00	0.00
	Zibotentan (ZD4054)	0.00	0.00	0.00
	SP600125	0.00	0.00	0.00
	Ponatinib (AP24534)	0.00	0.00	0.00
	BIX 02188	0.03	0.92	0.89
	RO4929097	0.00	0.00	0.00
	Curcumin	0.00	0.00	0.00
	Sodium butyrate	0.00	0.00	0.00
	GANT61	0.00	0.00	0.00
	Aurothiomalate	0.00	0.00	0.00
	(OSI-906)	0.11	1.00	0.89
	Pelitinib (EKB-569)	0.01	0.62	0.63

Notes: Avg. Err., average of the 44 leave-one-out errors; Error = |Pred. Sens. − Exp. Sens.|; Exp. Sens., experimental sensitivity; Pred. Sens., predicted sensitivity [2].

Table 9.8 Leave-One-Out Error Table for Osteosarcoma Primary Culture Sy

Avg. Err.	Drug	Error	Pred. Sens.	Exp. Sens.
0.056	Veliparib (ABT-888)	0.00	0.00	0.00
	Selumetinib (AZD6244)	0.00	0.00	0.00
	Bortezomib	0.00	0.00	0.00
	Bosutinib (SKI-606)	0.00	0.59	0.59
	Dasatinib	0.01	0.63	0.62
	Erlotinib	0.00	0.00	0.00
	Panobinostat (LBH-589)	0.05	0.86	0.80
	Pazopanib (GW-786034)	0.00	0.00	0.00
	PI-103	0.00	0.00	0.00
	Sorafenib	0.00	0.00	0.00
	Vorinostat (SAHA)	0.11	0.60	0.71
	Obatoclax (GX15-070)	0.00	0.00	0.00
	Crizotinib (PF-2341066)	0.00	0.59	0.59
	MK-2206	0.04	0.53	0.58
	Vismodegib (GDC-0449)	0.00	0.00	0.00
	Alisertib (MLN8237)	0.00	0.00	0.00
	SNS-032 (BMS-387032)	0.06	0.62	0.69
	Carfilzomib	0.38	0.62	1.00
	Imatinib	0.00	0.00	0.00
	BIX 01294	0.20	1.00	0.82
	BMS-754807	0.01	0.53	0.54
	SJ-172550	0.00	0.00	0.00
	Barasertib (AZD1152-HQPA)	0.00	0.00	0.00
	Ruxolitinib (INCB018424)	0.00	1.00	1.00
	Cediranib (AZD2171)	0.44	0.30	0.75
	Lapatinib	0.00	0.00	0.00
	Sunitinib	0.02	0.60	0.58
	Trichostatin A	0.20	0.80	1.00
	Tozasertib (VX-680)	0.00	1.00	1.00
	Enzastaurin	0.00	0.00	0.00
	PD0332991	0.05	0.62	0.67
	Valproate	0.00	0.00	0.00
	Resveratrol	0.00	0.00	0.00
	Zibotentan (ZD4054)	0.00	0.00	0.00
	SP600125	0.00	0.00	0.00
	Ponatinib (AP24534)	0.00	0.00	0.00
	BIX 02188	0.00	0.00	0.00
	RO4929097	0.00	0.00	0.00
	Curcumin	0.00	0.00	0.00
	Sodium butyrate	0.00	0.00	0.00
	GANT61	0.38	0.31	0.69
	Aurothiomalate	0.00	0.00	0.00
	(OSI-906)	0.47	0.53	1.00
	Pelitinib (EKB-569)	0.00	0.00	0.00

Notes: Avg. Err., average of the 44 leave-one-out errors; Error = |Pred. Sens. − Exp. Sens.|;
Exp. Sens., experimental sensitivity; Pred. Sens., predicted sensitivity [2].

Table 9.9 Cross-Validation Results

Cell Culture	Average MAE
Bailey	0.080
Charley	0.087
Cora	0.083
Sy	0.072

Notes: The 10-fold cross-validation error estimates were calculated for five runs and the average MAE across all the drugs is shown [2].

similarity measure Λ as:

$$\Lambda(D_1, D_2) = \frac{\sum_{i=1}^{n} \min(E_1(i), E_2(i)) * V_1(i) * V_2(i)}{\sum_{i=1}^{n} \max(E_1(i), E_2(i))} \tag{9.9}$$

Note that $\Lambda(D_1, D_1) = 1$ and similarity between drugs with no overlapping targets is 0. If two drugs have 50% targets overlapping with same EC_{50}s, then the similarity measure is 0.5. The summary of the similarities between the drugs is shown in Table 9.10. Note that except two drugs Rapamycin and Temsirolimus that have a similarity measure of 0.989, all other drugs have significantly lower similarities with each other. The maximum similarity between two different drugs (other than Rapamycin and Temsirolimus) is 0.169. This shows that any two drugs in the drug screen are not significantly overlapping and the prediction algorithm is still able to predict the response.

The low error rate illustrates the accuracy and effectiveness of this new method of modeling and sensitivity prediction. Furthermore, these error rates are significantly lower than those of other existing sensitivity prediction methodologies. Consistent with the analysis in [31], the sensitivity prediction rates improve dramatically when incorporating more information about drug-protein interaction. To more effectively compare the results generated via the TIM framework with the results in [8], we also present the correlation coefficients between the predicted and experimental drug sensitivity values in Table 9.11. The correlation coefficients for predicted and experimentally generated sensitivities for 24 drugs and more than 500 cell lines range from 0.1 to 0.8 when genomic characterizations are used to predict the drug sensitivities in the CCLE study [8]. In comparison, the PTIM approach based on sensitivity data on training set of drugs and drug-protein interaction information produced correlation coefficients >0.92 (Table 9.11) for both LOO and 10-fold CV approaches for error estimation.

It should be noted that the sensitivity prediction is performed in a continuous manner, not discretely, and thus effective dosage levels can be inferred from the predictions made from the TIM. This shows that the TIM framework is capable of predicting the sensitivity to anticancer targeted drugs outside the training set, and

Table 9.10 Drug Similarity Results [2]

Max Similarity (Excluding Itself)	Drug
0.000	Veliparib (ABT-888)
0.001	Selumetinib (AZD6244)
0.006	Bortezomib
0.043	Bosutinib (SKI-606)
0.072	Dasatinib
0.100	Erlotinib
0.120	Panobinostat (LBH-589)
0.103	Pazopanib (GW-786034)
0.077	PI-103
0.989	Rapamycin (sirolimus)
0.154	Sorafenib
0.989	Temsirolimus (CCI-779)
0.115	Vorinostat (SAHA)
0.000	Obatoclax (GX15-070)
0.043	Crizotinib (PF-2341066)
0.000	MK-2206
0.028	Vismodegib (GDC-0449)
0.006	Alisertib (MLN8237)
0.050	SNS-032 (BMS-387032)
0.006	Carfilzomib
0.077	Imatinib
0.000	BIX 01294
0.088	BMS-754807
0.000	SJ-172550
0.154	Barasertib (AZD1152-HQPA)
0.006	Ruxolitinib (INCB018424)
0.010	Cediranib (AZD2171)
0.053	Lapatinib
0.104	Sunitinib
0.120	Trichostatin A
0.169	Tozasertib (VX-680)
0.003	Enzastaurin
0.000	PD0332991
0.071	Resveratrol
0.000	Zibotentan (ZD4054)
0.088	SP600125
0.010	Ponatinib (AP24534)
0.010	BIX 02188
0.000	RO4929097
0.041	Curcumin
0.028	GANT61
0.000	Aurothiomalate
0.000	OSI-906
0.169	Pelitinib (EKB-569)

Table 9.11 Correlation Coefficients of Predicted Sensitivities vs Experimental Sensitivities

Cell Line	ρ LOO[a]	ρ 10-Fold CV[b]
Bailey	0.97	0.92
Charley	0.97	0.95
Cora	0.98	0.94
Sy	0.94	0.92

[a]Average Pearson correlation coefficient for test samples selected based on Leave-One-Out approach.
[b]Average Pearson correlation coefficient for test samples selected based on 10-fold cross-validation approach [2].

as such is viable as a basis for a solution to the complicated problem of sensitivity prediction.

9.9 DISCUSSION

The framework described in this chapter presents an alternative approach for precision medicine. Each individual tumor is treated as a system and the response of the system to various known perturbations, along with observation of the internal state of the system provided by genomic measurements is utilized to generate a Boolean-based model of the system. The framework shows promise in terms of high accuracy prediction of sensitivity to new drugs and can be utilized to design personalized combination therapies. The utilization of the PTIM framework to design synergistic combination therapy and its validation are described in details in Chapter 11. The framework can also be utilized to generate visual representation of the underlying tumor proliferation network which is appealing to a practitioner for understanding the inferred model. Note that the inferred model represents a steady-state model of the system and the generated TIM circuit fails to provide the directionality of the various inferred blocks. Chapter 10 considers the extension of this framework by incorporating additional data to infer the directionality of the target blocks and subsequently generating dynamic Boolean Network and Stochastic Markov Chain representations of the tumor proliferation system.

Software availability: The software to infer PTIMs and TIM circuits from drug perturbation data is available at http://www.myweb.ttu.edu/rpal/Softwares/PTIM1.zip. The software is written in Matlab and a Graphical User Interface is provided. A stand-alone compiled version is available for use by researchers without access to Matlab license. A descriptive user guide for the software can be downloaded from http://www.myweb.ttu.edu/rpal/Softwares/PTIMUserGuide.pdf.

REFERENCES

[1] R. Pal, N. Berlow, A kinase inhibition map approach for tumor sensitivity prediction and combination therapy design for targeted drugs, in: Pacific Symposium on Biocomputing, 2012, PMID: 22174290, pp. 351–362, http://psb.stanford.edu/psb-online/proceedings/psb12/pal.pdf.

[2] N. Berlow, L.E. Davis, E.L. Cantor, B. Seguin, C. Keller, R. Pal, A new approach for prediction of tumor sensitivity to targeted drugs based on functional data, BMC Bioinform. 14 (2013) 239.

[3] N. Berlow, S. Haider, Q. Wan, M. Geltzeiler, L.E. Davis, C. Keller, R. Pal, An integrated approach to anti-cancer drugs sensitivity prediction, IEEE/ACM Trans. Comput. Biol. Bioinform. (2014), doi:10.1155/2014/873436.

[4] C.S. Grasso, Y. Tang, N. Truffaux, N.E. Berlow, L. Liu, M. Debily, M.J. Quist, L.E. Davis, E.C. Huang, P.J. Woo, A. Ponnuswami, S. Chen, T. Johung, W. Sun, M. Kogiso, Y. Du, Q. Lin, Y. Huang, M. Hutt-Cabezas, K.E. Warren, L.L. Dret, P.S. Meltzer, H. Mao, M. Quezado, D.G. van Vuurden, J. Abraham, M. Fouladi, M.N. Svalina, N. Wang, C. Hawkins, J. Nazarian, M.M. Alonso, E. Raabe, E. Hulleman, P.T. Spellman, X. Li, C. Keller, R. Pal, J. Grill, M. Monje, Functionally-defined therapeutic targets in diffuse intrinsic pontine glioma, Nat. Med. 21 (2015) 555–559.

[5] M.L. Sos, K. Michel, T. Zander, J. Weiss, P. Frommolt, M. Peifer, D. Li, R. Ullrich, M. Koker, F. Fischer, T. Shimamura, D. Rauh, C. Mermel, S. Fischer, I. Stückrath, S. Heynck, R. Beroukhim, W. Lin, W. Winckler, K. Shah, T. LaFramboise, W.F. Moriarty, M. Hanna, L. Tolosi, J. Rahnenführer, R. Verhaak, D. Chiang, G. Getz, M. Hellmich, J. Wolf, L. Girard, M. Peyton, B.A. Weir, T.-H.H. Chen, H. Greulich, J. Barretina, G.I. Shapiro, L.A. Garraway, A.F. Gazdar, J.D. Minna, M. Meyerson, K.-K.K. Wong, R.K. Thomas, Predicting drug susceptibility of non-small cell lung cancers based on genetic lesions, J. Clin. Investig. 119 (6) (2009) 1727–1740.

[6] J.K. Lee, D.M. Havaleshko, H. Cho, J.N. Weinstein, E.P. Kaldjian, J. Karpovich, A. Grimshaw, D. Theodorescu, A strategy for predicting the chemosensitivity of human cancers and its application to drug discovery, Proc. Natl. Acad. Sci. USA 104 (32) (2007) 13086–13091.

[7] A. Mitsos, I.N. Melas, P. Siminelakis, A.D. Chairakaki, J. Saez-Rodriguez, L.G. Alexopoulos, Identifying drug effects via pathway alterations using an integer linear programming optimization formulation on phosphoproteomic data, PLoS Comput. Biol. 5 (12) (2009) e1000591.

[8] J. Barretina, et al., The Cancer Cell Line Encyclopedia enables predictive modelling of anticancer drug sensitivity, Nature 483 (7391) (2012) 603–607.

[9] J.C. Costello, et al., A community effort to assess and improve drug sensitivity prediction algorithms, Nat. Biotechnol. (2014), doi:10.1038/nbt.2877.

[10] Q. Wan, R. Pal, An ensemble based top performing approach for NCI-DREAM drug sensitivity prediction challenge, PLoS ONE 9 (6) (2014) e101183.

[11] M.W. Karaman, S. Herrgard, D.K. Treiber, P. Gallant, C.E. Atteridge, B.T. Campbell, K.W. Chan, P. Ciceri, M.I. Davis, P.T. Edeen, R. Faraoni, M. Floyd, J.P. Hunt, D.J. Lockhart, Z.V. Milanov, M.J. Morrison, G. Pallares, H.K. Patel, S. Pritchard, L.M. Wodicka, P.P. Zarrinkar, A quantitative analysis of kinase inhibitor selectivity, Nat. Biotechnol. 26 (1) (2008) 127–132.

[12] B.B. Hasinoff, D. Patel, The lack of target specificity of small molecule anticancer kinase inhibitors is correlated with their ability to damage myocytes in vitro, Toxicol. Appl. Pharmacol. 249 (2010) 132–139.

[13] R. Kurzrock, M. Markman, Targeted Cancer Therapy (Current Clinical Oncology), Humana Press, Totowa, NJ, USA, 2008.

[14] S. Haider, N. Berlow, R. Pal, L. Davis, C. Keller, Combination therapy design for targeted therapeutics from a drug–protein interaction perspective, in: 2012 IEEE International Workshop on Genomic Signal Processing and Statistics (GENSIPS), 2012, pp. 58–61, http://dx.doi.org/10.1109/GENSIPS.2012.6507726.

[15] M. Kuhn, M. Campillos, I. Letunic, L.J. Jensen, P. Bork, A side effect resource to capture phenotypic effects of drugs, Mol. Syst. Biol. 6 (2010) 343.

[16] A. Gaulton, L.J. Bellis, A.P. Bento, J. Chambers, M. Davies, A. Hersey, Y. Light, S. McGlinchey, D. Michalovich, B. Al-Lazikani, J.P. Overington, ChEMBL: a large-scale bioactivity database for drug discovery, Nucleic Acids Res. 40 (D1) (2011) gkr777–D1107.

[17] N. Berlow, L. Davis, C. Keller, R. Pal, Inference of dynamic biological networks based on responses to drug perturbations, EURASIP J. Bioinform. Syst. Biol. 14 (2014), http://dx.doi.org/10.1186/s13637-014-0014-1.

[18] M. Kanehisa, S. Goto, KEGG: Kyoto Encycolpedia of genes and genomes, Nucleic Acids Res. 28 (2000) 27–30.

[19] P.P. Zarrinkar, R.N. Gunawardane, M.D. Cramer, M.F. Gardner, D. Brigham, B. Belli, M.W. Karaman, K.W. Pratz, G. Pallares, Q. Chao, K.G. Sprankle, H.K. Patel, M. Levis, R.C. Armstrong, J. James, S.S. Bhagwat, AC220 is a uniquely potent and selective inhibitor of FLT3 for the treatment of acute myeloid leukemia (AML), Blood 114 (14) (2009) 2984–2992.

[20] A. Andreassen, T. Oyjord, E. Hovig, R. Holm, V. Florenes, et al., p53 abnormalities in different subtypes of human sarcomas, Cancer Res. 53 (3) (1993) 468–471.

[21] M. Kansara, D.M. Thomas, Molecular pathogenesis of osteosarcoma, DNA Cell Biol. 26 (1) (2007) 1–18.

[22] M. Overholtzer, P.H. Rao, R. Favis, X.-Y. Lu, M.B. Elowitz, F. Barany, M. Ladanyi, R. Gorlick, A.J. Levine, The presence of p53 mutations in human osteosarcomas correlates with high levels of genomic instability, Proc. Natl. Acad. Sci. USA 100 (20) (2003) 11547–11552.

[23] M.S. Benassi, L. Molendini, G. Gamberi, P. Ragazzini, M.R. Sollazzo, M. Merli, J. Asp, G. Magagnoli, A. Balladelli, F. Bertoni, P. Picci, Alteration of pRb/p16/cdk4 regulation in human osteosarcoma, Int. J. Cancer 84 (5) (1999) 489–493.

[24] A. Besson, V.W. Yong, Involvement of p21Waf1 Cip1 in protein kinase C alpha induced cell cycle progression, Mol. Cell. Biol. 20 (13) (2000) 4580–4590.

[25] Y. Shapovalov, D. Benavidez, D. Zuch, R.A. Eliseev, Proteasome inhibition with bortezomib suppresses growth and induces apoptosis in osteosarcoma, Int. J. Cancer 127 (1) (2010) 67–76.

[26] M. Scheffner, Ubiquitin, E6-AP, and their role in p53 inactivation, Pharmacol. Ther. 78 (3) (1998) 129–139.

[27] W. Zhang, J.C. Lee, S. Kumar, M. Gowen, ERK pathway mediates the activation of Cdk2 in IGF1 induced proliferation of human osteosarcoma MG63 cells, J. Bone Miner. Res. 14 (4) (1999) 528–535.

[28] E. Choy, F. Hornicek, L. MacConaill, D. Harmon, Z. Tariq, L. Garraway, Z. Duan, High-throughput genotyping in osteosarcoma identifies multiple mutations in phosphoinositide-3-kinase and other oncogenes, Cancer 118 (11) (2012) 2905–2914.

[29] D. Morgensztern, H.L. McLeod, PI3K/Akt/mTOR pathway as a target for cancer therapy, Anticancer Drugs 16 (8) (2005) 797–803.

[30] P. Robin, I. Boulven, C. Desmyter, S. Harbon, D. Leiber, ET1 stimulates ERK signaling pathway through sequential activation of PKC and Src in rat myometrial cells, Am. J. Physiol. Cell Physiol. 283 (1) (2002) C251–C260.

[31] R. Pal, N. Berlow, S. Haider, Anticancer drug sensitivity analysis: an integrated approach applied to Erlotinib sensitivity prediction in the CCLE database, in: 2012 IEEE International Workshop on Genomic Signal Processing and Statistics (GENSIPS), 2012, pp. 9–12, http://dx.doi.org/10.1109/GENSIPS.2012.6507714.

Inference of dynamic biological networks based on perturbation data

10

CHAPTER OUTLINE

10.1 INTRODUCTION

The Target Inhibition Map (TIM) model described in Chapter 9 is able to predict the steady-state behavior of target inhibitor combinations but does not provide us with the dynamics of the model or the directionality (upstream or downstream) of the inferred target blocks. In this chapter, we analyze the generation of possible dynamic models satisfying the steady-state model representation. We first show that the TIM [1,2] approach can generate blocks of targets that are connected in series to form a pathway but the directionality of the blocks are unknown. Subsequently, we establish that a directional pathway can be converted to a deterministic Boolean network (BN) [3] model. The discrete representation of the TIM as a directional pathway allows us to select a minimal number of sequential inhibition experiments for inferring the actual dynamic model. To incorporate the continuous sensitivity behavior following drug inhibition, we consider the inverse problem of generation of Markov chains that satisfies for every target inhibition condition: the steady-state probability of nontumorous state is equal to the normalized sensitivity. The set of dynamic models producing the static TIM can be utilized for robustness analysis of the combination therapy design and design of time-dependent combination therapies. The approach presented in this chapter extends the static design to incorporate possible dynamics.

10.2 DISCRETE DETERMINISTIC DYNAMIC MODEL INFERENCE

To arrive at potential discrete deterministic dynamical models, we consider the likely directional pathways that can generate the inferred TIM and map the directional pathways to deterministic BN models. The TIM can be used to locate the feasible mutation patterns and constrain the search space of the dynamic models generating the TIM. As an example, consider the TIM for three targets K_1, K_2, K_3 shown in Fig. 10.1. The map in Fig. 10.1 shows that inhibition of K_3 alone can inhibit the tumor, or inhibition of both K_1 and K_2 can inhibit the tumor. The current setting of the TIM approach will consider only those targets that are functionally relevant in cell death in a new cancer sample. These targets are often upregulated in cancer, either due to their own mutations or activations by some other enzymes (from now on, we will call such activations by enzyme(s) not considered in the final TIM *latent activations*).

Mutation or external activation of K_2 or K_1 alone cannot result in the TIM of Fig. 10.1, otherwise the inhibition of K_2 or K_1 should have been able to block the tumor. Thus feasible mutations or latent activation patterns are reduced to the following 5 sets of combinations: $\{K_1, K_2\}, \{K_1, K_3\}, \{K_2, K_3\}, \{K_3\}, \{K_1, K_2, K_3\}$ out of 8 possible combinations. For each mutation or latent activation pattern, we can arrive at possible directional pathways producing the required steady-state TIM output. For instance, Fig. 10.2 shows two directional pathway possibilities configurations 1 and 2 for mutation or activation patterns $\{K_1, K_2\}$ and $\{K_3\}$, respectively. The pathways in Fig. 10.2 show possible tumor survival circuits. In this model, if a left-to-right tumor survival pathway exists, the cancer survives. If the path is stopped, the tumor cells stop growing or involute.

$k_1\,k_2$ / k_3	0 0	0 1	1 1	1 0
0	0	0	1	0
1	1	1	1	1

FIG. 10.1

TIM for mutations in K_1 and K_2 [4].

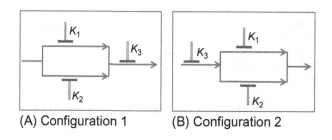

(A) Configuration 1   (B) Configuration 2

FIG. 10.2

Possible directional pathways based on the TIM in Fig. 10.1 [4].

10.2.1 OPTIMAL SET OF EXPERIMENTS TO INFER THE DIRECTIONAL PATHWAY STRUCTURE

In this section, we analyze the minimum number of expression measurement experiments required to decipher the pathway directionality once the steady-state structure (TIM) has been inferred. Knowledge of target expressions can be used to narrow down the possible directional pathways. For instance, expressed K_1 following inhibition of K_3 for our earlier example will denote the feasibility of the directional pathway of Fig. 10.2A and removing the possibility of the directional pathway shown in Fig. 10.2B. Note that latent activations and functionally irrelevant mutations may restrict the usefulness of mutation status in restricting the pathway search space. In the following paragraphs, we will consider a general pathway obtained from a TIM having the structure shown in Fig. 10.3, but with unknown directionalities of the blocks and target positions. We will consider that the pathway has L blocks in series $(B_1, B_2, \ldots, B_L)$ and each block B_i has a_i parallel path segments with each segment j containing b_j^i targets $(K_{1,1}^i, K_{1,2}^i, \ldots, K_{1,b_j^i}^i)$. The total number of targets in the general map is $N_K = \sum_{i=1}^{L} \sum_{j=1}^{a_i} b_j^i$.

Assuming that the N_K targets are distinct, the maximum number of distinct discrete dynamic models satisfying the structure is $L! \prod_{i=1}^{L} \prod_{j=1}^{a_i} (b_j^i)!$ If Fig. 10.3 represents a possible directional orientation, the only targets that will have initial activations for the target inhibition combination $K_{1,1}^1, K_{2,1}^1, \ldots, K_{a_1,1}^1$ to be effective are $K_{1,1}^1, K_{2,1}^1, \ldots, K_{a_1,1}^1$. For our analysis, we are assuming that we can inhibit specific targets of our choice and we can measure the steady-state target expression following application of the target inhibitions. We can locate the directionality of the blocks B_1 to B_L with respect to each other (downstream or upstream) with the worst-case scenario of $L - 1$ steady-state measurements. The expected number of experiments required to detect the directionality of L serial blocks is $\frac{2L-1}{3}$ for $L \geq 2$. To infer the directionality of targets in each parallel line of the block, one target from each line up to a maximum of $a_i - 1$ lines will be inhibited for each block B_i. If we consider a single block B_i, each experiment can detect the location of $a_i - 1$ targets, thus the total number of experiments required to decipher the

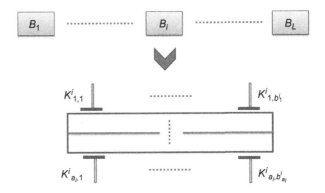

FIG. 10.3

A general abstract pathway resulting from a TIM [4].

possible directionalities (upstream or downstream) of the targets in the block B_i is $\leq \max\left(\max_{j\in S_i} b^i_j - 2, \left\lceil \frac{\sum_{j\in S_i} b^i_j - a_i}{a_i - 1}\right\rceil - 1\right)$ where $S_i = \{1, \ldots, a_i\}$. Thus, for the overall map, the worst-case number of experiments N^w_E required to decipher the directionalities of all the targets is upper-bounded by Berlow et al. [5]

$$N^w_E \leq \max_{i\in S}\left\{\max\left(\max_{j\in S_i} b^i_j - 2, \left\lceil \frac{\sum_{j\in S_i} b^i_j - a_i}{a_i - 1}\right\rceil - 1\right)\right\} + L - 1 \tag{10.1}$$

where $S = \{1, \ldots, L\}$. The expected number of experiments N^a_E required to decipher the directionalities of all the targets is upper-bounded by

$$N^a_E \leq \max_{i\in S}\left\{\max\left(\max_{j\in S_i} \frac{2b^i_j - 4}{3}, \left\lceil \frac{\sum_{j\in S_i} 2b^i_j - a_i}{3(a_i - 1)}\right\rceil - 1\right)\right\} + \frac{2L - 1}{3} \tag{10.2}$$

10.2.1.1 Simulation results on optimal experimental steps

For our simulation results, we consider a pathway derived from targeted drug perturbation experiments carried out at Keller Laboratory at Oregon Health and Science University on canine osteosarcoma cell cultures. Sixty targeted cancer drugs were tested on cell cultures and a TIM was generated based on the viability data using the approach provided in [1,2]. For our simulation results, we will consider one of the plausible directional pathways derived from the TIM to be the actual pathway, and estimate the number of target expression measurements required to arrive at it if the directional information is not known. The directional pathway assumed to be the actual pathway is shown in Fig. 10.4 consisting of 13 targets. If we compare Fig. 10.4 with the general pathway in Fig. 10.3, the number of serial blocks $L = 6$. Similarly, $a_1 = 4, a_2 = 1, a_3 = 1, a_4 = 2, a_5 = 1, a_6 = 2$ and

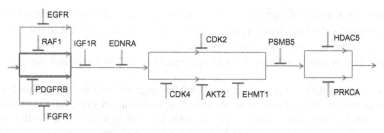

FIG. 10.4

Pathway derived from perturbation experiments [4].

$b_j^i = 1$ for all i and j except $b_2^4 = 3$. Because there is only one serial block with $b_j^i > 2$ we can reduce Eq. (10.1) to $N_E^w \leq \max_{i \in S, j \in S_i} b_j^i - 2 + L - 1 = 6$ and Eq. (10.2) to $N_E^a \leq \max_{i \in S, j \in S_i} \frac{2b_j^i - 4}{3} + \frac{2L-1}{3} = 4.33$. To compare these numbers with simulation results, we conducted 10,000 simulation runs to detect the pathway shown in Fig. 10.4, starting from random inhibition of serial blocks. The distribution of the number of steady-state experiments required to detect the directional pathway is shown in Fig. 10.5. We note that the maximum number of experiments required was 6, as given by N_E^w in Eq. (10.1) and the expectation of the distribution is 4.33 which is the same as the bound on N_E^a given by Eq. (10.2).

10.2.2 DETERMINISTIC DYNAMICAL MODEL FROM DIRECTIONAL PATHWAY

To generate a BN model of a directional pathway, we will first consider the starting mutations or latent activations. The number of states in the BN will be 2^{n+1} for

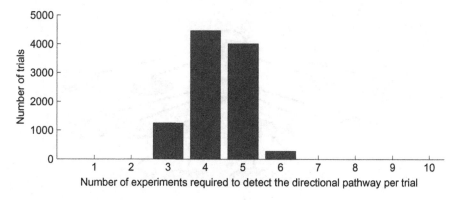

FIG. 10.5

Distribution of number of target expression measurements [4].

n targets. Each state will have $n+1$ bits with first n bits referring to the discrete state of the n targets, and the least significant bit (LSB) will correspond to the binarized phenotype, that is, tumor (1) or normal (0).

The rules of state transition for this special class of BNs are as follows [5]:

Rule a: A target state at time $t+1$ becomes 1 if any immediate upstream neighbor has state 1 at time t for *OR* relationships or all immediate upstream neighbors have state 1 at time t for *AND* relationships. Note that the examples have *OR* type of relations as they are the most commonly found relations in biological pathways (based on illustrated pathways in KEGG).

Rule b: For the BN without any drug, the targets that are mutated or have latent activations will transition to state 1 within one time step.

Rule c: For a target with no inherent mutation or latent activation, the state will become 0 at time $t+1$ if the immediate upstream activators of the target have state 0 at time t.

The BN construction from directional pathways mentioned above is described for targets acting as oncogenes (activation causing cancers), but it can also be extended to tumor suppressors (inhibition causing cancers) by considering the inverse state of the tumor suppressor in the above framework.

We illustrate the BN construction algorithm using the example of the pathway shown in Fig. 10.2A. The downstream target K_3 can be activated by either of the upstream activated targets K_1 or K_2. The corresponding BN transition diagram for this pathway is shown in Fig. 10.6. For instance, if we consider the state 1001 at time t, it denotes K_2, K_3 being inactive and K_1 being active, and the phenotype being tumorous. Based on the directional pathway in Fig. 10.2A, tumor proliferation is caused by activated K_3 and thus the phenotype will change to nontumorous (i.e., 0) at $t+1$. The activated K_1 will activate K_3 at time $t+1$ and K_2 will also be activated in the absence of continued inhibition, as we assumed that mutation or latent activations activate both K_1 and K_2. Thus the next state at time $t+1$ will be 1110. Note that we

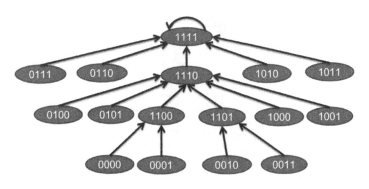

FIG. 10.6

State transitions of the BN for the directional pathway in Fig. 10.2A [4].

are considering that the effect of one application of the drug remains for one time step and thus the targets K_1 and K_2 revert back to 1 if the drug is not continued in the next time step. If the drug effect continues for multitime steps, then 1001 will transition to 1010. Note that some transitions may appear like the tumor state is oscillating in the transient phase, such as the path $0010 \rightarrow 1101 \rightarrow 1110 \rightarrow 1111$. The reason is that the network can only be in the starting state 0010 where K_1 and K_2 are inactivated through application of some external interventions and not through normal transitions as the network has K_1 and K_2 mutated. Scenarios following application of drugs can produce alternating tumor proliferation and inactivation states in the transient phase.

10.2.3 ALTERED BN FOLLOWING TARGET INHIBITION

The BN in Fig. 10.6 can also be represented by a 16×16 transition matrix P representing the state transitions. To generate the dynamic model after inhibition of s-specific targets $I = \{K_1, K_2, \ldots, K_s\}$ (by application of targeted drugs), the transition $i \rightarrow j$ in the untreated system will be converted to $i \rightarrow z$ in the treated system where z is j with targets I set to 0. Each target inhibition combination can be considered as multiplying the initial transition matrix P by an intervention matrix T_c. Each row of T_c contains only one nonzero element of 1 based on how the inhibition alters the state. If we consider n targets, n T_cs in combination can produce a total of 2^n possible transformation matrices $T_1, T_2, \ldots, T_{2^n}$. The TIM denotes the state of the LSB of the attractor for the 2^n transition matrices $PT_1, PT_2, \ldots, PT_{2^n}$ starting from initial state $11 \ldots 1$ (i.e., all targets considered in the TIM and tumor are activated). For instance, if we consider that our drug inhibits the targets K_1 and K_2 (i.e., set $S_1 = \{K_1, K_2\}$), the discrete dynamic model following application of the drug is shown in Fig. 10.7. The intervention matrix corresponding to the inhibition of K_3 is shown in Table 10.1. The transition $i \rightarrow j$ is 1 only when inhibition of the first and second bits of i results in j.

We should note that the equilibrium state of the network 0000 has 0 for the tumor state. This is because the tumor is activated by K_3 and inhibition of K_1 and K_2 blocks activation of K_3 and thus should eradicate the tumor. On the other hand, because both

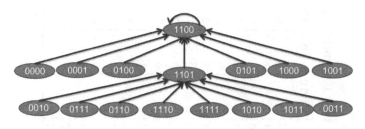

FIG. 10.7

BN state transition following inhibition of targets K_1 and K_2 [4].

Table 10.1 Inhibition Matrix T_c for Inhibition of K_1 and K_2 [4]

	0000	0001	0010	0011	0100	0101	0110	0111	1000	1001	1010	1011	1100	1101	1110	1111
0000	1	0	0	0	0	0	0	0	0	0	0	0	0	0	0	0
0001	0	1	0	0	0	0	0	0	0	0	0	0	0	0	0	0
0010	0	0	1	0	0	0	0	0	0	0	0	0	0	0	0	0
0011	0	0	0	1	0	0	0	0	0	0	0	0	0	0	0	0
0100	1	0	0	0	0	0	0	0	0	0	0	0	0	0	0	0
0101	0	1	0	0	0	0	0	0	0	0	0	0	0	0	0	0
0110	0	0	1	0	0	0	0	0	0	0	0	0	0	0	0	0
0111	0	0	0	1	0	0	0	0	0	0	0	0	0	0	0	0
1000	1	0	0	0	0	0	0	0	0	0	0	0	0	0	0	0
1001	0	1	0	0	0	0	0	0	0	0	0	0	0	0	0	0
1010	0	0	1	0	0	0	0	0	0	0	0	0	0	0	0	0
1011	0	0	0	1	0	0	0	0	0	0	0	0	0	0	0	0
1100	1	0	0	0	0	0	0	0	0	0	0	0	0	0	0	0
1101	0	1	0	0	0	0	0	0	0	0	0	0	0	0	0	0
1110	0	0	1	0	0	0	0	0	0	0	0	0	0	0	0	0
1111	0	0	0	1	0	0	0	0	0	0	0	0	0	0	0	0

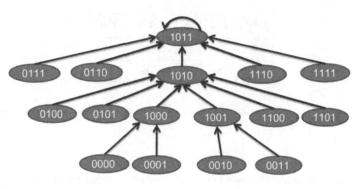

FIG. 10.8

BN state transitions following inhibition of target K_2 [4].

K_1 and K_2 can cause tumor through activation of intermediate K_3, inhibition of only one of K_1 and K_2 will not block the tumor. The BN following inhibition of K_2 is shown in Fig. 10.8 where the attractor 1011 denotes a tumorous phenotype.

10.3 DISCRETE STOCHASTIC DYNAMIC MODEL INFERENCE

The analysis so far has considered deterministic discrete binary states for the targets and tumor phenotype. A stochastic modeling approach will be preferred when we want to take into consideration that tumor phenotype (measured in terms of tumor size reduction, IC_{50}, or cell cycle arrest) is a continuous variable. We have extended our TIM approach to Probabilistic Target Inhibition Map (PTIM), where the PTIM provides continuous sensitivity prediction values between 0 and 1 for all possible kinase inhibition combinations [1,2]. From a stochastic dynamical model perspective, we can consider the sensitivity prediction value provided by the PTIM as the steady-state probability of the tumor phenotype being 0. (A similar approach with deterministic differential equation models for modeling the tumor sensitivity was considered in [6] and experimental data were assumed to reflect the steady-state values.) For instance, if we consider that a Markov chain of 16 states explains our dynamical model for the pathway shown in Fig. 10.9, the entry $PTIM(i, j)$ will reflect the steady-state probability for the LSB = 0 for the model with target inhibitions i, j. For instance, p_5 reflects the sensitivity with target inhibition K_1 and K_3.

In this chapter, the discrete stochastic dynamic behavior will be modeled by a Markov chain where the states of the Markov chain contain information on the protein expressions of the targets and the tumor status. Note that a detailed stochastic master equation model is a continuous time Markov chain and can be approximated by a discrete time Markov chain based on a suitable time step [7]. Also, BNs can be

k_3 \ $k_1 k_2$	0 0	0 1	1 1	1 0
0	p_0	p_2	p_6	p_4
1	p_1	p_3	p_7	p_5

FIG. 10.9

A probabilistic TIM [4].

incorporated as Markov chains where each row of the transition probability matrix contains a single 1 with remaining all entries being 0.

For the subsequent analysis, we will consider that we have n binarized targets in our model and the states of the Markov chain will be $0 \dots 0$ to $1 \dots 1$ where the LSB will denote the state of the tumor (1 denoting tumor proliferation and 0 denoting tumor reduction) and the remaining n bits denote the state of the n targets. The set of states of the Markov chain, denoted by the set $\mathcal{I}$, is of size $N = 2^{n+1}$. Let P denote the $N \times N$ transition probability matrix. Let $f_c(j)$ denote the value of the state j following Boolean intervention equal to the inverse binary value of decimal c, that is, under intervention $f_3(x) = x$ AND (0011) denoting intervention of 1st and 2nd targets and $f_5(x) = x$ AND (0101) denoting intervention of 1st and 3rd targets. Let $\mathcal{I}_c$ denote the possible set of states following application of intervention c, that is, $\mathcal{I}_c$ contains only the states i s.t. $f_c(i) = i$. For the above example, $\mathcal{I}_3 = \{0000, 0001, 0010, 0011\}$. Let $S_{c,i}$ denote the set of states j for which $i = f_c(j)$. Here, $S_{3,0} = \{0000, 1000, 0100, 1100\}$, the set of states which, under inhibition $f_3(\cdot)$, transition into state i.

The targeted drugs usually inhibit a set of target proteins and modeling such a behavior can be approached in one of the two following ways:

(A_i) If the targeted drugs inhibit the set of proteins I, the dynamics of the system under drug delivery can be considered as a new Markov chain with transition probability matrix P_2 where the jth row of P_2 is same as the ith row of P, where j is i with targets I set to 0. For instance, in a four-target system where I is targets 1 and 2, rows 0000, 1000, 0100, and 1100 of P_2 will be the same as row 0000 of P. This approach refers to resetting the system to the state obtained by applying the drug and letting it evolve from there. Note that the above-described system will still show nonzero transition probabilities to states where the target set I may have nonzero values.

(A_{ii}) If we have transition probability $P(i,j)$ of moving from state i to state j in the uncontrolled system, for the new system with transition probability matrix P_3, we will add the transition probability $P(i,j)$ to $P(i,z)$ where z is j with targets I set to 0. This encompasses the behavior that if the system transitions from i to j, j has been turned to z by the intervention.

The following theorem proves that the aggregated steady-state probability distribution for both the approaches is equal.

Theorem 10.1. *Let π_2 and π_3 denote the stationary probability distributions of P_2 and P_3. If π_2^* denotes the aggregation of states after intervention C, that is, $\pi_2^*(i) = \sum_{i_1 \in S_{i,c}} \pi_2(i_1)$ for $i \in \mathcal{I}_C$ and $\pi_2^*(i) = 0$ for $i \notin \mathcal{I}_C$, then π_2^* also satisfies the stationary probability distribution equations for P_3, that is, $\pi_2^* = \pi_2^* P_3$. If P is ergodic, then $\pi_2^* = \pi_3$ [4].*

Proof. Let $f_c(\cdot)$ be a Boolean intervention function. We have $\forall i, j \in \mathcal{I}$

$$P_2(i,j) = P(f_c(i),j) \tag{10.3}$$

and $P_3(i,j) = \sum_{k \in S_{c,j}} P(i,k)$. The stationary distribution for P_2 will satisfy

$$\pi_2(i) = \sum_{j \in \mathcal{I}} \pi_2(j) P_2(j,i)$$

$$= \sum_{z \in \mathcal{I}_C} P(z,i) \sum_{k \in S_{c,z}} \pi_2(k) \tag{10.4}$$

Similarly, the stationary distribution for P_3 will satisfy $\pi_3(i) = 0$ for $i \notin \mathcal{I}_C$ and for $i \in \mathcal{I}_C$:

$$\pi_3(i) = \sum_{j \in \mathcal{I}} \pi_3(j) P_3(j,i)$$

$$= \sum_{z \in \mathcal{I}_C} \pi_3(z) P_3(z,i)$$

$$= \sum_{z \in \mathcal{I}_C} \pi_3(z) \sum_{k \in S_i} P(z,k) \tag{10.5}$$

If π_2^* denotes the aggregation of states, that is, $\pi_2^*(i) = \sum_{i_1 \in S_{c,i}} \pi_2(i_1)$ for $i \in \mathcal{I}_C$, then we have for $i \in \mathcal{I}_C$:

$$\pi_2^*(i) = \sum_{k \in S_{c,i}} \pi_2(k) = \sum_{k \in S_{c,i}} \sum_{z \in \mathcal{I}_C} P(z,k) \sum_{j \in S_z} \pi_2(j)$$

$$= \sum_{z \in \mathcal{I}_C} \sum_{k \in s_{c,i}} P(z,k) \sum_{j \in S_{c,z}} \pi_2(j)$$

$$= \sum_{z \in \mathcal{I}_C} \pi_2^*(z) \sum_{k \in S_{c,i}} P(z,k) \tag{10.6}$$

Comparing Eqs. (10.5), (10.6), we note that π_2^* also satisfies the stationary probability distribution equation for P_3, that is, $\pi_2^* P_3 = \pi_2^*$. $\qquad\square$

We will model the target intervention based on perspective A_i. We next analyze whether every PTIM can be represented by a Markov chain. Theorem 10.2 shows that there always exists a Markov chain construction that can satisfy the PTIM steady-state sensitivities.

Theorem 10.2. *For any given PTIM, $\exists$ at least one Markov chain satisfying the PTIM [4].*

Proof. Consider a PTIM with n targets $K_1, K_2, \ldots, K_n$ and thus 2^n PTIM entries $p_0, p_1, \ldots, p_{2^n-1}$, where p_i denotes the PTIM-predicted steady-state probability of tumor reduction when the active targets in the binary representation of i are inhibited. Denote the treatments corresponding to each p_i as g_i. A trivial Markov chain satisfying the PTIM can be generated as follows: $\forall i \in [0, \ldots, 2^n-1]$ we can generate an unique pair of $n+1$ dimensional states $D_1 = 2(2^n-i-1)$ and $D_2 = 2(2^n-i-1)+1$. D_1 and D_2 differ only in the last bit indicating tumor proliferation status. Here, LSB of $D_1 = 0$ and LSB of $D_2 = 1$. The first n bits of the binary representation of D_1 and D_2 are 0 where the representation of i has value 1. Consider a $2^{n+1} \times 2^{n+1}$ Markov chain with transition probability matrix P. $\forall i \in [0, \ldots, 2^n-1]$, let us assign probabilities as $P(D_1, D_1) = p_i$, $P(D_1, D_2) = 1 - p_i$, $P(D_2, D_1) = p_i$, and $P(D_2, D_2) = 1 - p_i$. This particular Markov chain will satisfy our given PTIM. Because there are 2^n closed classes of 2 states each, the stationary probability for inhibiting i can be calculated from considering the steady-state probabilities of the Markov chain

$$\begin{vmatrix} p_i & 1 - p_i \\ p_i & 1 - p_i \end{vmatrix}$$

which is p_i for the state with tumor $= 0$ and $1 - p_i$ for the state with tumor $= 1$. $\quad\square$

10.3.1 GENERATION OF MARKOV CHAINS BASED ON PATHWAY CONSTRAINTS

In this section, we will discuss two algorithms to generate Markov chains satisfying the PTIM steady-state sensitivities, while incorporating the directional pathway structures as emphasized in Section 10.2.

Each target inhibition combination can be considered as multiplying a matrix T_c to the initial Markov chain P. Each row of T_c contains only one nonzero element of 1 based on how the inhibition alters the state. If we consider n targets, n T_cs in combination can produce a total of 2^n possible transformation matrices $T_1, T_2, \ldots, T_{2^n}$. The PTIM denotes the stationary state probability of the LSB $= 0$ for the 2^n Markov chains $PT_1, PT_2, \ldots, PT_{2^n}$ starting from initial state $11\ldots1$ (i.e., all kinases considered in the PTIM and tumor are activated). The transition probability matrix has $2^{n+1} \times 2^{n+1}$ variables to be inferred and the number of equations available is 2^n. To narrow down the constraints, we will consider the possible BNs that can be generated for each set of possible mutations or outside activations of the thresholded PTIM. Each BN corresponding to a different mutation or initial activation pattern can provide information on possible alterations producing the required PTIM.

10.3.1.1 Algorithm 10.1

The first algorithm to generate Markov chains satisfying the PTIM sensitivities is presented in Algorithm 10.1.

ALGORITHM 10.1 ALGORITHM TO GENERATE A MARKOV CHAIN P FROM PTIM Ψ

Step 1: Convert the PTIM Ψ to a TIM ψ using a threshold of α.

Step 2: Based on the genetic mutation information or sequential protein expression measurements, generate the BN Ω corresponding to the TIM ψ using approach of Section 10.2.

Step 3: If we have n targets, the TIM has n levels 0 to n representing the number of target inhibitions. Consider each level, starting from level n. For the inhibition $B_{n,1} = [1\ 1 \ldots 1]$, if PTIM $\Psi(B_{n,1}) < 1$ then we should consider a latent variable that may be responsible for tumor growth. Thus the dynamic model should allow a transition from $0\ 0 \ldots 0\ 0$ to $0\ 0 \ldots 0\ 1$ and $0\ 0 \ldots 0\ 1$ to $0\ 0 \ldots 0\ 1$. The probabilities of this transition should be equal to $1 - \Psi(B_{n,1})$.

Step 4: Consider level $n - 1$. There are n possibilities of inhibition at this level $011111..1$, $1011...1$, ..., $11..10$ denoted by $B_{n-1,1}, B_{n-1,2}, \ldots B_{n-1,n}$, respectively. For the inhibition $B_{n-1,i}$, removing inhibition of target i has opened up another tumor proliferation pathway with a steady-state mass of $1 - \Psi(B_{n-1,i})$. To capture this behavior, we will assign the following transition probability $0\ 0 \ldots 0\ 1\ 0 \ldots 0 \rightarrow 0\ 0 \ldots 0\ 1\ 0 \ldots 1 = 1 - \Psi(B_{n-1,i})$ and $0\ 0 \ldots 0\ 1\ 0 \ldots 1 \rightarrow 0\ 0 \ldots 0\ 1\ 0 \ldots 1 = 1 - \Psi(B_{n-1,i})$. The 1 is in position i.

Step 5: The next step is to consider level $n - 2$. There are $n(n-1)/2$ possibilities of inhibition at this level $001111..1$, $01011..1.,,11.00$ denoted by $B_{n-2,[1,1]}, B_{n-2,[1,2]}, \ldots, B_{n-2,[n-1,n]}$, respectively. For inhibition $B_{n-2,[i,j]}$ in this level, it means that removing inhibition of the targets i and j has opened up another tumor proliferation pathway with a steady-state mass of $1 - \Psi(B_{n-2,[i,j]})$. To capture this constraint, we will assign the following transition probability $00..10..1..00 \rightarrow 00..10..1..01 = 1 - \Psi(B_{n-2,[i,j]})$ and $00..10..1..01 \rightarrow 00..10..1..01 = 1 - \Psi(B_{n-2,[i,j]})$. Note that any of these transitions will not affect the inhibitions of its supersets in levels $n - 1$ and n. This is because state $B_{n-2,[i,j]}$ will not be reached if one of its supersets in levels $n - 1$ or n is inhibited.

Steps 6 to $n + 3$: Repeat the above process till level 0.

Step $n + 4$: Finally, we have to consider the cases where activation cannot be sustained based on our initial mutation assumptions. As an example, let us consider that $p_1, p_3, p_7, p_5, p_6 \geq 0.5$ and $p_0, p_2, p_4 = 0$ for the PTIM in Fig. 10.9. Using a threshold of 0.5, we will arrive at the TIM of Fig. 10.1 that has Fig. 10.6 as the deterministic dynamic model assuming K_1 and K_2 as initial mutations. Thus, for the case of inhibition of K_1 and K_2, the system may not return to 0010 and 0011 once it leaves the states. An approach to tackle this is to allow transition back from 0000 to 0010 and the transition probability is based on the value of p_6 and p_7.

A simulation example for the application of Algorithm 10.1 is shown next based on the PTIM in Table 10.2. If we consider a threshold of $\alpha = 0.5$ and assuming K_1 and K_2 as initial mutations, the estimated Boolean Network is as shown in Fig. 10.6. Note that the threshold α is selected based on the minimum sensitivity considered significant from the perspective of intervention. Because a drug is often considered effective if the concentration to reduce the tumor volume by 50% is within approved dosage, we considered a threshold of 0.5 for normalized sensitivity to

Table 10.2 Example PTIM [4]

	0 0	0 1	1 1	1 0
0	0	0	0.8	0
1	0.55	0.65	0.9	0.7

denote effectiveness. The threshold should be decreased if we want to incorporate low sensitivity inhibitions in our modeling. To achieve the probabilities shown in Table 10.2, we apply steps 3–7 of Algorithm 10.1 to generate the Markov chain shown in Table 10.3. Note that the Markov chain shown in Table 10.3 is not ergodic and thus the stationary distribution may depend on the starting state. To make the Markov chain ergodic, we can add a small perturbation probability to the Markov chain [8]. The corresponding steady-state sensitivities generated by the Markov chain for a perturbation probability $p = 0.001$ is shown in Table 10.4 which closely reflects the PTIM steady-state sensitivities shown in Table 10.2.

10.3.1.2 Algorithm 10.2

Another perspective on this issue is based on considering that the tumor is heterogeneous and the observed PTIM response is the aggregate effect of inhibition on multiple clones. The dynamics of each clone can be represented by a BN and there is a small probability q of one clone converting to another clone. Thus the overall system can be represented by a context-sensitive probabilistic BN with perturbation probability p and network transition probability q [9]. The algorithm to generate a context-sensitive PBN satisfying the observed PTIM behavior is presented in Algorithm 10.2. Note that based on collapsed steady-state probabilities of context-sensitive PBNs [9], Algorithm 10.2 will always achieve the desired PTIM response within an error of ϵ when p and q are selected to be small.

ALGORITHM 10.2 ALGORITHM TO GENERATE A MARKOV CHAIN P FROM PTIM Ψ BASED ON CONTEXT-SENSITIVE PBN APPROACH [4]

Let ϵ denote the minimum change in PTIM values that needs to be differentiated.
Initialize $L_{max} = \epsilon$, $L_{last} = 0$, $count = 0$
while $L_{max} < 1$ **do**
    Let L_{min} = minimum among the PTIM values $> L_{max}$
    if $L_{min} \neq \emptyset$ **then**
        Let $L_{max} = \min(1, L_{min} + \epsilon)$
        Binarize the PTIM using L_{min} as the threshold.
        $count = count + 1$
        This provides BN $count$ with selection probability $L_{max} - L_{last}$.
        $L_{last} = L_{max}$
    else
        Increase selection probability of BN $count$ by $1 - L_{max}$
        $L_{max} = 1$
    end if
end while

Table 10.3 Example Markov Chain Transition Probability Matrix [4]

	0000	0001	0010	0011	0100	0101	0110	0111	1000	1001	1010	1011	1100	1101	1110	1111
0000	0	0.1	0.12	0	0	0	0	0	0	0	0	0	0.78	0	0	0
0001	0	0.1	0	0	0	0	0	0	0	0	0	0	0.9	0	0	0
0010	0	0	0	0.2	0	0	0	0	0	0	0	0	0	0.8	0	0
0011	0	0	0	0.2	0	0	0	0	0	0	0	0	0	0.8	0	0
0100	0	0	0	0	0	0.3	0	0	0	0	0	0	0	0	0.7	0
0101	0	0	0	0	0	0.3	0	0	0	0	0	0	0	0	0.7	0
0110	0	0	0	0	0	0	0	0	0	0	0	0	0	0	0	1
0111	0	0	0	0	0	0	0	0	0	0	0	0	0	0	0	1
1000	0	0	0	0	0	0	0	0	0	0.35	0	0	0	0	0.65	0
1001	0	0	0	0	0	0	0	0	0	0.35	0	0	0	0	0.65	0
1010	0	0	0	0	0	0	0	0	0	0	0	0	0	0	0	1
1011	0	0	0	0	0	0	0	0	0	0	0	0	0	0	0	1
1100	0	0	0	0	0	0	0	0	0	0	0	0	0	0.45	0.55	0
1101	0	0	0	0	0	0	0	0	0	0	0	0	0	0.45	0.55	0
1110	0	0	0	0	0	0	0	0	0	0	0	0	0	0	0	1
1111	0	0	0	0	0	0	0	0	0	0	0	0	0	0	0	1

Table 10.4 Simulated PTIM From Markov Chain [4]

	0 0	0 1	1 1	1 0
0	0.002003	0.002994	0.800463	0.002995
1	0.549716	0.649251	0.89785	0.698992

As an example of application of Algorithm 10.2, let us consider the PTIM shown in Table 10.5. Based on Algorithm 10.2 with $\epsilon = 0.05$, we will have three individual Boolean networks BN_1, BN_2, and BN_3 with selection probabilities of 0.65, 0.25, and 0.1, respectively. The TIMs corresponding to the BNs are shown in Table 10.6. The BNs satisfying the TIMs in Table 10.6 are shown in Figs. 10.10–10.12. Using a $p = 0.001$ and $q = 0.001$, we arrive at the simulated PTIM shown in Table 10.7 which closely reflects the starting PTIM shown in Table 10.5.

Note that the dynamical models allow us to generate further insights into possible outcomes with sequential application of drugs. For instance, if we consider the previous example with the inferred context-sensitive PBN generating the PTIM shown in Table 10.5 and continuously apply a drug D_1 that inhibits K_2 and K_3, we achieve a sensitivity of 0.9. Similarly, continuous application of a drug D_2 that inhibits K_1 and K_3 will generate a sensitivity of 0.9. However, if we alternate the application of D_1 and D_2, we achieve a sensitivity of 0.94. It shows that alternate inhibition of these pathways allows us to lower the steady-state mass of tumorous

Table 10.5 Example PTIM 2 [4]

	0 0	0 1	1 1	1 0
0	0.02	0.01	0.98	0.03
1	0.65	0.89	1	0.9

Table 10.6 TIMs for BN_1, BN_2, and BN_3, Respectively [4]

	0 0	0 1	1 1	1 0
0	0	0	1	0
1	1	1	1	1
	0 0	0 1	1 1	1 0
0	0	0	1	0
1	0	1	1	1
	0 0	0 1	1 1	1 0
0	0	0	1	0
1	0	0	1	0

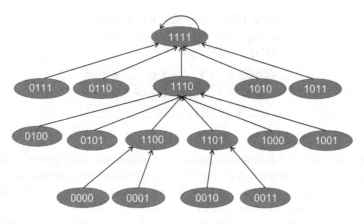

FIG. 10.10

State Diagram for BN_1 in Table 10.6 [4].

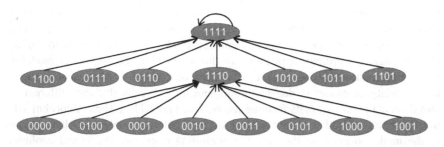

FIG. 10.11

State Diagram for BN_2 in Table 10.6 [4].

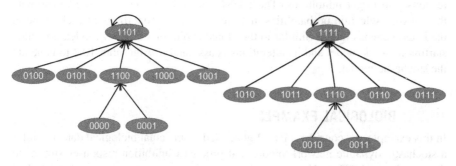

FIG. 10.12

State Diagram for BN_3 in Table 10.6 [4].

Table 10.7 Simulated PTIM Based on Algorithm 10.2 With $p = 0.001$ and $q = 0.001$ [4]

	0 0	0 1	1 1	1 0
0	0.0017	0.0029	0.9964	0.0029
1	0.6495	0.8982	0.9981	0.8982

states. On the other hand, different sequence of inhibitions can negatively affect the final sensitivity. For instance, if a drug D_3 that inhibits K_1 and K_2 and another drug D_4 that inhibits K_3 is applied alternatively, we achieve a sensitivity of 0.50. Note that D_3 alone produces a sensitivity of 0.99 and D_4 produces a sensitivity of 0.65. This shows that stopping the inhibition of D_3 or D_4 at every alternate step causes the tumor to grow back again. For instance, if no inhibition is applied at every alternate time step, we achieve a sensitivity of 0.49 for D_3 and 0.01 for D_4.

In this section we presented two algorithms for generation of Markovian models that have inhibition profiles (termed model generated PTIM) similar to our starting PTIM. The motivation behind the two algorithms is based on two widely accepted evolution models of cancer (Cancer Stem Cell model and Clonal Evolution model [10]) because the primary application of this study is in the context of modeling tumor proliferation pathways. A cancer stem cell model assumes that observed heterogeneity in cancer is due to tumorigenic cancer cells that can differentiate into diverse progeny of cells forming the bulk of tumor [10]. Thus Algorithm 10.1 tries to capture this idea of starting with a single network model and altering parts of the model to generate the observed inhibition response. The clonal evolution of cancer model assumes that a tumor can consist of multiple clones without hierarchical organization [10]. Thus Algorithm 10.2 considers the inhibition response to be based on diverse multiple clones (modeled as separate BNs) with different responses to target inhibitions. The PTIM sensitivity values are used to estimate the network selection probabilities that are similar to proportions of each clone in the heterogeneous tumor. Similar to the clonal evolution of cancer model, no single starting network model and its alterations is assumed in Algorithm 10.2 to generate the stochastic model.

10.3.2 BIOLOGICAL EXAMPLE

In this example, we consider a PTIM generated from actual biological data and infer a stochastic dynamic network model that produces inhibition responses similar to the experimental PTIM. We consider a canine osteosarcoma tumor sample perturbed with 60 targeted drugs, with unique target inhibition profiles to generate steady-state cell viability values [1]. Note that available time series data for perturbation studies are mostly for single gene knockouts/knockdowns [11] which are unable to provide sufficient information to estimate the cell viability response for all possible target inhibition combinations. Thus due to the absence of time series data and ground truth dynamic networks for drug inhibition studies, our model design criteria are to generate dynamic models that can create the experimentally inferred PTIM, while satisfying structural constraints of cancer pathways.

The PTIM generated from experimental 60-drug screen data and satisfying biological constraints [1] for canine tumor sample Sy is shown in Table 10.8. There are 6 target kinases (IGF1R, PSMB5, TGFBR2, AKT2, EGFR, HDAC1) in this model and the 64 entries in Table 10.8 refer to the $2^6 = 64$ possible target inhibitions of the kinases. For instance, 2nd row and 7th column entry of 0.76 refers to sensitivity of 0.76 when the tumor culture is inhibited by IGFR1, TGFBR2, and HDAC1.

Considering the overall idea of generation of context-sensitive PBNs, we arrive at the TIM shown in Table 10.9 using a threshold of 0.3. One of the possible directional pathways that will produce the TIM of Table 10.9 is shown in Fig. 10.13. Note that there can be multiple other possible directional pathway combinations that can produce the above TIM and we are selecting only one of them with assumed mutation in PSMB5. Further biological data, such as gene mutation and expression data and analysis presented in Section 10.2.1 can be used to narrow down the possible combinations.

Subsequently, to select the next level of differences in sensitivities, we considered a threshold of 0.55, which introduces three more possible combinations that fail to stop proliferation (i.e., binarized sensitivity of 0). The TIM is shown in Table 10.10 and a corresponding directional pathway that produces the TIM is shown in Fig. 10.14. Note that the pathway in Fig. 10.14 requires inhibition of multiple targets as compared to the previous pathway in Fig. 10.13 for stopping tumor proliferation. The first three kinases are the same for the two pathways, but the next possibilities are combinations of two kinases, rather than single kinase inhibitions.

We next consider a threshold of 0.8 that differentiates the cluster of sensitivity values $\{0.84, 0.84, 0.88\}$ from the remaining values. The TIM for this threshold is shown in Table 10.11 and a corresponding directional pathway that produces the TIM is shown in Fig. 10.15. The directional pathway is more constrained than the previous pathways in having blocks of targets that require more number of inhibitions to stop tumor proliferation.

Note that the thresholds can be selected in various ways. For instance, we considered equal intervals of 0.25 following the starting threshold of 0.3 resulting in thresholds of 0.3, 0.55, and 0.8. Another approach can use unequal increment thresholds to maintain sensitivity clusters. Because the experiments conducted to generate the sensitivity information can contain noise, it is preferable to ignore small sensitivity differences.

Once we had the three directional pathways, we used the directional pathway to BN approach of Section 10.2.2 to generate the Boolean networks BN_1, BN_2, and BN_3 corresponding to the directional pathways of Figs. 10.13, 10.14, and 10.15, respectively. Based on the limits of the thresholds, we assigned a selection probability of 0.5 for BN_1 $(0.25 < 0.5 < 0.55)$, 0.25 for BN_2 $(0.55 < 0.5 + 0.25 < 0.8)$, and remaining 0.25 for BN_3. Using a value of $p = 0.001$ and $q = 0.001$, we generated

Table 10.8 PTIM Generated From a 60-Drug Screen Data for Canine Tumor Sample Sy [1,4]

AKT2	EGFR	HDAC1		TGFBR2	PSMB5 TGFBR2	PSMB5	IGF1R PSMB5	IGF1R PSMB5 TGFBR2	IGF1R TGFBR2	IGF1R
		HDAC1	0.11	0.38	1	1	1	1	0.68	0.60
		HDAC1	0.55	0.64	1	1	1	1	0.76	0.68
	EGFR		0.64	0.64	1	1	1	1	0.76	0.76
	EGFR		0.17	0.52	1	1	1	1	0.68	0.68
AKT2	EGFR		0.58	0.64	1	1	1	1	0.76	0.76
AKT2	EGFR	HDAC1	0.73	0.76	1	1	1	1	0.88	0.84
AKT2		HDAC1	0.64	0.73	1	1	1	1	0.84	0.76
AKT2		HDAC1	0.47	0.57	1	1	1	1	0.76	0.68

Table 10.9 TIM Generated From the PTIM in Table 10.8 Using a Threshold of 0.3 [4]

			TGFBR2	PSMB5 / TGFBR2	PSMB5	IGF1R / PSMB5	IGF1R / PSMB5 / TGFBR2	IGF1R / TGFBR2	IGF1R
	HDAC1	0	1	1	1	1	1	1	1
EGFR	HDAC1	1	1	1	1	1	1	1	1
EGFR		1	1	1	1	1	1	1	1
EGFR		0	1	1	1	1	1	1	1
EGFR	HDAC1	1	1	1	1	1	1	1	1
AKT2	HDAC1	1	1	1	1	1	1	1	1
AKT2		1	1	1	1	1	1	1	1
AKT2									
AKT2									

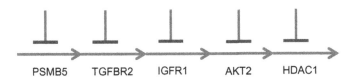

FIG. 10.13

Directional pathway satisfying TIM of Table 10.9 [4].

a context-sensitive PBN and calculated the PTIM for the model by generating the steady-state probabilities of tumor state = 0 for each target inhibition combination. The generated PTIM for the designed model is shown in Table 10.12 (up to two decimal digits). The model generated PTIM is similar to our initial experimental PTIM shown in Table 10.8. The mean and maximum absolute errors of the entries between the experimental and model generated PTIM are 0.043 and 0.2, respectively, which is low considering that only three BNs were used to generate the context sensitive PBN. Further reduction in the differences between the experimental and model generated PTIM can possibly be achieved by increasing the number of BNs and optimizing the thresholds and network selection probabilities to reduce the mean error.

10.4 DISCUSSION

In this chapter, we analyzed the inference of dynamical models from static TIM models. We showed that the inferred blocks from the TIM approach could be converted to directional pathways based on different mutation scenarios, and subsequently converted to dynamic BN models. In terms of stochastic model inference, we presented two algorithms where (i) the first technique was based on altering the BN generated from binarizing the PTIM based on a single threshold and (ii) the second approach considered generation of multiple BNs based on different thresholds and integrating them in the form of a context-sensitive PBN. We provided examples to show the application of the algorithms to generate Markovian models whose steady-state inhibition profiles are close to the experimental PTIMs.

Note that the inference algorithms presented here are primarily focused on dynamic models of tumor proliferation. The number of targets considered is small as they are a subset of the targets of targeted drugs (usually tyrosine kinase inhibitors) that are required to faithfully capture the tumor proliferation of a particular system without overfitting. Consequently, any properties of large-scale genetic regulatory networks [11,12], such as adherence to power law [13], were not incorporated in these studies. The incorporation of characteristics of large-scale networks in inference of dynamic models from PTIMs remains an open problem. Furthermore, the current algorithms can be refined by analyzing mutation and time series data to restrict the possible directional pathways.

Table 10.10 TIM Generated From the PTIM in Table 10.8 Using a Threshold of 0.55 [4]

				TGFBR2	PSMB5 / TGFBR2	PSMB5	IGF1R / PSMB5	IGF1R / PSMB5 / TGFBR2	IGF1R / TGFBR2	IGF1R
AKT2	EGFR		0	0	1	1	1	1	1	1
AKT2	EGFR	HDAC1	1	1	1	1	1	1	1	1
AKT2	EGFR	HDAC1	1	1	1	1	1	1	1	1
AKT2	EGFR		0	0	1	1	1	1	1	1
	EGFR	HDAC1	1	1	1	1	1	1	1	1
	EGFR	HDAC1	1	1	1	1	1	1	1	1
AKT2			0	1	1	1	1	1	1	1

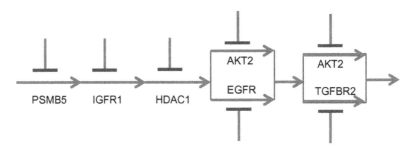

FIG. 10.14

Directional pathway satisfying TIM of Table 10.10 [4].

Table 10.11 TIM Generated From the PTIM in Table 10.8 Using a Threshold of 0.8 [4]

							IGF1R	IGF1R	IGF1R	IGF1R
				PSMB5	PSMB5	PSMB5	PSMB5			
			TGFBR2	TGFBR2					TGFBR2	TGFBR2
			0	0	1	1	1	1	0	0
		HDAC1	0	0	1	1	1	1	0	0
	EGFR	HDAC1	0	0	1	1	1	1	0	0
	EGFR		0	0	1	1	1	1	0	0
AKT2	EGFR		0	0	1	1	1	1	0	0
AKT2	EGFR	HDAC1	0	0	1	1	1	1	1	1
AKT2		HDAC1	0	0	1	1	1	1	1	0
AKT2			0	0	1	1	1	1	0	0

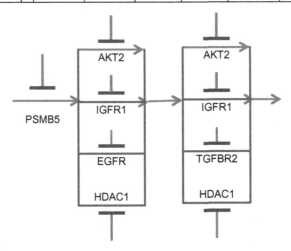

FIG. 10.15

Directional pathway satisfying TIM of Table 10.11 [4].

Table 10.12 PTIM Generated From Context-Sensitive Probabilistic Boolean Network Model Based on Algorithm 10.2 [4]

				TGFBR2	PSMB5 TGFBR2	PSMB5	IGF1R PSMB5	IGF1R PSMB5 TGFBR2	IGF1R TGFBR2	IGF1R
			0.00	0.50	0.99	0.99	0.99	1.00	0.75	0.75
		HDAC1	0.75	0.75	1.00	0.99	0.99	1.00	0.75	0.75
	EGFR	HDAC1	0.75	0.75	1.00	0.99	0.99	1.00	0.75	0.75
	EGFR		0.00	0.50	0.99	0.99	0.99	1.00	0.75	0.75
AKT2	EGFR		0.75	0.75	1.00	0.99	0.99	1.00	0.75	0.75
AKT2	EGFR	HDAC1	0.75	0.75	1.00	1.00	1.00	1.00	1.00	0.75
AKT2		HDAC1	0.75	0.75	1.00	0.99	0.99	1.00	1.00	1.00
AKT2			0.50	0.75	1.00	0.99	0.99	1.00	0.75	0.75

REFERENCES

[1] N. Berlow, L.E. Davis, E.L. Cantor, B. Seguin, C. Keller, R. Pal, A new approach for prediction of tumor sensitivity to targeted drugs based on functional data, BMC Bioinform. 14 (2013) 239.

[2] R. Pal, N. Berlow, A kinase inhibition map approach for tumor sensitivity prediction and combination therapy design for targeted drugs, in: Pacific Symposium on Biocomputing, 2012, pp. 351–362, PMID: 22174290, http://psb.stanford.edu/psb-online/proceedings/psb12/pal.pdf.

[3] S.A. Kauffman, The Origins of Order: Self-Organization and Selection in Evolution, Oxford University Press, New York, 1993.

[4] N. Berlow, L. Davis, C. Keller, R. Pal, Inference of dynamic biological networks based on responses to drug perturbations, EURASIP J. Bioinform. Syst. Biol. 14 (2014), http://dx.doi.org/10.1186/s13637-014-0014-1.

[5] N. Berlow, R. Pal, L. Davis, C. Keller, Analyzing pathway design from drug perturbation experiments, in: 2012 IEEE Statistical Signal Processing Workshop (SSP), 2012, pp. 552–555, http://dx.doi.org/10.1109/SSP.2012.6319757.

[6] S. Nelander, W. Wang, B. Nilsson, Q.-B. She, C. Pratilas, N. Rosen, P. Gennemark, C. Sander, Models from experiments: combinatorial drug perturbations of cancer cells, Mol. Syst. Biol. 4 (1) (2008) 216.

[7] R. Pal, S. Bhattacharya, Characterizing the effect of coarse-scale PBN modeling on dynamics and intervention performance of genetic regulatory networks represented by stochastic master equation models, IEEE Trans. Signal Process. 58 (2010) 3341–3351.

[8] R. Pal, A. Datta, M.L. Bittner, E.R. Dougherty, Intervention in context-sensitive probabilistic Boolean networks, Bioinformatics 21 (2005) 1211–1218.

[9] R. Pal, Context-sensitive probabilistic Boolean networks: steady state properties, reduction and steady state approximation, IEEE Trans. Signal Process. 58 (2010) 879–890.

[10] M. Shackleton, E. Quintana, E.R. Fearon, S.J. Morrison, Heterogeneity in cancer: cancer stem cells versus clonal evolution, Cell 138 (5) (2009) 822–829.

[11] R.J. Prill, D. Marbach, J. Saez-Rodriguez, P.K. Sorger, et al., Towards a rigorous assessment of systems biology models: the DREAM3 challenges, PLoS ONE 5 (2) (2010) e9202.

[12] H. De Jong, Modeling and simulation of genetic regulatory systems: a literature review, J. Comput. Biol. 9 (2001) 67–103.

[13] T. Zhou, Y.-L. Wang, Causal relationship inference for a large-scale cellular network, Bioinformatics 26 (16) (2010) 2020–2028.

Combination therapeutics

11

11.1 INTRODUCTION

Multiple drugs and treatment options have been developed for cancer in recent years, but the numerous aberrations in molecular pathways that can produce cancer necessitate the use of drug combinations, as compared to single drug for treatment of individual cancers. Furthermore, use of combination drugs to target multiple pathways that individually can reduce proliferation can assist in avoiding drug resistance. A general consensus is building up in the medical community for the use

Predictive Modeling of Drug Sensitivity. http://dx.doi.org/10.1016/B978-0-12-805274-7.00011-7

of drug combinations in treating diseases such as cancer that are caused by a complex interplay of molecular and environmental factors. The use of drug combinations opens up the potential treatment opportunities, but bestows a combinatorial and experimental challenge for arriving at the optimal solution.

The prohibitively high number of potential combination drugs rules out exhaustive testing of all drug combinations. For instance, if we consider n individual drugs and combinations of m drugs with k concentrations for each drug, there are potentially $\Lambda(n, m, k) = k^m \binom{n}{m}$ combinations. The number of combinations can be enormous for reasonable values of n, m, k such as $\Lambda(500, 3, 2) = 165,668,000 \simeq 165$ million. Thus a systematic approach to evaluate the potential of different drug combinations is required. Also desirable is an approach to evaluate if a drug combination is more efficacious than individual drugs.

The selection of a drug combination for application to a patient can be approached from various perspectives. Some applicable cancer drug combinations are based on empirical information where a commonly used standard of care drug is combined with a newer drug with the expectation of increasing efficacy and avoiding resistance. Examples of such an approach include *trastuzumab* (targeted drug affecting HER2) being applied in combination with *paclitaxel* (chemotherapeutic drug) for breast cancer treatment, and *cetuximab* (targeted EGFR inhibitor) being applied in combination with *irinotecan* (chemotherapeutic drug) for metastatic colorectal cancer therapy. Based on empirical evidence of efficaciousness in animal models along with sensitivity of individual drugs, clinical trials of drug combinations are frequently designed [1]. However, these clinical- and empirical-based techniques can fail to provide an unbiased optimal combination for an individual patient, along with the inability to guide the reasoning behind the success or failure of a drug combination.

Alternative directions include the use of individual patient tumor culture to design an optimal drug combination. In this chapter, the drug combination designs for an individual tumor culture are considered from two standpoints: (a) *Model-Based* combination design, where a mathematical model for drug sensitivity prediction is designed from the tumor culture, and a drug combination effective on the estimated model is designed. We utilize the TIM framework as the predictive model and set cover- and hill climbing-based techniques are used to arrive at the desired drug combinations. (b) *Model-Free* combination drug design for the scenario where limited information is available to infer a model. The chapter presents a stochastic search approach to arrive at an efficacious drug combination that requires limited experimental iterations.

This chapter is organized as follows: Section 11.2 covers approaches to analyze the effectiveness of a drug combination from experimental data. Section 11.3 considers model-based combination therapy design approaches, whereas Section 11.4 considers model-free combination therapy design techniques.

11.2 ANALYZING DRUG COMBINATIONS

The measure of success of a drug combination is dependent on the medium on which the combination is to be tested. If the drug combination is designed for a model, the success measure will be dependent on how the model predicts efficacy and toxicity, such as discrete or continuous point estimates or prediction of the entire dose-response curve. Because these measures are tied to the applicable model, we will discuss the measure for effectiveness of drug combinations for model-based design in the section on *Model-Based Drug Combination Design*.

If we consider the effectiveness of a drug combination for in vivo human patients or animal models, we usually consider the increase in life span with the application of drug combinations, as compared to individual drugs or no drugs. For this purpose, Kaplan-Meier (KM) curves [2] are often designed for survival data where the two-dimensional KM graphs have survival time span represented in the X-axis and percentage of population surviving on the Y-axis.

11.2.1 KAPLAN-MEIER SURVIVAL CURVES

To explain KM survival curves, let us consider an example study conducted on a cohort of 10 mice with xenografted tumor to evaluate the survival efficacy. Let the survival time for the mice in the cohort be given in Table 11.1.

The observed time of deaths can be ordered as $t_1 \leq t_2 \leq t_3 \leq \cdots$ and let $d_1, d_2, d_3, \ldots$ denote the number of deaths that occur at each of these times and let $n_1, n_2, n_3, \ldots$ denote the number of mice remaining in the cohort. Let $S(t)$ denote the probability that mice from the population belonging to the samples will survive time t. The KM estimate for this survival probability is calculated as

$$\hat{S}(t) = \prod_{t_i < t} \frac{n_i - d}{n_i} \tag{11.1}$$

The KM estimation based on survival data in Table 11.1 is shown in Table 11.2 and plotted as a curve in Fig. 11.1.

The effectiveness of a drug combination is expected to be measured by creating four cohorts of (a) mice not given any drug; (b) mice treated with drug 1; (c) mice treated with drug 2; and (d) mice treated with a combination of drugs 1 and 2. A successful synergistic drug combination is expected to produce KM plots similar

Table 11.1 Example Table Showing Life Spans for a Cohort of 10 Mice

Mice	1	2	3	4	5	6	7	8	9	10
Survival time (in months)	2	3	3	3	3	5	5	5	10	12

Table 11.2 Kaplan-Meier Estimate for Survival Data Shown in Table 11.1

t_i	2	3	5	10	12	>12
d_i	0	1	4	3	1	1
n_i	10	9	5	2	1	0
$\hat{S}(t)$	1	0.9	0.5	0.2	0.1	0

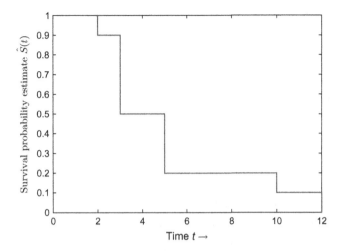

FIG. 11.1

Kaplan-Meier survival plot for data shown in Table 11.1.

to ones shown in Fig. 11.2. To quantify the behavior, the statistical significance of the difference in life spans for the drug combination as compared to groups (a), (b), and (c) can be computed.

11.2.2 COMBINATION INDEX

A measure for effectiveness of a drug combination for in vitro tumor culture or cell line experimental studies will consider the combination and individual drug response curves to estimate synergy of the combination. We will discuss the most commonly used approach to measure combination synergy based on in vitro combination studies in this section.

As described in [3], a test of a combination measure can be based on the ability to predict the effect of combining the same drug. To explain further, if we run tests on drug A and drug B, the combination measure should estimate it as additive when drug A = drug B. The fractional product method is applicable when the drug effects are independent (nonmutually exclusive agents) and the dose-effect relationship follows Michaelis-Menten relationship (hill slope of 1 in sigmoidal curve). When the assumptions are valid, the effect of the combination of drug A (amount C_A where f_A is the fraction of the unaffected to total activity for the drug) and drug B (amount C_B where f_B is the fraction of the unaffected to total activity for the drug) is given by $f_{AB} = f_A f_B$ where f_{AB} is the fraction of the unaffected to total activity for the drug combination. A more general approach for all forms of drug-effect response curves is given by the median-effect relationship where $f_a/f_u = (D/EC_{50})^m$ where m is the hill slope of the sigmoidal curve, f_a denotes the fraction affected, f_u denotes the fraction

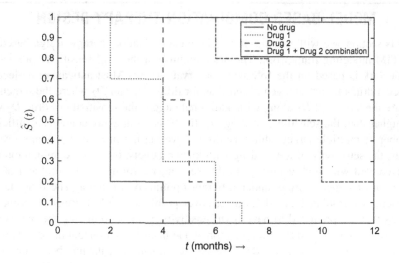

FIG. 11.2

Kaplan-Meier survival plot for hypothetical data of four cohorts of untreated mice, mice treated with drug 1, mice treated with drug 2, and mice treated with a combination of drugs 1 and 2.

unaffected, EC_{50} denotes the concentration of the drug to reach 50% effect, and D denotes the drug concentration.

The multiple drug-effect equation when drugs share a similar mode of actions (as denoted by effects are mutually exclusive in pharmacology literature)

$$\left(\frac{f_a(A,B)}{f_u(A,B)}\right)^{1/m} = \left(\frac{f_a(A)}{f_u(A)}\right)^{1/m} + \left(\frac{f_a(B)}{f_u(B)}\right)^{1/m} = \frac{D(A)}{EC_{50}(A)} + \frac{D(B)}{EC_{50}(B)} \qquad (11.2)$$

The multiple drug-effect equation when drugs have independent mode of actions (as denoted by effects are nonmutually exclusive in pharmacology literature)

$$\left(\frac{f_a(A,B)}{f_u(A,B)}\right)^{1/m} = \frac{D(A)}{EC_{50}(A)} + \frac{D(B)}{EC_{50}(B)} + \frac{D(A)D(B)}{EC_{50}(A)EC_{50}(B)} \qquad (11.3)$$

Combination index (CI) was defined by Chou and Talalay [4] as

$$CI = \frac{D(A)}{D_X(A)} + \frac{D(B)}{D_X(B)} = \frac{D(A)}{EC_{50}(A)(f_a/(1-f_a))^{1/m_A}} + \frac{D(B)}{EC_{50}(B)(f_a/(1-f_a))^{1/m_B}} \qquad (11.4)$$

where $CI < 1, = 1$, and > 1 indicate synergism, additive effect, and antagonism, respectively. If A and B have a combined effect of $X\%$, then $D_X(A)$ and $D_X(B)$ denote the concentrations of drugs A and B that induce effect $X\%$ when acting alone, respectively.

11.3 MODEL-BASED COMBINATION THERAPY DESIGN

In this section, we will consider model-based combination therapy design based on the TIM modeling framework. The basic premise of the TIM sensitivity prediction framework is based on the following observations: (a) Most anticancer molecular targeted drugs target oncogenes and thus for drugs D_1 and D_2 where the targets of D_2 are the targets of D_1 along with additional targets, the sensitivity of drug D_2 will be higher than the sensitivity of drug D_1. (b) If the drug assay contains a sufficient number of multiple target inhibitory drugs covering the important proteins several times, the sensitivity of a new drug with known targets, or a drug combination can be predicted with high accuracy. In this section, we formulate the problem of the selection of targeted drugs under different optimization criteria. The drug design problem is formulated as [5] (a) a set cover problem to select drugs covering the targets that are predicted to have high sensitivity, while minimizing the number of nonessential targets and (b) minimization of a joint cost function that incorporates the sensitivity of a combination and the toxicity as measured by the number of inhibited targets. The formulation a is solved using a greedy set cover algorithm and the optimization in b is tackled using a hill climbing search. The objective of combination therapy can vary depending on the condition of the patient, such as the ability to endure off-target effects and chances of drug resistance development. Thus a second optimization criterion of avoiding resistance is considered by targeting mutually exclusive pathways that by themselves can inhibit the tumor. The expectation is that if resistance develops to one of the blocked pathways, the other inhibited pathways will still be able to avoid tumor growth. We formulate the resistance optimization problem as locating the cover of a set that has mutually exclusive subsets with high sensitivity.

11.3.1 MATHEMATICAL FORMULATION

Let us assume that there are T protein targets denoted by $k_1, k_2, \ldots, k_T$ for each cell line or tumor culture. Each targeted drug inhibits a specific set of targets. The amount of inhibition can be measured by dissociation constants (K_d) or EC_{50}. If we discretize the inhibitions based on a threshold, there can be a total of $D = 2^T - 1$ different drugs that can be generated theoretically. However, the library of available drugs is quite limited. For instance, if 20 important proteins for cancers such as BRAF, AKT, EGFR, etc. are considered, that is, $T = 20$, the total number of possible inhibitions is $2^{20} = 1,048,576$, but the number of available drugs will be in the range of <1000. Note that if we consider combinations of the available drugs, then the possibilities increase tremendously. Thus we will assume that among D possible drugs, we have only m drugs available to us. Let the known multitarget inhibiting sets for these drugs be denoted by $S_1, S_2, \ldots, S_m$ which are obtained from drug inhibition studies [6–8]. We consider that the sensitivities of $m_1 \leq m$ drugs are known (using a drug assay). Our goal is to select a drug or drug combination from the m drugs that will have high sensitivity, while maintaining low toxicity. The toxicity will be proportional to the number of targets inhibited by the drug. For drug combinations, toxicity will be proportional to the number of the targets in the union of the set of targets.

For the T targets, each drug can be considered to be in one of the $T+1$ levels (0 to T) denoting the number of targets being inhibited by the drug. An example is shown in Fig. 11.3 for $T = 3$. Here the experimentally tested drugs will be nodes in the graph and the sensitivity of a target set combination will be between the maximum of the sensitivities of the subsets and the minimum of the sensitivities of the supersets. For instance, if we are trying to find the sensitivity of $k_1 k_3$ and we know that $S(k_1) = 0.5$, $S(k_3) = 0.6$, and $S(k_1 k_2 k_3) = 0.9$, then $\max(0.5, 0.6) \leq S(k_1 k_3) \leq \min(0.9)$. In such a scenario, we do not have to generate the sensitivities of the 2^T combinations at the beginning, but we can generate the predictions for each interested combination when required.

Each drug $D_i(i \in 1, 2, \ldots, D)$ is a T length vector with 0 everywhere except 1s at places that are targeted. $|D_i|$ denotes the number of targets inhibited by the drug D_i. To arrive at the sensitivity of a target combination that has not been experimentally tested, we will fit a curve between the maximum of subset sensitivities and minimum of superset sensitivities. Let $L_{\min}$ denote the level of a drug having minimum sensitivity $\lambda_{\sup/\min}$ among the supersets, $L_{\max}$ denote the level of a drug achieving maximum sensitivity $\lambda_{\sub/\max}$ among the subsets, L_{D_i} denote the level of the drug D_i. So, the sensitivity of a drug D_i can be approximated by λ_{D_i} as shown next:

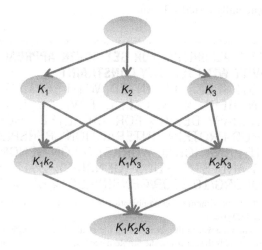

FIG. 11.3

Levels of drugs.

(©2012 IEEE. Reprinted, with permission, from S. Haider, N. Berlow, R. Pal, L. Davis, C. Keller, Combination therapy design for targeted therapeutics from a drug-protein interaction perspective, in: 2012 IEEE International Workshop on Genomic Signal Processing and Statistics (GENSIPS), 2012, pp. 58–61, http://dx.doi.org/10.1109/GENSIPS.2012.6507726.)

$$\lambda_{D_i} = \lambda_{\text{sub/max}} + \frac{(L_{D_i} - L_{\min})(\lambda_{\text{sup/min}} - \lambda_{\text{sub/max}})}{L_{\min} - L_{\max}} \quad (11.5)$$

Note that Eq. (11.5) provides a basic linear unsupervised fitting. We can use supervised learning to fit higher-order curves.

The objective is to maximize λ_D while maintaining toxicity less than a number Z. Here toxicity of each drug or drug combination is equal to the number of targets among T targets inhibited by the drug or drug combination. Thus our maximum sensitivity is upper-bounded by the maximum sensitivity achieved at level Z. This is because level $Z + 1$ will not satisfy the toxicity condition and the union of collection of drugs satisfying the toxicity condition can maximally reach level Z.

The above-described sensitivity maximization problem will be solved using *set cover approach* and a marginally modified version of the problem will be solved using *hill climb approach*.

11.3.2 SET COVER APPROACH TO MAXIMIZE SENSITIVITY WITH CONSTRAINT ON TOXICITY

The set cover approach for maximizing sensitivity with constraint on toxicity is described in Algorithm 11.1. The computational complexity at step (a) of Algorithm 11.1 can be reduced by considering that any set in level Z that has a superset with experimental value less than the maximum of the experimental sensitivities for set size $\leq$ Z can be ignored, as that solution will be worse than the already experimentally tested solution.

ALGORITHM 11.1 ALGORITHM FOR SET COVER APPROACH TO MAXIMIZE SENSITIVITY WITH TOXICITY CONSTRAINT

(©2012 IEEE. REPRINTED, WITH PERMISSION, FROM S. HAIDER, N. BERLOW, R. PAL, L. DAVIS, C. KELLER, COMBINATION THERAPY DESIGN FOR TARGETED THERAPEUTICS FROM A DRUG-PROTEIN INTERACTION PERSPECTIVE, IN: 2012 IEEE INTERNATIONAL WORKSHOP ON GENOMIC SIGNAL PROCESSING AND STATISTICS (GENSIPS), 2012, PP. 58–61, HTTP://DX.DOI.ORG/10.1109/GENSIPS.2012.6507726)

Step (a): Locate the maximum target combination at level Z. Let that maximum set be D_Z with sensitivity $\lambda(D_Z) = y$.

Step (b): Try to find a cover of the set D_Z using the available drugs while minimizing the sum of toxicities. If Ξ is the set cover, the problem becomes *Minimize $|S_U|$ where S_U is the union of all the sets in Ξ subject to $|S_U| \geq 1$*

Step (c): If a set cover cannot be found in step (b), then either reduce Z by one and go to step (a), or consider the next highest sensitivity at level Z and go to step (b).

The greedy set cover solution [9] shown in Algorithm 11.2 can be used to solve step (b) of Algorithm 11.1. For Algorithm 11.2, the weights w_i is assumed to be equal for all drugs, that is, $w_1 = w_2 = \cdots = w_m = 1$.

ALGORITHM 11.2 SET COVER GREEDY ALGORITHM

(©2012 IEEE. REPRINTED, WITH PERMISSION, FROM S. HAIDER, N. BERLOW, R. PAL, L. DAVIS, C. KELLER, COMBINATION THERAPY DESIGN FOR TARGETED THERAPEUTICS FROM A DRUG-PROTEIN INTERACTION PERSPECTIVE, IN: 2012 IEEE INTERNATIONAL WORKSHOP ON GENOMIC SIGNAL PROCESSING AND STATISTICS (GENSIPS), 2012, PP. 58–61, HTTP://DX.DOI.ORG/10.1109/GENSIPS.2012.6507726)

Input: A set system $(\Gamma, \mathcal{S})$ with $\bigcup_{i \in \mathcal{S}} S_i = \Gamma$, weights $w_i \in \mathbb{R}_+$
Output: A set cover $\mathcal{R}$ with $\bigcup_{i \in \Pi} S_i = \Gamma$
Initialize $\mathcal{R} = \emptyset$ and $W = \emptyset$
while $W \neq \Gamma$ **do**
    choose an $i \in \mathcal{S} \setminus \mathcal{R}$ for which $S_i \setminus W \neq \emptyset$ and $\frac{w_i}{|S_i \setminus W|}$ is minimum.
    $\mathcal{R} = \mathcal{R} \cup \{i\}$ and $W = W \cup S_i$.
end while

11.3.3 HILL CLIMB APPROACH TO MAXIMIZE SENSITIVITY WITH CONSTRAINT ON TOXICITY

In this section, a single cost function that integrates the objectives of maximizing sensitivity and minimizing toxicity is considered. We describe the cost function as follows: $C(S_U) = \frac{1}{Z}|S_U| + p$ where S_U is the union of the sets in combination and $p = 1 - \lambda(S_U)$ denotes the proportion of tumor cells still alive as compared to no control scenario. We can apply hill climbing technique to solve this optimization problem as was considered in the Medicinal Algorithmic Combinatorial Screen (MACS) design approach [10]. Algorithm 11.3 describes the steps of this approach.

ALGORITHM 11.3 HILL CLIMBING APPROACH

(©2012 IEEE. REPRINTED, WITH PERMISSION, FROM S. HAIDER, N. BERLOW, R. PAL, L. DAVIS, C. KELLER, COMBINATION THERAPY DESIGN FOR TARGETED THERAPEUTICS FROM A DRUG-PROTEIN INTERACTION PERSPECTIVE, IN: 2012 IEEE INTERNATIONAL WORKSHOP ON GENOMIC SIGNAL PROCESSING AND STATISTICS (GENSIPS), 2012, PP. 58–61, HTTP://DX.DOI.ORG/10.1109/GENSIPS.2012.6507726)

Step (a): Select R random drug combinations and calculate the cost function for each combination.
Step (b): Select the top combination (lowest cost).
Step (c): Add each of the remaining drugs to the top combination from step (b) one at a time to create new combinations.
Step (d): Remove one drug at a time from the top combination of step (b) to create further new combinations.
Step (e): Find the new lowest cost combination from the combinations in steps (c) and (d).
Keep repeating steps (b)–(e) until no benefit in cost is found from the new combinations.

As compared to set cover approach, we do not have to calculate all the sensitivities at level Z.

Computational complexity: The computational complexity of the set cover approach consists of the calculations involved in steps (a) and (b) of Algorithm 11.1. The step (a) has a complexity of $\mathcal{O}(\binom{T}{Z})$ and step (b) has a complexity of $\mathcal{O}(mZ)$. The complexity of the hill climb approach is $\mathcal{O}(mI)$ where I is the number of iterations required by the hill climb search. In comparison, the exhaustive search has a computational complexity of $\mathcal{O}(2^m)$.

11.3.4 RESISTANCE CONTROL PROBLEM

Tumors often develop resistance to drugs. In other words, if a tumor is restricted by blocking proliferation pathway P_1, another pathway P_2 might become active to support proliferation. Thus, for our goal of avoiding resistance, we would like to inhibit more than one independent blocking pathway such that for the scenario when resistance to one of the blocking pathway develops, the other independent pathway(s) can still keep the tumor under check. In other words, we want to select a set of targets that can be divided into two or more nonintersecting sets such that the sensitivity of each set is higher than a threshold T.

Consequently, the control problem can be formulated as selection of k drugs $D_{a(1)}, D_{a(2)}, \ldots, D_{a(k)}$ from m drugs $D_1, D_2, \ldots, D_m$ such that $S_U = \cup_{i=1}^{k} S_{a(i)}$ can be divided into $c(c \geq 2)$ nonintersecting subsets $B_1, \ldots, B_c$ ($B_i \cap B_j = \phi$ for $i, j \in 1, \ldots, c$) under the constraint $S(B_i) > T$ for $i \in 1, \ldots, c$ and $\left| \cup_{i=1}^{k} S_{a(i)} \right| < Z$. A possible approach to solve the resistance control problem is illustrated in Algorithm 11.4.

ALGORITHM 11.4 ALGORITHM FOR SOLVING RESISTANCE CONTROL PROBLEM

(©2012 IEEE. REPRINTED, WITH PERMISSION, FROM S. HAIDER, N. BERLOW, R. PAL, L. DAVIS, C. KELLER, COMBINATION THERAPY DESIGN FOR TARGETED THERAPEUTICS FROM A DRUG-PROTEIN INTERACTION PERSPECTIVE, IN: 2012 IEEE INTERNATIONAL WORKSHOP ON GENOMIC SIGNAL PROCESSING AND STATISTICS (GENSIPS), 2012, PP. 58–61, HTTP://DX.DOI.ORG/10.1109/GENSIPS.2012.6507726)

Step (a): Look for top sensitivity sets at level Z

Step (b): For a set S with high sensitivity at level Z, try to find out if S can be broken into mutually exclusive c subsets, each having sensitivity greater than threshold T. This can be done in the following way:

(i): Let $k = \lfloor Z/c \rfloor$.

(ii): Search for all subsets of S of size k.

if None of the subsets satisfy *Sensitivity* $> T$ condition **then**

    Go back to step (a) and select a new set

else

    Go back to step (i) with new $Z = Z - c$ and repeat until a total of c subsets are found

end if

Step (*c*): If step (*b*) is successful, then try to find a cover for set *S* with the existing drugs using Algorithm 11.2.

11.3.5 APPLICATIONS

In this section, we discuss the performance of the discussed algorithms when applied to synthetic models and models generated from experimental osteosarcoma tumor culture data.

Synthetic models: We randomly generated sensitivity data satisfying biological constraints for different numbers of targets and available drugs. Table 11.3 summarizes the performance results for eight of these experiments for set cover, hill climb, and exhaustive search. Note that exhaustive search is not feasible when number of drugs, *m*, is large. We have shown the exhaustive search results when $m \leq 15$. The selected drugs are denoted by the decimal equivalent of their binary representations. For instance, drug "01001000" means that it can inhibit 4th and 7th target and this drug is represented by the decimal equivalent of "01001000," that is, 72. The constraint on toxicity *Z* varies from 4 to 8. The column on mutually exclusive sets denotes whether the solution contains two or more mutually exclusive sets with individual sensitivity greater than threshold 0.5. The computation time in seconds using Matlab R2011b on Windows 64-bit machine with a Quad Core Intel Xeon Processor E5507 and 6 GB RAM is also reported in the table. The optimization criteria for the approaches are marginally different: set cover tries to maximize sensitivity while satisfying the toxicity constraint of $\leq Z$, whereas hill climb and exhaustive search try to minimize the cost = 1 − sensitivity + toxicity/Z. Thus, as the results show, the cost is usually low for hill climb as compared to set cover, whereas the sensitivity is higher for set cover as compared to hill climb. The results also show that the set cover and hill climb approaches can attain the global minimum cost in some cases and the cost of the suboptimal approaches always remains within two times that of exhaustive search, thus illustrating their effectiveness.

Models based on experimental data: We have used experimental data generated from an osteosarcoma cell line cultured from a large-breed dog named Bailey. The tumor culture underwent a small-molecule inhibitor drug screen developed at the Keller Laboratory at Oregon Health and Sciences University. The tumor culture was treated with a set of 60 targeted drugs and the drugs with known targets and nonzero response were considered for our modeling. Different numbers of target sizes (10, 12, and 15) were considered for generating the models. For the resistance control problem pertaining to set cover approach, we checked if the highest sensitivity target set at level *Z* can be divided into two nonintersecting subsets (both having sensitivity greater than a threshold). It should be noted that we can sometimes find two nonintersecting subsets, but the cover among the existing drugs may not exist.

Table 11.3 Performance Comparison for *Set Cover*, *Hill Climb*, and *Exhaustive Search* for Synthetic Models

Data	Set 1	Set 2	Set 3	Set 4	Set 5	Set 6	Set 7	Set 8
Number of targets	8	8	10	10	12	12	12	15
Number of available drugs	15	50	10	12	8	30	75	100
Z	4	4	5	5	6	6	6	7
Threshold	0.5	0.5	0.5	0.5	0.5	0.5	0.5	0.5
Set cover								
Selected drugs	[101 36]	[224..160]	[784 2]	[14 5]	[2883]	[582..514]	[822]	[4353]
Cost	1.01	1	1.06	0.81	1.01	1	1	0.43
Sensitivity	0.99	1	0.94	0.99	1	1	1	1
Run-time	0.3747	0.4437	1.1237	1.3633	0.7646	5.413	5.36	753.20
Mutually Exclusive sets	Found	N/A	N/A	N/A	Found	Found	N/A	N/A
Hill climb								
Selected drugs	[204..72]	[64 66]	[785]	[5 10]	[164]	[849]	[5 2577]	[22563]
Cost	1.05	0.62	0.86	0.81	0.78	0.85	0.97	0.93
Sensitivity	0.95	0.88	0.94	0.99	0.72	0.98	0.86	0.93
Run-time	0.3741	0.6742	0.3226	0.3822	0.4765	1.9723	2.0416	18.14
Exhaustive								
Selected drugs	[36]	N/A	[785]	[10 14]	[404]	N/A	N/A	N/A
Cost	0.59	N/A	0.86	0.61	0.71	N/A	N/A	N/A
Sensitivity	0.91	N/A	0.94	0.99	0.96	N/A	N/A	N/A
Run-time	3.2491	N/A	0.3305	0.8146	0.6061	N/A	N/A	N/A

©2012 IEEE. Reprinted, with permission, from S. Haider, N. Berlow, R. Pal, L. Davis, C. Keller, Combination therapy design for targeted therapeutics from a drug-protein interaction perspective, in: 2012 IEEE International Workshop on Genomic Signal Processing and Statistics (GENSIPS), 2012, pp. 58–61, http://dx.doi.org/10.1109/GENSIPS.2012.6507726.

The performance of the two techniques is reported in Table 11.4. The results show that both the approaches have been able to achieve smaller costs. Note that the cost ($\frac{1}{Z}|S_U| + p$) can attain a maximum value of $15/7 + 1 = 3.14$ for $Z = 7$ and $T = 15$.

11.3.5.1 Diffuse intrinsic pontine glioma case study

In this case study, we consider the modeling of diffuse intrinsic pontine glioma (DIPG) cell lines and tumor cultures and validation of predicted drug combinations [11]. This study was a collaborative effort among multiple laboratories in the

Table 11.4 Combination Drug Design Performance Using *Set Cover* and *Hill Climb* Approaches for Canine Osteosarcoma Models

Data	Bailey₁₀	Bailey₁₂	Bailey₁₅
Number of targets	10	12	15
Number of available drugs	12	16	20
Z	5	6	7
Threshold	0.5	0.5	0.5
Set cover			
Selected drugs	[192..256]	[12..2048]	[12..4096]
Cost	1	1	1
Sensitivity	1	1	1
Run-time	0.7397	5.4729	302.80
Mutually Exclusive sets	Found	Found	Found
Hill climb			
Selected drugs	[2..192]	[4..768]	[4..72]
Cost	0.6	0.67	0.43
Sensitivity	1	1	1
Run-time	0.7220	1.4354	10.6903

United States and Europe [11]. The genetic characterizations in the form of RNA-seq expressions and functional characterization in the form of responses across 60 drugs with known target inhibition profiles were generated for 13 DIPG tumor cell lines. A TIM model representing the average of these 13 DIPG tumor cell lines were generated. A discretized version of the TIM model including the seven most relevant blocks is shown in Fig. 11.4. Each block represents potential effective target inhibition combinations, for example, inhibiting HSP90 and HSPB (block 2), is predicted be an effective treatment. The scores within each block represent the predicted efficacy of the block in terms of the expected sensitivity following inhibition of that blocks gene targets.

Fig. 11.5 shows the heatmap of the ratio of the gene expression (RNA-seq) levels of the tumor samples to their matched normal or averaged group normal samples. The majority of targets are highly differentiated, even on samples not included in the TIM analysis. This may indicate a dependence on some of these overexpressed pathways for the cancerous samples. For example, drug *cediranib* (PDGFRB is one of its targets) is sensitive for sample VU-DIPG.A which may be due to the overexpression

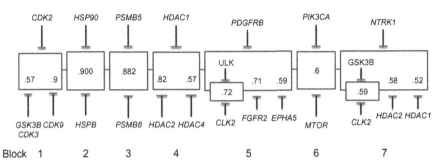

FIG. 11.4

Tumor inhibition circuit for the overall set of DIPG tumor sample drug responses. The circuit was derived from a set of 13 DIPG RNA-seq-matched drug screen responses, and is a model for a general DIPG sample [11].

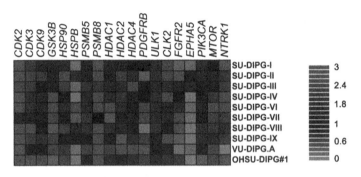

FIG. 11.5

Heatmap of differential expression between tumor and control samples for protein targets identified by the DIPG TIM models [11].

of PDGFRB (a target in the fifth TIM block). On the other hand, SU-DIPG-VI, which is not sensitive to *cediranib*, has limited PDGFRB expression.

We next considered the combination of drugs *BKM120* and *RAD001*. Figs. 11.6–11.8 show the cell viability at different combination concentrations of *BKM120* and *RAD001* for six cell lines. To illustrate the synergy achieved by the predicted drug combinations, we calculated the combination index using *CalcuSyn* software as shown in Fig. 11.9. We note that the combination index is less than 1 for nearly all combinations which validates the synergy that was predicted by the TIM model in Fig. 11.4.

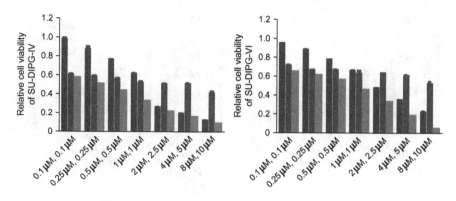

FIG. 11.6

Relative cell viability of SU-DIPG-IV and SU-DIPG-VI cell lines with application of BKM120, RAD001, and combination of BKM120 + RAD001. DIPG cells were seeded in 96-well plates and treated with the indicated two drugs individually or in combination at the indicated concentrations for 72 h in at least triplicate. Cell viabilities were then assessed using Celltiter Glo assay relative to 0.1% DMSO control. Data shown as mean, SD [11].

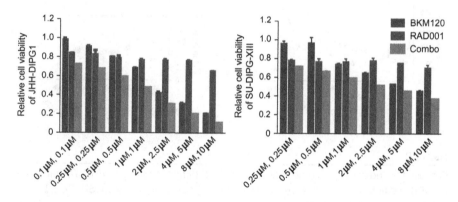

FIG. 11.7

Relative cell viability of JHH-DIPG-I and SU-DIPG-XIII cell lines with application of BKM120, RAD001, and combination of BKM120 + RAD001. DIPG cells were seeded in 96-well plates and treated with the indicated two drugs individually or in combination at the indicated concentrations for 72 h in at least triplicate. Cell viabilities were then assessed using Celltiter Glo assay relative to 0.1% DMSO control. Data shown as mean, SD [11].

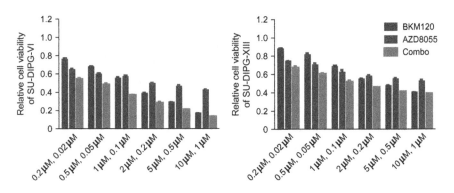

FIG. 11.8

Relative cell viability of SU-DIPG-VI and SU-DIPG-XIII cell lines with application of BKM120, RAD001, and combination of BKM120 + RAD001. DIPG cells were seeded in 96-well plates and treated with the indicated two drugs individually or in combination at the indicated concentrations for 72 h in at least triplicate. Cell viabilities were then assessed using Celltiter Glo assay relative to 0.1% DMSO control. Data shown as mean, SD [11].

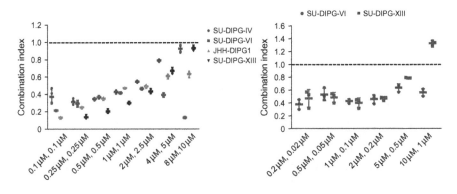

FIG. 11.9

Combination index (CI) for combination of *BKM120* (targets PI3Ki) and *RAD001* (targets mTORi) at different concentrations for six cell lines. Data shown as mean, SD. CI < 1 is considered to be synergistic [11].

11.4 MODEL-FREE COMBINATION THERAPY DESIGN

The primary concern with the combination approach is the enormous increase in the possible candidate concentrations that need to be experimentally tested. One possible solution can be detailed modeling of the cellular system and design of the combination therapy based on analytical optimization and simulation which was considered in the previous section. However, the kind of detailed model that captures the synergy or antagonism of drugs at different levels can require a significant amount of experimentation to infer the model parameters. Furthermore, this kind of approach may only work for molecularly targeted drugs where the specific drug targets are known, but modeling chemotherapeutic drug synergies can often be difficult.

For generation of optimal drug cocktails, systematic empirical approaches tested on in vitro patient tumor cultures are often considered. Some existing approaches include (i) systematic screening of combinations [12–14], which require numerous test combinations; (ii) MACS based on laboratory drug screen for multiple drug combinations guided by sequential search using a fitness function [10], and (iii) deterministic and stochastic optimized search algorithms [15–18].

The systematic search approach has to be focused on locating the global maximum instead of getting stuck in a local maximum. Furthermore, the optimization algorithm needs to be effective in search spaces without existing prior knowledge, and easily adaptable to higher dimensional systems. Because the knowledge of the drug sensitivity distribution is unknown, the algorithm should be effective over a number of unrelated search spaces. A common problem related to the stochastic search algorithms in literature is the normalization issue mentioned in [18]. Some of the search algorithms like Gur Game require proper normalization of the search space without having any prior information about it [16]. In order to overcome this problem, effective algorithms should be adaptive to nonnormalized search spaces. In this section, we present a diversified stochastic search algorithm (termed DSS) that does not require prior normalization of the search space, and can find optimum drug concentrations efficiently. At this point, it is important to emphasize that the primary objective of the algorithm is to minimize the number of experimental steps and not the computational time. The critical problem is the cost of the experiments that are necessary to find the most efficient combination, because the cost of biological experiments is significantly higher than the cost of computation in terms of time and money. As the number of biological experiments necessary to find the global maximum is equal to the number of steps, the objective is to minimize the experimental steps in order to reduce the cost of the overall process. The presented algorithm is able to significantly decrease the number of steps necessary to find the maximum. The iterative algorithm is based on estimating the sensitivity surface and incorporating the response of previous experimental iterations. By generating an estimate of the sensitivity distribution based on currently available response data, we are able to make larger moves in the search space, as compared to smaller steps for gradient-based approaches. The proposed algorithm is composed of two parts: (a) the first part consists of generating a rudimentary knowledge of the search space, (b) in the second part, we utilize the crude knowledge of the sensitivity distribution to run a focused search to locate the exact maximums, or to run a diverse search to gather further information on the sensitivity distribution.

11.4.1 DIVERSIFIED STOCHASTIC SEARCH ALGORITHM

The primary objective of the search algorithm is to locate the global maximum in minimum number of iteration steps. Numerous approaches can be considered for this purpose and the method considered here is based on a combination of stochastic and deterministic approaches. We expect that the efficiency in terms of average number of iteration steps can be increased by large jumps over the search space, rather than using a traditional step-by-step gradient descent approach. The algorithm consists of two parts: an initial parallel part and a subsequent iterative segment. The objective of the initial part is to generate a rudimentary idea of the search space. The objective of the iteration part is twofold: (a) it tries to find the exact maximum using the currently available knowledge and (b) it searches the space further to add new knowledge, that is, attempts to find new hills that the previous iterations could not locate.

Step-by-step schema of the search algorithm is described as follows [19]:

- Step 1: Generation of Latin Hypercube Numbers (LHNs).

  (i) In this step, m points in the given grid are selected for drug response experiments. We first generate m points in the continuous search space based on an LHN generation approach with the criterion of maximizing the minimum distance between these points. This approach assists in distributing the points homogeneously in the search space, such that the maximum possible distance between a given target point and the nearest point whose coordinates are represented by LHN will be minimum. Consequently, we map these points to the nearest grid points and term these mapped points approximate Latin hypercube points. We considered this continuous-discrete grid mapping to compare DSS results with previous studies that utilized a grid structure for the search space.

  (ii) In this step, experiments are conducted to determine the efficiency of the m drug combinations determined by the approximate Latin hypercube points.

- Step 2: Iterative segment

  (i) Normalize the experimental drug efficacy results to numbers between 0 and 1. Then the $(n-1)$th power of the normalized drug efficacies are considered where n denotes the number of drugs. The power step emphasizes the hills of the distribution—and the value $n-1$ is termed *Power used for the inputs.*

  (ii) Estimate the drug efficacies of the unknown grid points using the sensitivity surface estimation algorithm. The details of the surface estimation algorithm are explained in subsequent sections. At the end of this procedure, we have estimates for the efficacies of each and every point on the search grid. The grid points are classified into two groups: known points from experimental data and estimated points based on interpolation and extrapolation.

  (iii) Decide on the objective of the iteration step based on a probabilistic approach. For our case, the algorithm follows *path a* to find the exact maximum based on previous knowledge with a 0.3 probability and follows *path b* with a 0.7 probability to explore the search space with a diversified approach.

(iv) Path a (Focused Search)

   (a) The main idea of the focused search is to experimentally search the estimated maximums generated following the surface estimation mapping. The algorithm also tries to avoid focusing on an individual local maximum by exploring geographically apart multiple estimated local maximums. To achieve this purpose, we employ a tracking algorithm to label the local maximums and avoid prolonged emphasis on individual maximum points. The individual steps of the Focused Search part are described as follows:

   (b) Sort the grid sensitivities (both experimental and estimated sensitivities) from higher to lower sensitivity values.

   (c) Check if the location corresponding to the highest sensitivity is an experimental point or an estimated point. If it is an estimated point, generate the experimental sensitivity for this grid location.

   (d) If the highest sensitivity point is an experimental point, check the second highest point. If the second highest point is an estimated point, generate the experimental value for this grid location.

   (e) If the second highest point is also an experimental point, generate the gradient from the second highest point based on the mapped surface. If the upward path from the second highest point leads up to the highest point, label both the points as 1, which implies that they belong to the same hill. Otherwise, label the highest point as 1 and label the second highest point as 2, which indicates that they belong to different hills.

   (f) Repeat this procedure until an estimated point is located. Meanwhile, keep labeling the experimental points with respect to the hill they belong to and the order of the point on the hill (e.g., 3rd highest point in hill 2, etc.).

   (g) If a hill's highest ξ points are experimental points, then label the hill as discovered, which indicates that we have enough information on this hill and collecting information on other hills might be more beneficial.

   (h) If the search continues until 1% of grid points without finding a suitable candidate for experimentation, halt the search. Locate all the considered points that are inside a sphere of volume $1/500$ of the whole search space with the center being the highest sensitivity point. Assign a value of "0" for the sensitivities of all points inside this sphere (maintain the record of their actual values in another place). Then go to the beginning of step 2.

(v) Path b (Diverse Search)

   (a) The aim of the diverse search is to explore the space to locate new possible candidate hills that were not discovered in the previous searches.

   (b) Assume that the surface generated by the experimental and estimated points is a probability distribution function (PDF).

(c) Generate points by sampling this distribution. For the generation process, we use Gibbs sampling algorithm. The number of points generated by Gibbs algorithm is termed *Number of points to generate Gibbs sampling*. Because the points are generated from sampling the PDF, the points are denser around the hills and less dense at locations where the efficacy estimate is close to 0.

(d) Randomly select one of the generated points as the candidate point and generate its sensitivity experimentally.

Sensitivity surface estimation algorithm

The sensitivity surface estimation algorithm considered here is established on the n-dimensional application of the penalized least square regression analysis based on discrete cosine transform (LS-DTC) proposed by Garcia [20,21]. The code is generated to compute missing values in datasets. The estimation algorithm contains a parameter s, termed *smoothing parameter* that determines the smoothness of the output. For our case, the smoothness parameter is adjusted to a small value so that the result of surface estimation goes through the actual experimental points. The core of the algorithm is based on minimizing the equation: $F(y) = wRSS + s * P(y)$, where wRSS corresponds to weighted residual sum of squares and P is the penalty function related to the smoothness of the output. wRSS can be written explicitly as $\|W^{1/2} \cdot (\hat{y} - y)\|^2$ where y represents the actual data with missing values and $\hat{y}$ provides the estimate of the data without any missing values. W is a diagonal matrix, whose entries represent the reliability of the points and can take values between [0, 1]. For our case, the missing values represent unknown points and are assigned a value of 0 and the experimental points are reliable points which are assigned a value of 1. The solution to $\hat{y}$ that minimizes $F(y)$ can be generated based on an iterative process starting from an arbitrary initial point $\hat{y}_0$.

Choice of parameters

The implementation of the DSS algorithm includes several parameters that can affect the performance of the search process. In this section we present the guiding principles behind the selection of the parameters based on the dimensionality and the total number of grid points in the search space.

Latin hypercube numbers (LHNs): The LHNs denote the number of points that will be tested in step 1 of the algorithm. These points are supposed to provide an initial estimate of the search space. Based on simulations and theoretical analysis, we observe that increasing the number of LHNs provides limited benefit in terms of reaching the maximum sensitivity combination after a certain point. On the other hand keeping this number too low will cause the program to start the second step with limited knowledge and to search low sensitivity locations. Thus there is an optimum number of LHNs to maximize the benefit of the algorithm. Although this optimum number depends on the search space; our simulations for four surfaces with two different LHNs (10 and 40) illustrate that the DSS algorithm provides better results than ARU algorithm [18] for a fairly large interval of LHNs.

Latin hypercube iterations: The LHNs are distributed homogeneously through an iterative algorithm. The iterations maximize the minimum distance between the points. It is desirable to have a higher number of iterations but after a point, the benefits of increasing the iterations become negligible. For our simulations, we selected a threshold point following which the increase in the maximum-minimum distance is negligible.

Number of iterations to generate the sensitivity surface estimate: This parameter is related to a sensitivity surface estimation algorithm and describes the number of iterations used to find a smooth surface passing through the given points in high-dimensional space. A higher value for this parameter will provide a smoothed surface (that still passes through the experimental points), but will carry a high computational time cost. Furthermore, the benefits of increasing the iterations become negligible after a threshold and the output surface become more stable. For our examples, we have fixed this number to 100.

Probability of focused search: Probability of focused search denotes the probability that the search algorithm follows *Path a*. *Path a* attempts to discover the exact local maximum of a hill, and *Path b* attempts to learn new hills. For all our simulations, this parameter has been assigned a value of 0.3.

Power used by inputs: This parameter attempts to emphasize the hills. After normalizing the experimental points, we take the $(n-1)$th power of the values so that the high peaks are emphasized as compared to dips or grid points with average values in the estimated surface. Thus the probability of point selection around hills is increased during the Gibbs sampling process.

Number of points generated by Gibbs sampling: This parameter describes the number of points generated by Gibbs sampling in *Path b* of step 2 of the algorithm. More points provide a better representation of the estimated surface. After a level, the number of points is sufficient to represent the probability distribution and the benefits of increasing the iterations become negligible. We achieve better sampling by increasing this parameter. This parameter is required to be large for problems in higher dimensions or problems containing a huge number of grid points. For our examples, if the number of grid points is below 7500, this parameter has been assigned a value equal to twice the number of grid points. Otherwise, we have fixed the number of Gibbs sampling points to 15,000.

Clustering-related parameters: The clustering concept is introduced to avoid the search being stuck in one dominant hill.

Cluster threshold ξ: This denotes the maximum number of experimental points in an individual hill. Further exploration of the hill is paused once this value is reached. For our examples, if the number of drugs (dimensions) n is less than 5, ξ is assigned a value of $2n-1$. Otherwise, the parameter is fixed at 7.

Cluster break: This parameter denotes the maximum number of high efficacy point estimates in a single hill. If this condition is reached, we assign a value of 0 sensitivity for points around the known top of the hill. This parameter is considered to be around 1% of the total grid points.

Cluster distance: This parameter represents the radius of the sphere around the hill top for which any grid point within the sphere is assigned a value of 0. The cluster distance is selected such that the volume of the sphere is 0.2% of the total volume. The parameter considers that the algorithm has no knowledge of the hills that are narrower than the 0.2% of the total search space.

THEORETICAL ANALYSIS OF THE STOCHASTIC SEARCH ALGORITHM

In this section we will attempt to theoretically analyze the distance of the point with the global sensitivity maximum from the points that are tested by DSS algorithm. We will consider that each drug is discretized from 0 to T levels and that we are considering n drugs. Thus any drug cocktail can be represented by a n length vector $V = \{V(1), V(2), \ldots, V(n)\}$, where $V(i) \in \{0, 1, 2, \ldots, T\}$ for $i \in \{0, 1, \ldots, n\}$. Thus the search space of drug cocktails (denoted by Ω) is of size $(T + 1)^n$ and represents points in an n-dimensional hypercube of length T. Let V_{max} denote the drug cocktail with the maximum sensitivity among the $(T + 1)^n$ points. The mapping from the drug cocktail to sensitivity will be denoted by the function $f: \Omega \to [0\ 1]$, that is, the maximum sensitivity will be given by $f(V_{max})$. We will assume that if the distance of the test point (V) from the point with the global maximum (V_{max}) is small, the sensitivity will be close to the global maximum, that is, a small $|V_{max}V|$ will imply a small $|f(V_{max}) - f(V)|$. We will primarily analyze the L_1 norm of $|V_{max}V|$.

Note that $|V_{max}V|_1 = \sum_{i=1}^{n} |V_{max}(i)V(i)|$. The first m points in our algorithm are chosen randomly in the search space and thus we will consider that $V(i)$ has a uniform distribution between 0 and T. V_{max} can also be situated in any portion of the search space and thus we will consider V_{max} to have a uniform distribution between 0 and T. Thus the probability mass function of the random variable $Z = V(i) - V_{max}(i)$ will be given by

$$f_Z(z) = \begin{cases} \frac{T+1-|z|}{(T+1)^2} & z = \{-T, -T+1, \ldots, T-1, T\} \\ 0 & \text{otherwise} \end{cases} \tag{11.6}$$

Subsequently, the PMF of the random variable $W = |Z|$ will be given by

$$f_W(w) = \begin{cases} \frac{2(T+1-w)}{(T+1)^2} & w = \{0, \ldots, T-1, T\} \\ \frac{1}{T+1} & w = 0 \\ 0 & \text{otherwise} \end{cases} \tag{11.7}$$

The random variable R_1 denoting the L_1 norm $|V_{max}V|_1 = \sum_{i=1}^{n} |V_{max}(i)V(i)|$ will be a sum of n random variables with PMF given by Eq. (11.7). The distribution for the sum of any two random variables consists of the convolution of the individual distributions of the random variables. Thus the probability distribution of R_1 can be calculated by convolving f_W n times. The distribution of R_1 for $T = 10$ and $n = \{1, \ldots, 15\}$ is shown in Fig. 11.10.

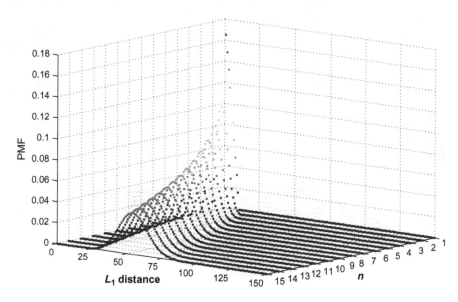

FIG. 11.10

Distribution of random variable R_1 (denoting L_1 distance from optimal point) for $T = 10$ and different values of n [19].

At the beginning of our algorithm, we are selecting m points in random. Thus the nearest neighbor distance from the optimal point will be given by the random variable R_2 that denotes the minimum of m random variables $X_1, X_2, \ldots, X_m$ selected independently based on the distribution of R_1. Thus the cumulative distribution function (CDF) of R_2 is given by Bertsekas and Tsitsikilis [22]

$$P(R_2 \leq x) = 1 - P(X_1 > x, \ldots, X_m > x)$$
$$= 1 - P(X_1 > x) * \cdots * P(X_m > x)$$
$$= 1(1 - CDF_{R_1}(x))^m \qquad (11.8)$$

The PMF of R_2 given by $PMF_{R_2}(x) = CDF_{R_2}(x) - CDF_{R_2}(x - 1)$ for $i = 1, 2, \ldots, nT$ and $PMF_{R_2}(0) = CDF_{R_2}(0)$ is shown in Fig. 11.11.

For example, the expected minimum distance from the optimal point for $m = 40$, $T = 10$ is 6.86 for $n = 5$. The mean $\mu(n, T, m)$ and variance $\sigma(n, T, m)^2$ of the minimum distance from the optimal point for different values of n, T, and m are shown in Table 11.5. Note that if there are k optimal points in diverse locations, the mean $\mu_k(n, T, m)$ and variance $\sigma_k(n, T, m)^2$ of the minimum distance of the selected points from any of the optimal points are given by: $\mu_k(n, T, m) = \mu(n, T, k * m)$ and $\sigma_k(n, T, m)^2 = \sigma(n, T, k * m)^2$. This is because when there are k optimal points, the minimum distance will consist of the minimum of $m \times k$ distances (m distances from each optimal point). If there are multiple optimal points in one hill with small

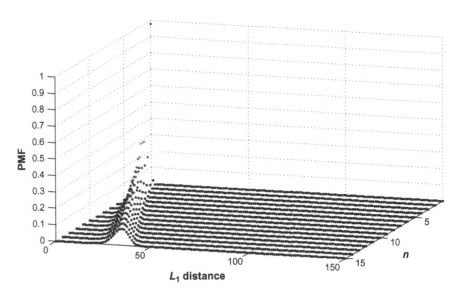

FIG. 11.11

Distribution of random variable R_2 (denoting the minimal L_1 distance from the optimal point for $m = 40$) for $T = 10$ and different values of n [19].

distances between each other, they will be considered as one single optimal point for the minimum distance analysis.

As Table 11.5 shows, the L_1 distance will increase with increasing n and T and following the step 1 of the algorithm, our point with highest experimental sensitivity may not be close to the optimal point, but rather may belong to another hill with a local optima. However, based on the nearest neighbor L_1 distances, we would expect to have at least one point close to the optimal point in the top k optimal points. Thus, if we keep selecting ρ points for further experimentation from around the top k experimental points sequentially, we expect that on an average $\rho_1 = \rho/k$ points will be selected around the optimal point.

Consider that the L_1 distance from the optimal point was given by the random variable R_2 and if a point is selected randomly between the experimental point and the randomly selected point, the subsequent nearest neighbor distance from the optimal point will be given by the random variable $R_3 = R_2 * G_1$ where G_1 is a uniform random variable between 0 and 1. The distance in each dimension will reduce by a number selected based on a uniform random variable and consequently, we will approximate the L_1 distance (sum of n such distances) to reduce by a number selected based on a uniform random variable. Thus, after ρ_1 points have been selected sequentially around the optimal point, the distance to the optimal point will be given by the random variable $R_{\rho_1} = R_2 * G_1 * G_2 * \cdots * G_{\rho_1}$. The PDF of the multiplication of ρ_1 random variables with uniform distribution between [0, 1] is given by Dettmann and Georgiou [23]

Table 11.5 Expectation and Variance of the Minimum Distances From the Optimal Point for Various Values of n, T, and m [19]

n	T	m	Mean	Variance
5	5	20	4.15	1.77
5	5	40	3.42	1.29
5	5	60	3.04	1.09
10	5	20	11.35	4.29
10	5	40	10.20	3.32
10	5	60	9.59	2.90
15	5	20	19.16	6.88
15	5	40	17.69	5.43
15	5	60	16.92	4.79
5	10	20	8.15	5.25
5	10	40	6.86	3.71
5	10	60	6.21	3.06
10	10	20	21.75	13.35
10	10	40	19.70	10.21
10	10	60	18.62	8.83
15	10	20	36.45	21.76
15	10	40	33.83	17.02
15	10	60	32.45	14.94

$$f_{G_1 * \cdots * G_{\rho_1}}(x) = \frac{(\ln(1/x))^{n-1}}{(n-1)!} \tag{11.9}$$

Thus, if the expected distance from the optimal point after the initial selection of m points is D and we select ρ_1 points sequentially between the optimal point and its current nearest neighbor, the expected nearest neighbor distance from the optimal point will be $D/2^{\rho_1}$.

As an example, if $n = 10$, $T = 10$, $m = 40$ we have $D = 19.7$ from Table 11.5. The expected L_1 distance from the optimal point at the end of $40 + 20 = 60$ iterative steps will be $19.7/2^6 = 0.3078$ assuming a single hill and a 0.3 probability for the focused search (*Path a* of step 2 of the algorithm). Based on the focused search probability, at the end of 60 iterations, we expect to have $(60 - 40) * 0.3 = 6$ points selected around the optimal point.

11.4.2 ALGORITHM PERFORMANCE EVALUATION

In this section, we present the performance of DSS algorithm for nine different examples. The number of drugs considered in the examples are 2, 3, 4, and 5 with 21, 11, 11, and 11 discretized concentration levels, respectively, resulting in search space sizes of 21^2, 11^3, 11^4, and 11^5 for the synthetic examples and search grid sizes of 9^2 and 10^2 for the experimentally generated examples with 2 drugs and number of discretization levels of 9 and 10. The results compare DSS with another stochastic search algorithm for drug cocktail generation (termed ARU algorithm) [18] which was shown to outperform earlier approaches [16,17]. Similar to comparisons in [18], two parameters are primarily considered (a) *Cost*: Average number of steps required to reach within 95% of the maximum sensitivity and (b) *Success rate*: percentage of times that the search algorithm reaches 95% of the maximum sensitivity within a fixed number of steps. The performance of DSS approach, as compared to ARU algorithm for the nine examples, is summarized in Table 11.6. The results indicate that DSS achieve 100% success rate for all 9 examples (13 different evaluations) whereas ARU has slightly lower success rate in 2 of these examples. The primary benefit of our approach is the lower average number of iterations to reach within 95% of the maximum. For all the examples considered, DSS require significantly lower average number of iterations to reach within 95% of the maximum. Note that the standard deviation of the number of iterations required to reach within 95% of the maximum is relatively small as compared to the difference in average iterations between ARU and DSS. For instance, the first example in two dimensions requires an average of 15.96 iterations for DSS approach as compared to 46.2 iterations for ARU approach. The ARU algorithm has earlier been shown to outperform other existing algorithms such as Gurgame and its variants. Please refer to Tables 1 and 2 of [18] for the detailed comparison results of ARU with Gurgame. This strongly illustrates that DSS algorithm is able to generate high sensitivity drug combinations in lower number of average iterations than existing approaches.

To further illustrate the significance of DSS approach, let us consider one of the experimental example results. The experimental data on lung cancer contain the sensitivity for $10^2 = 100$ drug concentration combinations where each drug is assumed to have 1 of 10 discrete concentrations. These data have been utilized to study the efficacy of the proposed algorithm. For instance, an exhaustive search approach will experimentally test the sensitivity of each of these 100 concentrations and select the one with the highest sensitivity and thus it will require 100 experimental steps. On the other hand the stochastic search algorithms such as ARU and DSS will start with random drug concentration combinations and try to sequentially select drug concentrations that will provide an improved knowledge of the sensitivity surface over these two drugs. As Table 11.6 shows, ARU will require an average of 12.4 sequential steps to reach a drug combination that has sensitivity within 95% of the maximum sensitivity, whereas the proposed DSS will require an average of 5.97 sequential steps to reach within 95% of the maximum sensitivity. Thus DSS will reach within 95% of the maximum sensitivity on an average of 5.97 experimental steps, whereas ARU will require 12.4 experimental steps and exhaustive search will require 100 experimental steps. Note that since ARU and DSS are stochastic approaches, the number of sequential steps required can vary with each experimental run and the numbers 12.4 and 5.97 represent the mean of multiple experimental runs.

Table 11.6 Summary of the Results for Seven Synthetic and Two Experiment-Based Examples

	Number of Points With $\geq 0.95 \times$ Max$_{efficacy}$	Search Grid Size	ARU [18] Cost	ARU [18] Success Rate (%)	DSS Cost	DSS STD	DSS Worst Case	DSS Success Rate (%)	Initial LHC Points
Synthetic examples									
2 DeJong	2	21^2	46.2	99	16.0	7.99	48	100	5
3a	4	11^3	74	100	24.7	11.73	72	100	10
3b	1	11^3	79.4	100	52.9	32.20	149	100	10
4a	1	11^4	136.8	100	65.3	14.11	106	100	40
4a	1	11^4	136.8	100	50.7	21.81	159	100	10
4b	12	11^4	91.6	100	52.7	8.80	85	100	40
4b	12	11^4	91.6	100	28.3	9.17	57	100	10
5a	4	11^5	80.6	100	79.3	23.25	157	100	40
5a	4	11^5	80.6	100	61.8	27.58	176	100	10
5b	8	11^5	216.8	100	159.5	90.51	402	100	40
5b	8	11^5	216.8	100	194.2	150.15	647	100	10
Experiment-based examples									
Bacterial inhibition [14]	34	9^2	4.8	100	1.85	0.78	3	100	3
Lung cancer [12]	7	10^2	12.4	98	5.97	4.74	23	100	3

Notes: The cost of DSS algorithm is significantly lower than ARU algorithm [18] which has been shown to be efficient than other existing algorithms. Furthermore, the success rate of DSS algorithm is better than the ARU algorithm [19].

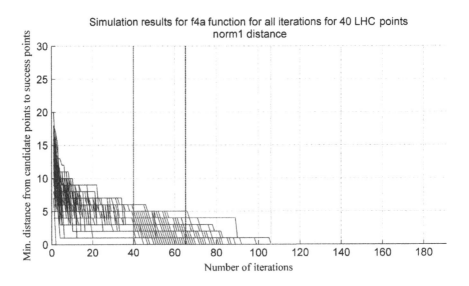

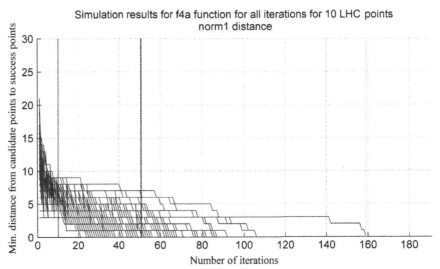

FIG. 11.12

Minimum distance to optimal points for function f4a. (The surface is described by the four-dimensional response function $z = x1 * \exp(-(x1^2 + x2^2 + x3^2 + x4^2))$.) We simulate the f4a function 100 times with two different LHC numbers. The *first figure* represents the simulation with LHC equal to 40 and the *second figure* represents the simulation LHC equal to 10. The analyzed function has 1 optimal point. The *lines* represent the minimum norm 1 distance between the optimal points and DSS selected points. The *left vertical-dotted line* represents the end of step 1, that is, Latin Hypercube Numbers, which is equal to 40th iteration in first simulation set and 10th iteration in the second simulation set; and the *middle vertical-dotted line* represents the average value of iterations (cost of proposed algorithm) required to find one of the points with $\geq 0.95 \times$ Max$_{efficacy}$ (=79.25). The *rightmost vertical-dotted line* represents the worst situation out of 100 different runs [19].

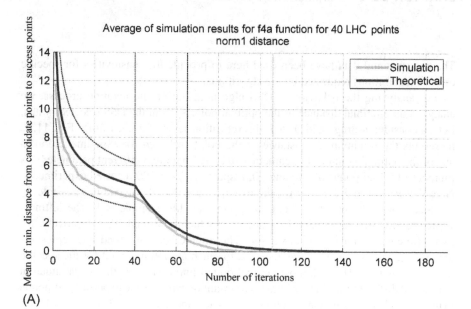

(A)

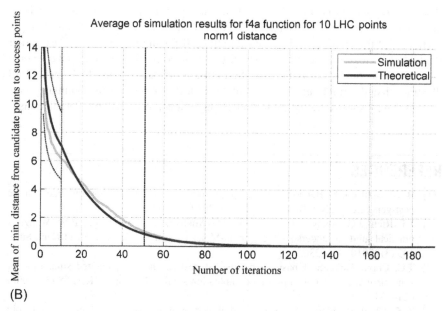

(B)

FIG. 11.13

Simulation average and theoretical expected minimum distance to optimal points for function f4a. (The surface is described by the four-dimensional response function $z = x1 * \exp(-(x1^2 + x2^2 + x3^2 + x4^2))$).) The *dark solid line* represents the theoretical values for L_1 distance and the *dashed line* represents the error margins ($\mu \pm \sigma$) for the analytically calculated values for step 1 of the iteration. (A) The simulation with LHC equal to 40 and (B) the simulation LHC equal to 10. The *left vertical-dotted line* represents the end of step 1, that is, Latin Hypercube Numbers, which is equal to 40th iteration; and the *middle vertical-dotted line* represents the average value of iterations (cost of proposed algorithm) required to find one of the points with $\geq 0.95\times$ Max$_{\text{efficacy}}$. The *rightmost vertical line* represents the worst situation out of 100 different runs [19].

The experimental data have been used here to provide the sensitivities for specific drug concentrations requested by the algorithms.

For analyzing the behavior of DSS algorithm during the iteration process, we analyzed the minimal distance of the optimal point(s) from the DSS selected points. Let us consider n drugs and 0 to T discretization levels for each drug. Fig. 11.12 represents the minimum L_1 distance of the points selected for experimentation to any of the optimal point(s) for synthetic example 4 with two different parameter sets (number of initial points are 40 and 10, respectively) for 100 repeated experiments. Note that $n = 4$ and $T = 10$ for the example and thus the maximum possible L_1 distance is 40. The number of optimal points for this example is 1. The leftmost vertical-dotted line represents the value of m which is 40 and 10, respectively, for two different solutions of this example. The middle vertical-dotted line represents the average number of iterations required to reach an optimal point for the specific response function. The rightmost vertical-dotted line represents the worst situation out of 100 different runs. The average minimum distance of the experimental points to the optimal point is shown as solid lighter shade line in Fig. 11.13. The solid darker shade line represents the analytically calculated expected minimum L_1 distance. Note that there is a change in the shape of the darker shade curve after the end of step 1 (iteration 40 for the Fig. 11.13A and iteration 10 for the Fig. 11.13B). The darker shade dotted curve denotes the analytically calculated $\mu \pm \sigma$ where μ and σ denote the mean and standard deviation for the minimum distance. Figs. 11.12 and 11.13 illustrate that the minimum distance of the selected points to the optimal points decreases with successive iterations and closely matches the analytical predictions.

REFERENCES

[1] B. Al-Lazikani, U. Banerji, P. Workman, Combinatorial drug therapy for cancer in the post-genomic era, Nat. Biotechnol. 30 (7) (2012) 679–692.

[2] J.T. Rich, J.G. Neely, R.C. Paniello, C.C. Voelker, B. Nussenbaum, E.W. Wang, A practical guide to understanding Kaplan-Meier curves, Otolaryngol. Head Neck Surg. 143 (3) (2010) 331–336.

[3] T.C. Chou, Theoretical basis, experimental design, and computerized simulation of synergism and antagonism in drug combination studies, Pharmacol. Rev. 58 (3) (2006) 621–681.

[4] T.-C. Chou, P. Talalay, Analysis of combined drug effects: a new look at a very old problem, Trends Pharmacol. Sci. 4 (1983) 450–454.

[5] S. Haider, N. Berlow, R. Pal, L. Davis, C. Keller, Combination therapy design for targeted therapeutics from a drug-protein interaction perspective, in: 2012 IEEE International Workshop on Genomic Signal Processing and Statistics (GENSIPS), 2012, pp. 58–61, http://dx.doi.org/10.1109/GENSIPS.2012.6507726.

[6] M.W. Karaman, S. Herrgard, D.K. Treiber, P. Gallant, C.E. Atteridge, B.T. Campbell, K.W. Chan, P. Ciceri, M.I. Davis, P.T. Edeen, R. Faraoni, M. Floyd, J.P. Hunt, D.J. Lockhart, Z.V. Milanov, M.J. Morrison, G. Pallares, H.K. Patel, S. Pritchard, L.M. Wodicka, P.P. Zarrinkar, A quantitative analysis of kinase inhibitor selectivity, Nat. Biotechnol. 26 (1) (2008) 127–132.

[7] P.P. Zarrinkar, R.N. Gunawardane, M.D. Cramer, M.F. Gardner, D. Brigham, B. Belli, M.W. Karaman, K.W. Pratz, G. Pallares, Q. Chao, K.G. Sprankle, H.K. Patel, M. Levis, R.C. Armstrong, J. James, S.S. Bhagwat, AC220 is a uniquely potent and selective inhibitor of FLT3 for the treatment of acute myeloid leukemia (AML), Blood 114 (14) (2009) 2984–2992.

[8] Library containing dissociation constants of drugs, http://pubchem.ncbi.nlm.nih.gov/.

[9] V. Vazirani, Approximation Algorithms, Springer, Berlin Heidelberg, 2004.

[10] R.G. Zinner, B.L. Barrett, E. Popova, P. Damien, A.Y. Volgin, et al., Algorithmic guided screening of drug combinations of arbitrary size for activity against cancer cells, Mol. Cancer Ther. 8 (2009) 521–532.

[11] C.S. Grasso, Y. Tang, N. Truffaux, N.E. Berlow, L. Liu, M. Debily, M.J. Quist, L.E. Davis, E.C. Huang, P.J. Woo, A. Ponnuswami, S. Chen, T. Johung, W. Sun, M. Kogiso, Y. Du, Q. Lin, Y. Huang, M. Hutt-Cabezas, K.E. Warren, L.L. Dret, P.S. Meltzer, H. Mao, M. Quezado, D.G. van Vuurden, J. Abraham, M. Fouladi, M.N. Svalina, N. Wang, C. Hawkins, J. Nazarian, M.M. Alonso, E. Raabe, E. Hulleman, P.T. Spellman, X. Li, C. Keller, R. Pal, J. Grill, M. Monje, Functionally-defined therapeutic targets in diffuse intrinsic pontine glioma, Nat. Med. 21 (2015) 555–559.

[12] A.A. Borisy, P.J. Elliott, N.W. Hurst, M.S. Lee, J. Lehár, E.R. Price, G. Serbedzija, G.R. Zimmermann, M.A. Foley, B.R. Stockwell, C.T. Keith, Systematic discovery of multicomponent therapeutics, Proc. Natl. Acad. Sci. U. S. A. 100 (13) (2003) 7977–7982.

[13] M. Wadman, The right combination, Nature 439 (2006) 390–401.

[14] G.R. Zimmermann, J. Lehár, C.T. Keith, Multi-target therapeutics: when the whole is greater than the sum of the parts, Drug Discov. Today 12 (1–2) (2007) 34–42.

[15] D. Calzolari, S. Bruschi, L. Coquin, J. Schofield, J.D. Feala, J.C. Reed, A.D. McCulloch, G. Paternostro, Search algorithms as a framework for the optimization of drug combinations, PLoS Comput. Biol. 4 (12) (2008) e1000249.

[16] P.K. Wong, F. Yu, A. Shahangian, G. Cheng, R. Sun, C.-M. Ho, Closed-loop control of cellular functions using combinatory drugs guided by a stochastic search algorithm, Proc. Natl. Acad. Sci. U. S. A. 105 (13) (2008) 5105–5110.

[17] B.J. Yoon, Enhanced stochastic optimization algorithm for finding effective multi-target therapeutics, BMC Bioinform. 12 (Suppl. 1) (2011) S18.

[18] M. Kim, B.-J. Yoon, Adaptive reference update (ARU) algorithm. A stochastic search algorithm for efficient optimization of multi-drug cocktails, BMC Genomics 13 (6) (2012) 1–15.

[19] M.U. Caglar, R. Pal, A diverse stochastic search algorithm for combination therapeutics, BioMed Res. Int. (2014) 9.

[20] D. Garcia, Robust smoothing of gridded data in one and higher dimensions with missing values, Comput. Stat. Data Anal. 54 (4) (2010) 1167–1178.

[21] G. Wang, D. Garcia, Y. Liu, R. de Jeu, A.J. Dolman, A three-dimensional gap filling method for large geophysical datasets: application to global satellite soil moisture observations, Environ. Model. Softw. 30 (2012) 139–142.

[22] D.P. Bertsekas, J.N. Tsitsiklis, Introduction to Probability, second ed., Athena Scientific, Belmont, MA, USA, 2008.

[23] C.P. Dettmann, O. Georgiou, Product of n independent uniform random variables, Stat. Prob. Lett. 79 (2009) 2501–2503.

Online resources

CHAPTER OUTLINE

The goal of this chapter is to provide a collection of commonly used online resources for drug sensitivity prediction modeling and analysis. Resources for various steps of drug sensitivity modeling and application, such as online characterization databases, predictive modeling tools, drug synergy estimation, prediction challenges, and regulatory information, are presented next.

12.1 PATHWAY DATABASES

KEGG: KEGG (http://www.genome.jp/kegg/) is a database resource for understanding high-level functions and utilities of the biological system, such as the cell, the organism, and the ecosystem, from molecular-level information, especially large-scale molecular datasets generated by genome sequencing and other high-throughput experimental technologies. KEGG also provides commonly known pathways for various types of cancers [1].

Predictive Modeling of Drug Sensitivity. http://dx.doi.org/10.1016/B978-0-12-805274-7.00012-9

STRING Db: STRING (http://string-db.org/) is a database of known and predicted protein interactions. The interactions include direct (physical) and indirect (functional) associations derived from (a) text mining, (b) experiments, (c) databases, (d) coexpression, (e) neighborhood, (f) gene fusion, and (g) cooccurrence. STRING also provides a visual network representation of the interactions [2]. The STRING database can be used for biological network enrichment analysis of inferred genomic features, as was considered in [3].

GO: Gene Ontonology (geneontology.org) defines concepts/classes used to describe gene function, and relationships between these concepts. The database can be used to conduct enrichment analysis and decipher molecular activities of gene products and their involvement in biological processes.

GeneCards: GeneCards (www.genecards.org) is a searchable, integrative database that provides comprehensive, user-friendly information on all annotated and predicted human genes. It automatically integrates gene-centric data from around 125 web sources, including genomic, transcriptomic, proteomic, genetic, clinical, and functional information.

IntAct: IntAct Molecular Interaction Database (www.ebi.ac.uk/intact/) provides a freely available, open source database system and analysis tools for molecular interaction data. All interactions are derived from literature curation or direct user submissions and are freely available.

12.2 DRUG-PROTEIN INTERACTION AND PROTEIN STRUCTURE DATABASES

Binding DB: BindingDB (https://www.bindingdb.org/) is a public, web-accessible database of measured binding affinities, focusing chiefly on the interaction of proteins considered to be drug targets with small, drug-like molecules. BindingDB contains binding data, for protein targets and small molecules [4].

STITCH: chemical proteins interactions: STITCH (http://stitch.embl.de/) is a resource to explore known and predicted interactions of chemicals and proteins. Chemicals are linked to other chemicals and proteins by evidence derived from experiments, databases, and literature [5].

ChEMBL: ChEMBL (https://www.ebi.ac.uk/chembl/) is a database containing binding, functional, and ADMET information for a large number of drug-like bioactive compounds. These data are manually extracted from published literature and further curated to standardize and increase quality [6].

Drug Bank: Drug Bank (http://www.drugbank.ca/) is an open source database that contains a collection of chemical, pharmacological, pharmaceutical, and drug target data for a large collection of FDA approved and experimental drugs [7,8].

PubChem: PubChem (https://pubchem.ncbi.nlm.nih.gov/) is a database of chemical molecules and their activities against biological assays maintained by National

Center for Biotechnology Information (NCBI). PubChem can be utilized to extract drug-target inhibition profiles for various targeted drugs.

UniProt: Database (http://www.uniprot.org/) with a collection of functional information on proteins.

Protein Data Bank (PDB): Database (http://www.rcsb.org/pdb/home/home.do) containing 3D shapes of proteins, nucleic acids, and complex assemblies [9].

eMolecules: eMolecules (https://www.emolecules.com/info/plus/download-database) is a commercial database containing structural information based on simplified molecular-input line-entry system (SMILES) entries. **SMILES** can act as unique identifier in ASCII form for a specific chemical structure.

12.3 DRUG SENSITIVITY, GENETIC CHARACTERIZATION, AND FUNCTIONAL DATABASES

CCLE: The Cancer Cell Line Encyclopedia (CCLE) (http://www.broadinstitute.org/ccle/home) project is a collaboration between the Broad Institute, and the Novartis Institutes for Biomedical Research and its Genomics Institute of the Novartis Research Foundation to conduct a detailed genetic and pharmacologic characterization of a large panel of human cancer models, to develop integrated computational analyses that link distinct pharmacologic vulnerabilities to genomic patterns and to translate cell line integrative genomics into cancer patient stratification. The CCLE provides public access to genomic data, analysis, and visualization for about 1000 cell lines [10].

GDSC: The Genomics of Drug Sensitivity in Cancer (GDSC) database (http://www.cancerRxgene.org) is a publicly available resource for information on drug sensitivity in cancer cells and molecular markers of drug response. Data are freely available without restriction. GDSC currently contains drug sensitivity data for almost 75,000 experiments, describing response to 138 anticancer drugs across almost 700 cancer cell lines [11].

NCI 60: The NCI-60 Human Tumor Cell Lines Screen (https://dtp.cancer.gov/discovery_development/nci-60/) has been in existence for more than two decades. The screen was implemented in fully operational form in 1990 and utilizes 60 different human tumor cell lines to identify and characterize novel compounds with growth inhibition or killing of tumor cell lines. It is designed to screen up to 3000 small molecules (synthetic or purified natural products) per year for potential anticancer activity. The operation of this screen utilizes 60 different human tumor cell lines, representing leukemia, melanoma, and cancers of the lung, colon, brain, ovary, breast, prostate, and kidney.

Drug Sensitivity: https://wiki.nci.nih.gov/display/NCIDTPdata/NCI-60+Growth+Inhibition+Data.

Chemical Information: https://wiki.nci.nih.gov/display/NCIDTPdata/Chemical+Data.

Genetic Characterization Data including protein, mRNA, miRNA, DNA methylation, mutations, SNPs, enzyme activity, and metabolites (https://wiki.nci.nih.gov/display/NCIDTPdata/Molecular+Target+Data).

National Cancer Institute's Cancer Target Discovery and Development (CTD2) Network: Open-Access Data Portal that makes available raw data downloads from member centers, including all raw sensitivity and enrichment data and other supporting information related to cancer genetic and functional studies (https://ctd2.nci.nih.gov/dataPortal/).

Project Achilles: The project (http://www.broadinstitute.org/Achilles) uses genome-wide genetic perturbation reagents (shRNAs or Cas9/sgRNAs) to silence or knock-out individual genes and identify those genes that affect cell survival. Large-scale functional screening of cancer cell lines provides a complementary approach to those studies that aims to characterize the molecular alterations (e.g., mutations, copy number alterations) of primary tumors [12].

The Cancer Genome Atlas (TCGA): TCGA (http://cancergenome.nih.gov/) is a project funded by NIH that considers more than 500 cancer patient samples and generates multiple forms of genetic characterization data, including gene expression, copy number variation, single nucleotide polymorphism gentyping, DNA methylation profiling, microRNA profiling, and exon sequencing.

NIH LINCS: The Library of Integrated Network-based Cellular Signatures (LINCS) (http://www.lincsproject.org/) aims to create a network-based understanding of biology by cataloging changes in gene expression and other cellular processes that occur when cells are exposed to a variety of perturbing agents, and by using computational tools to integrate this diverse information into a comprehensive view of normal and disease states that can be applied for the development of new biomarkers and therapeutics.

12.4 DRUG TOXICITY

SIDER: Database (http://sideeffects.embl.de/) containing approved drugs and their recorded side effects [13].

Comparative Toxicogenomics Database (CTD): CTD (http://ctdbase.org/) is a robust, publicly available database that aims to advance understanding about how environmental exposures affect human health. It provides manually curated information about chemical-gene/protein interactions, chemical-disease, and gene-disease relationships. These data are integrated with functional and pathway data to aid in the development of hypotheses about the mechanisms underlying environmentally influenced diseases [14].

12.5 MISSING VALUE ESTIMATION

- *Matlab K-Nearest Neighbor-Based Missing Value Imputation*: Function *knnimpute* to impute missing data using nearest-neighbor method with various

options for the distance metric including Euclidean, Manhattan, Mahalanobis, and Hamming distance, or using similarity measure such as correlation (http://www.mathworks.com/help/bioinfo/ref/knnimpute.html).
- *Bayesian principal component analysis*-based estimation of missing values for gene expression profiles [15] (http://hawaii.aist-nara.ac.jp/~shige-o/tools/).

12.6 REGRESSION TOOLS

LASSO

- Least Angle Regression, Lasso, and Forward Stagewise (*lars* package) [16] in **R**
- *lasso* function in Matlab Statistics and Machine Learning Toolbox
- *L1Packv2* package in Mathematica [17]

Elastic Net regression

- Matlab *lasso* function
- *elasticnet* package in R [18]

Principal Component Regression and Partial Least Squares Regression

- *pls* package in **R**: Partial Least Squares and Principal Component Regression [19]
- *plsregress* and *pca* functions in Matlab Statistics and Machine Learning Toolbox

Support Vector Machine Regression

- Matlab provides a set of functions to train an SV regression model. They are included in *Statistics and Machine Learning Toolbox → Regression → Support Vector Machine Regression. Fitrsvm* function supports Linear, Gaussian, RBF, and Polynomial Kernel functions. The optimization can be solved using one of three algorithms: (a) sequential minimal optimization [20], (b) iterative single data algorithm [21], and (c) L1 soft-margin minimization by quadratic programming [22].
- *IBM ILOG CPLEX Optimization Studio*: The CPLEX studio provides state-of-the-art linear and quadratic programming solver that can be used to solve the optimization problem in SV regression.
- *kernlab* [23] contains multiple kernel-based algorithms in **R** including SV implementations.
- *libsvm* implementation in **C++** [24] (https://www.csie.ntu.edu.tw/~cjlin/libsvm/).

Adaboost and Bagging

- *adabag* package in R [25]
- *fitensemble* functions in Matlab Statistics and Machine Learning Toolbox

Random Forest

- **R** implementation *randomForest* [26]

- Matlab implementations (http://www.mathworks.com/matlabcentral/ fileexchange/31036-random-forest, https://code.google.com/archive/p/ randomforest-matlab/, https://github.com/sahaider/copulamrf)
- **C++** implementation (https://github.com/bjoern-andres/random-forest)

Probabilistic Random Forest

- Matlab implementation (https://github.com/razrahman/PRF_codes)

Multivariate Random Forest

- Matlab implementation (https://github.com/sahaider/copulamrf)
- **R** implementation *IntegratedMRF* [27]

12.7 TARGET INHIBITION MAPS

- Matlab implementation (http://www.myweb.ttu.edu/rpal/Softwares/PTIM1.zip). User guide (http://www.myweb.ttu.edu/rpal/Softwares/PTIMUserGuide.pdf).

12.8 ESTIMATING DRUG COMBINATION SYNERGY

Calcusyn: CalcuSyn (http://www.biosoft.com/w/calcusyn.htm) is a commercially available software for analyzing combined drug effects to quantify phenomena, such as synergism and antagonism using Median Effects Method designed by Chou and Talalay [28].

Synergy: Synergy (https://biostatistics.mdanderson.org/softwaredownload/SingleSoftware.aspx?Software_Id=18) can be used to determine the interaction of two drugs as synergy, additivity, or antagonism. The software coded in S-PLUS/R is described in [29].

12.9 SURVIVAL ANALYSIS

Kaplan-Meier Estimator

- **R** implementation *survival: Survival Analysis Package* [30]
- Matlab implementation *ecdf function* in Statistics and Machine Learning Toolbox
- **Phython** implementation: *Lifelines* package in Python

12.10 PREDICTION CHALLENGES

DREAM Challenges: The Dialogue for Reverse Engineering Assessments and Methods (DREAM) (http://dreamchallenges.org/) provides challenges based on fundamental questions about systems biology and translational medicine where multiple individuals and groups work on the same challenges with the hope of using the *wisdom of the crowd* to advance science.

Kaggle: Kaggle (https://www.kaggle.com/) run data science challenges in which people compete to solve problems posed by industry sponsors, and in exchange for sizeable cash prizes, the solutions to the challenges are owned by the sponsoring company.

12.11 REGULATORY INFORMATION

MedDRA: http://www.meddra.org/. Standardized medical dictionary for regulatory activities of approved drugs.

A to Z List of Cancer Drugs: http://www.cancer.gov/about-cancer/treatment/drugs. This list includes more than 200 cancer drug information summaries from NCI. The summaries provide consumer-friendly information about cancer drugs and drug combinations.

Bioresearch Monitoring Information System (BMIS): BMIS (http://www.accessdata.fda.gov/scripts/cder/bmis/) contains information that identifies clinical investigators, contract research organizations, and institutional review boards involved in the conduct of Investigational New Drug (IND) studies with human investigational drugs and therapeutic biologics.

REFERENCES

[1] M. Kanehisa, S. Goto, Y. Sato, M. Kawashima, M. Furumichi, M. Tanabe, Data, information, knowledge and principle: back to metabolism in KEGG, Nucleic Acids Res. 42 (Database issue) (2014) 199–205.

[2] D. Szklarczyk, A. Franceschini, S. Wyder, K. Forslund, D. Heller, J. Huerta-Cepas, M. Simonovic, A. Roth, A. Santos, K.P. Tsafou, M. Kuhn, P. Bork, L.J. Jensen, C. von Mering, STRING v10: protein-protein interaction networks, integrated over the tree of life, Nucleic Acids Res. 43 (Database issue) (2015) D447–D452.

[3] S. Haider, R. Rahman, S. Ghosh, R. Pal, A copula based approach for design of multivariate random forests for drug sensitivity prediction, PLoS ONE 10 (12) (2015) e0144490.

[4] T. Liu, Y. Lin, X. Wen, R.N. Jorissen, M.K. Gilson, BindingDB: a web-accessible database of experimentally determined protein-ligand binding affinities, Nucleic Acids Res. 35 (Database issue) (2007) 198–201.

[5] M. Kuhn, D. Szklarczyk, S. Pletscher-Frankild, T.H. Blicher, C. von Mering, L.J. Jensen, P. Bork, STITCH 4: integration of protein-chemical interactions with user data, Nucleic Acids Res. 42 (Database issue) (2014) D401–D407.

[6] A. Gaulton, L.J. Bellis, A.P. Bento, J. Chambers, M. Davies, A. Hersey, Y. Light, S. McGlinchey, D. Michalovich, B. Al-Lazikani, J.P. Overington, ChEMBL: a large-scale bioactivity database for drug discovery, Nucleic Acids Res. 40 (Database issue) (2012) D1100–D1107.

[7] V. Law, C. Knox, Y. Djoumbou, T. Jewison, A.C. Guo, Y. Liu, A. Maciejewski, D. Arndt, M. Wilson, V. Neveu, A. Tang, G. Gabriel, C. Ly, S. Adamjee, Z.T. Dame, B. Han, Y. Zhou, D.S. Wishart, DrugBank 4.0: shedding new light on drug metabolism, Nucleic Acids Res. 42 (Database issue) (2014) D1091–D1097.

[8] D.S. Wishart, C. Knox, A.C. Guo, S. Shrivastava, M. Hassanali, P. Stothard, Z. Chang, J. Woolsey, DrugBank: a comprehensive resource for in silico drug discovery and exploration, Nucleic Acids Res. 34 (Database issue) (2006) D668–D672.

[9] H.M. Berman, J. Westbrook, Z. Feng, G. Gilliland, T.N. Bhat, H. Weissig, I.N. Shindyalov, P.E. Bourne, The protein data bank, Nucleic Acids Res. 28 (1) (2000) 235–242.

[10] J. Barretina, et al., The cancer cell line encyclopedia enables predictive modelling of anticancer drug sensitivity, Nature 483 (7391) (2012) 603–607.

[11] W. Yang, et al., Genomics of Drug Sensitivity in Cancer (GDSC): a resource for therapeutic biomarker discovery in cancer cells, Nucleic Acids Res. 41 (D1) (2013) D955–D961.

[12] G.S. Cowley, B.A. Weir, F. Vazquez, P. Tamayo, J.A. Scott, S. Rusin, A. East-Seletsky, L.D. Ali, W.F. Gerath, S.E. Pantel, P.H. Lizotte, G. Jiang, J. Hsiao, A. Tsherniak, E. Dwinell, S. Aoyama, M. Okamoto, W. Harrington, E. Gelfand, T.M. Green, M.J. Tomko, S. Gopal, T.C. Wong, H. Li, S. Howell, N. Stransky, T. Liefeld, D. Jang, J. Bistline, B.H. Meyers, S.A. Armstrong, K.C. Anderson, K. Stegmaier, M. Reich, D. Pellman, J.S. Boehm, J.P. Mesirov, T.R. Golub, D.E. Root, W.C. Hahn, Parallel genome-scale loss of function screens in 216 cancer cell lines for the identification of context-specific genetic dependencies, Sci. Data 1 (2014) 140035.

[13] M. Kuhn, M. Campillos, I. Letunic, L.J. Jensen, P. Bork, A side effect resource to capture phenotypic effects of drugs, Mol. Syst. Biol. 6 (2010) 343.

[14] A.P. Davis, C.J. Grondin, K. Lennon-Hopkins, C. Saraceni-Richards, D. Sciaky, B.L. King, T.C. Wiegers, C.J. Mattingly, The Comparative Toxicogenomics Database's 10th year anniversary: update 2015, Nucleic Acids Res. 43 (Database issue) (2015) D914–D920.

[15] S. Oba, M.A. Sato, I. Takemasa, M. Monden, K. Matsubara, S. Ishii, A Bayesian missing value estimation method for gene expression profile data, Bioinformatics 19 (16) (2003) 2088–2096.

[16] T. Hastie, B. Efron, lars: Least Angle Regression, Lasso and Forward Stagewise, R package version 1.2, 2013.

[17] I. Loris, L1Packv2: a Mathematica package for minimizing an ℓ-penalized functional, Comput. Phys. Commun. 179 (12) (2008) 895–902.

[18] H. Zou, T. Hastie, elasticnet: Elastic-Net for Sparse Estimation and Sparse PCA, R package version 1.1, 2012.

[19] R. Wehrens, B.-H. Mevik, K.H. Liland, pls: Partial Least Squares and Principal Component Regression, R package version 2.5, 2015.

[20] J.C. Platt, Fast training of support vector machines using sequential minimal optimization, in: Advances in Kernel Methods—Support Vector Learning, MIT Press, Cambridge, MA, 1998.

[21] V. Kecman, T.-M. Huang, M. Vogt, Iterative single data algorithm for training kernel machines from huge data sets: theory and performance, in: L. Wang (Ed.), Support Vector Machines: Theory and Applications, Studies in Fuzziness and Soft Computing, vol. 177, Springer, Berlin, Heidelberg, 2005, pp. 255–274.

[22] B. Scholkopf, A.J. Smola, Learning with Kernels: Support Vector Machines, Regularization, Optimization, and Beyond, MIT Press, Cambridge, MA, 2001.

[23] A. Karatzoglou, A. Smola, K. Hornik, A. Zeileis, kernlab—an S4 package for kernel methods in R, J. Stat. Softw. 11 (9) (2004) 1–20.

[24] C.-C. Chang, C.-J. Lin, LIBSVM: a library for support vector machines, ACM Trans. Intell. Syst. Technol. 2 (2011) 27:1–27:27, software available at http://www.csie.ntu.edu.tw/~cjlin/libsvm.

[25] E. Alfaro, M. Gamez, N. Garcia, L. Guo, adabag: Applies Multiclass AdaBoost.M1, SAMME and Bagging, R package version 4.1, 2015.

[26] Fortran original by L. Breiman, A. Cutler, R port by A. Liaw, M. Wiener, randomForest: Breiman and Cutler's Random Forests for Classification and Regression, R package version 4.6-12, 2015.

[27] R. Rahman, R. Pal, IntegratedMRF: Integrated Prediction using Univariate and Multivariate Random Forests, R package version 1.0, 2016.

[28] T.-C. Chou, P. Talalay, Analysis of combined drug effects: a new look at a very old problem, Trends Pharmacol. Sci. 4 (1983) 450–454.

[29] J.J. Lee, M. Kong, G.D. Ayers, R. Lotan, Interaction index and different methods for determining drug interaction in combination therapy, J. Biopharm. Stat. 17 (3) (2007) 461–480.

[30] T.M. Therneau, T. Lumley, survival: Survival Analysis, R package version 2.38-3, 2015.

CHAPTER

Challenges

13

CHAPTER OUTLINE

13.1 IMPEDIMENTS TO THE DESIGN OF PREDICTIVE MODELS FOR PERSONALIZED MEDICINE

This chapter discusses several existing challenges that need to be addressed before personalized medicine becomes the standard of care in clinical oncology. We consider several research challenges, such as incorporating tumor heterogeneity, addressing data inconsistencies, improving accuracy of predictive models, and individualized toxicity prediction. We also discuss the collaborative constraints in this research area along with the ethical considerations that require attention.

13.2 TUMOR HETEROGENEITY

Tumor heterogeneity refers to the possibility of population of tumor cells showing differences in morphological and phenotypic profiles, such as differences in gene expression, metabolism, angiogenic, and proliferative potentials [1–4]. The predictive modeling approaches considered in this book primarily targeted *intertumor*

Predictive Modeling of Drug Sensitivity. http://dx.doi.org/10.1016/B978-0-12-805274-7.00013-0

325

heterogeneity (between tumors) as observed through dissimilarities in genomic and functional characterizations. Due to their distinct therapeutic responses, the *intratumor heterogeneity* (within tumors) can hinder personalized medicine strategies that depend on results from single biopsy samples [5]. A potential approach to tackle the problem of intratumor heterogeneity is to consider multiple tumor biopsies. A 2012 study [5] profiled multiple spatially separated samples obtained from primary renal carcinomas and associated metastatic sites and observed divergent behavior in terms of mutations and RNA expressions. The capability to measure genomic and functional profiles of multiple spatially separated biopsies (with approaches such as fine needle biopsy) can allow us to design therapeutic strategies that are optimal for multiple samples. The optimality can be considered from the perspective of increasing the average efficacy over all spatially separated samples, or maximizing the sensitivity of the worst case sample. We can also consider it from a multiobjective problem perspective where Pareto frontiers for different therapeutic options can be predicted and final decisions can be selected from the Pareto front based on average or worst case objective.

13.3 DATA INCONSISTENCIES

Cell line studies have played a prominent role in analyzing tumor pathways and response to various compounds irrespective of their limitations as models of patient tumors [6]. One of the techniques to evaluate the efficacy of cell line models consists of associating the genomic characterizations of commonly used cancer cell lines with genomic profiles of primary tumor samples. For instance, Domcke et al. [7] compared the copy-number changes, mutation, and mRNA expressions of commonly used ovarian cancer cell lines with ovarian cancer primary tumor samples and observed significant differences in the genomic profiles, but the authors were able to identify some rarely used cell lines that closely resembled the genetic characterization of patient tumors. The ability of cell lines to model tumors can also be assessed in terms of their ability to identify mechanisms of anticancer drug action and resistance [8]. For instance, the diverse set of 60 cancer cell lines NCI60 has been used as a panel to screen thousands of potential anticancer drugs since the late 1980s and has contributed to multiple advances in cancer chemotherapy [9].

A pair of recently published databases: Cancer Cell Line Encyclopedia (CCLE) [10] and Cancer Genome Project (CGP) [11] provided a collection of genomic characterization and drug response results for 947 cell lines and 24 drugs and 700 cell lines and 138 drugs, respectively. A study [12] published in the following year observed concordance in genomic characterizations, but discrepancies in drug response data for the two databases. They analyzed the gene expression profiles of 471 cell lines common to both the studies and observed high correlation (median correlation coefficient of 0.86) for identical cell lines from the two databases, as compared to different cell lines from the same database in spite of the use of different array platforms for gene expression measurements. Similar high correlation was observed for the presence of mutations in 64 genes in the 471 common cell lines.

However, the correlations between responses in the two databases for common cell lines were observed to be limited for the majority of the drugs. The Spearman's rank correlation coefficients were calculated for both area under the curve (AUC) and IC50 cell response measurements and observed to be <0.6 for the set of 15 common drugs [12]. Fig. 13.1 provides a overview of the analysis conducted in [12].

The authors in [12] have discussed potential sources of the discrepancies due to lack of standardization in experimental assays and analysis techniques. Another article in the same issue [13] argued that the low concordance was expected, as the pharmacological assay used by CGP measures metabolic activity in terms of reductase-enzyme product following 72 h of drug delivery, whereas the assay used by CCLE measures metabolic activity by evaluating energy-transfer molecule ATP levels after 7284 h of incubation.

Based on the previous studies, it is clearly evident that there is a one-to-one lack of concordance in the measured sensitivities between the two studies for the majority of the drugs. Thus one needs to be careful while using the available drug sensitivity databases as alterations in the cell lines, or differences in the assay and protocol used in different laboratories can produce discrepancies in the data. The majority

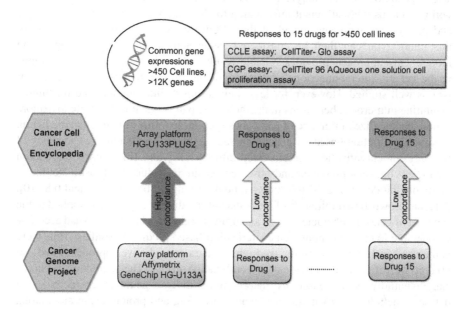

FIG. 13.1

Pictorial representation of the analysis conducted in [12] to evaluate concordance in CCLE and CGP databases.

of high-throughput genomic measurements have been standardized based on the work in the last two decades and thus we observe higher concordance in genomic measurements from different studies. However, high-throughput drug screen studies are yet to be properly normalized and thus inconsistencies are often observed in different studies. We can expect to design transformation mappings that can map the results of one database to the other database where the correlations between the studies are higher. It is expected that standardized approaches and environments for cell viability measurements, along with reduction in measurement noises will reduce the kind of inconsistencies observed in some current databases.

13.4 **PREDICTION ACCURACY LIMITATIONS**

The current state-of-the-art techniques for drug sensitivity prediction primarily generate a model based on the genomic characterizations and train it on drug responses to multiple samples to predict the response to a new sample [10,14,15]. The genomic characterizations are commonly observed under normal growth conditions and do not measure the genomic information following application of different perturbations (perturbations such as siRNAs to silence gene expression, targeted drugs, or other stimulus). Thus a single snapshot of the underlying system is observed and deciphering the interconnections based on a single snapshot remains difficult. The expectation is that analyzing numerous samples of the same form of cancer will provide us with sufficient information to elucidate the genetic interconnections and provide an aggregate model of the specific tumor type. Any new sample of this tumor type can then be considered as an activating part of this aggregate model. This approach provides reliable results when there are limited variations in the genetic interconnections of samples of the same tumor type and the tumor type is well studied. However, for less studied tumors like sarcoma and for tumors exhibiting numerous aberrations in the molecular pathways, it is desirable to explore a more personalized inference of pathway structure. In terms of numerical prediction accuracy, prediction based on genetic characterizations have shown (a) accuracy of 64–92% for classification for binarized efficacy [14] and (b) Pearson correlation coefficient between predicted and observed sensitivity values for Elastic Net-based predictive models using 10-fold cross-validation ranging between 0.1 and 0.8 [10]. A recent crowd-based initiative of more than 40 predictive approaches applied to the same set of genomic characterization and drug efficacy data [16,17] also had accuracy comparable to earlier genomic characterization-based prediction studies. Thus there is the need and scope for alternative higher accuracy predictive approaches that can predict the efficacy of individual or combination drugs. The prediction accuracy can potentially be increased by incorporating diverse genomic characterization datasets, including DNA mutations, gene expression, and protein expression, along with epigenomic characterizations. Furthermore, the tumor microenvironment as observed through metabolomics can provide valuable complementary information. Irrespective of these improvements, additional information on the time domain behavior of the tumor response or the response to multiple perturbations of the individual tumor can be highly crucial in improving prediction accuracy. We considered incorporating the steady-state tumor response to multiple perturbations for drug sensitivity modeling in this book that were able to improve prediction accuracy. Further research is ongoing on designing optimal time domain perturbation experiments, along with incorporating them in the predictive models for increased accuracy and robustness.

13.5 TOXICITY OF COMBINATION THERAPEUTICS

The majority of drug sensitivity modeling research considers the prediction of tumor cell viability to various drug or drug combinations without any explicit personalized predictive modeling of the toxicity of the drug combination. For application of the individual drugs, the toxicity is ignored based on the assumption that toxicity estimates were conducted during the FDA approval process. However, there can be individualized variations in toxicity based on a patients metabolism and morphological profile. The problem is further exacerbated when we consider combination drugs that have not been approved before as a combination. Due to the numerous potential drug combinations based on FDA-approved individual drugs, it will be hard to expect that experimental toxicity data on combinations of all drugs will be available in the near future. Thus it is imperative to incorporate predictive modeling of combination therapeutics in our therapy design approaches. One of the rudimentary ways to estimate drug toxicity is in considering it to be proportional to the number of target inhibitions [18–20]. The targets of a drug combination are considered to be the union of targets of the individual drugs as was considered in our combination therapy design based on target inhibition maps.

Toxicity estimates for drug combinations can also be considered based on existing side effects data for individual drugs, as available from publicly available databases such as Side Effect Resource (SIDER) [21], or information on the absorption, distribution, metabolism, excretion, and toxicity (ADMET) properties of different drugs such as ChEMBL [22]. Existing computational methods for drug toxicity prediction can be broadly categorized as *expert-driven systems* that depend on the knowledge of human experts and *data-driven methods* that depend on supervised learning from experimental data. Machine-learning approaches for toxicity prediction problems [23,24] utilize input features selected from drug combination chemical descriptors, drug-protein interactions, and individual drug side effects [25–28]. There are a few datasets available for validation of drug combination toxicity estimates [29,30], but are limited for detailed predictive modeling.

Most of these approaches are based on individual drug side effects and drug descriptors that are yet to map patient-specific traits (such as genomic characterizations) to drug toxicities. The challenge will be to design and validate personalized toxicity models that can accurately estimate the toxicity to a drug or drug combination for an individual patient.

13.6 COLLABORATIVE CONSTRAINTS

Successful personalized medicine research requires collaborative efforts between multiple disciplines, including medicine, biology, computer science, and engineering. The conventional structure of research restricted to individual disciplines lacks the optimal conditions for advancement of personalized medicine. The current research problems require a multidisciplinary approach where researchers from different disciplines spend time to understand the possibilities and limitations in each other's domains and take joint steps to mitigate the challenges. For instance, we often observe computational analysis of drug sensitivity datasets that were collected without any computational model in mind. A model-based experimental design can provide an optimal set of experiments for parameter inference. Additionally, understanding the molecular biology limitations of drug applications can restrict a computational researcher from designing intervention strategies that can never be employed in practice. Furthermore, increase in crowd-sourced science such as the DREAM challenges [16] will require appreciation of the effort provided by the teams that are not necessarily the top performers. A *winner-takes-all* scenario can restrict increased involvement in crowd-sourced science [31]. A 2014 study [32] on inequalities in science observed that the current system awards a disproportionate share of resources to a minority of researchers and institutions and scientists are increasingly likely to be judged by the numbers that they can generate in terms of publications, research funding, citations, research team size, awards, and memberships in prestigious academies, rather than their actual scientific contributions. The current structure provides limited incentives for sharing of results between multiple research groups with the aim of solving a bigger challenge with contributions from numerous researchers.

13.7 ETHICAL CONSIDERATIONS

Technology for editing human genomes has been in existence for some time, but their applications have been restricted due to ethical considerations on how they would be applied. For instance, *preimplant genetic diagnosis* (PGD), available since 1990, involves checking of embryos created through in vitro fertilization for specific hereditary conditions and avoiding passing of deleterious genetic traits from parents to offsprings [33]. Countries approving such techniques usually restrict the check to serious conditions and the procedure involves searching for embryos without the harmful genetic traits among the multiple in vitro fertilized embryos rather than altering the genes of one specific embryo. Ethical concerns arise when clinics try to modify nonlife threatening hereditary traits, such as physical appearance, intelligence, and agility to produce genetically advanced babies who can usher in a race for *designer babies* and a person's worth being measured in terms of their genotype. A controversial use of PGD has involved selecting the sex of a baby, which can result in skewed sex ratios in a population when conducted in large numbers due to cultural preferences for a specific sex.

The primary approaches considered in this book to tackle cancer involved the use of targeted or chemotherapeutic drugs, but technology can potentially allow us to alter the genes of an embryo based on what will provide the largest resistance to cancer. For instance, the observation that multiple copies of the TP53 gene in elephants enabling them to avoid cancer [34] can result in a plausible hypothesis that increasing the copies of TP53 genes during the conception of a human embryo can reduce the chances of cancer development in later stages of life. These ideas can augment the idea of *designer babies* that are resistant to multitude of genetic diseases. However, genetic modification-based preventive measures for avoiding cancer in humans at later stages can face numerous ethical considerations, such as clinical trials involving genetically modified children and hard to foresee adverse side effects. A study [35] published in 2015 reports the result of modification of nonviable human embryos to correct a mutation causing β thalassemia using clustered regularly interspaced short palindromic repeat (CRISPR)-associated system (Cas) gene editing approach. The publication renewed the discussion on the ethics and legality of human genetic modification [36].

To anticipate and address the ethical issues involved in genetic research, the National Human Genome Research Institute (NHGRI) has created the *Ethical, Legal, and Social Implications* (ELSI) Program [37] that funds conceptual and empirical research on the ethical, legal, and social implications of genomics. Such programs can assist in providing guidelines on the proper implementation of genetic research-based therapy.

REFERENCES

[1] A. Marusyk, K. Polyak, Tumor heterogeneity: causes and consequences, Biochim. Biophys. Acta 1805 (1) (2010) 105–117.
[2] A. Marusyk, V. Almendro, K. Polyak, Intra-tumour heterogeneity: a looking glass for cancer?, Nat. Rev. Cancer 12 (5) (2012) 323–334.
[3] M. Shackleton, E. Quintana, E.R. Fearon, S.J. Morrison, Heterogeneity in cancer: cancer stem cells versus clonal evolution, Cell 138 (5) (2009) 822–829.
[4] C. Swanton, R. Burrell, P.A. Futreal, Breast cancer genome heterogeneity: a challenge to personalised medicine?, Breast Cancer Res. 13 (1) (2011) 104.
[5] M. Gerlinger, et al., Intratumor heterogeneity and branched evolution revealed by multiregion sequencing, N. Engl. J. Med. 366 (10) (2012) 883–892.
[6] B. Borrell, How accurate are cancer cell lines?, Nature 463 (7283) (2010) 858.
[7] S. Domcke, R. Sinha, D.A. Levine, C. Sander, N. Schultz, Evaluating cell lines as tumour models by comparison of genomic profiles, Nat. Commun. 4 (5) (2013) 2126.
[8] J.-P. Gillet, A.M. Calcagno, S. Varma, M. Marino, L.J. Green, M.I. Vora, C. Patel, J.N. Orina, T.A. Eliseeva, V. Singal, R. Padmanabhan, B. Davidson, R. Ganapathi, A.K. Sood, B.R. Rueda, S.V. Ambudkar, M.M. Gottesman, Redefining the relevance of established cancer cell lines to the study of mechanisms of clinical anti-cancer drug resistance, Proc. Natl. Acad. Sci. U. S. A. 108 (46) (2011) 18708–18713.
[9] R.H. Shoemaker, The NCI60 human tumour cell line anticancer drug screen, Nat. Rev. Cancer 6 (10) (2006) 813–823.

[10] J. Barretina, et al., The cancer cell line encyclopedia enables predictive modelling of anticancer drug sensitivity, Nature 483 (7391) (2012) 603–607.

[11] W. Yang, et al., Genomics of Drug Sensitivity in Cancer (GDSC): a resource for therapeutic biomarker discovery in cancer cells, Nucleic Acids Res. 41 (D1) (2013) D955–D961.

[12] B. Haibe-Kains, N. El-Hachem, N.J. Birkbak, A.C. Jin, A.H. Beck, H.J.W.L. Aerts, J. Quackenbush, Inconsistency in large pharmacogenomic studies, Nature 504 (2013) 389–393.

[13] J.N. Weinstein, P.L. Lorenzi, Cancer: discrepancies in drug sensitivity, Nature 504 (7480) (2013) 381–383.

[14] J.E. Staunton, D.K. Slonim, H.A. Coller, P. Tamayo, M.J. Angelo, J. Park, U. Scherf, J.K. Lee, W.O. Reinhold, J.N. Weinstein, J.P. Mesirov, E.S. Lander, T.R. Golub, Chemosensitivity prediction by transcriptional profiling, Proc. Natl. Acad. Sci. U. S. A. 98 (2001) 10787–10792.

[15] M.L. Sos, K. Michel, T. Zander, J. Weiss, P. Frommolt, M. Peifer, D. Li, R. Ullrich, M. Koker, F. Fischer, T. Shimamura, D. Rauh, C. Mermel, S. Fischer, I. Stückrath, S. Heynck, R. Beroukhim, W. Lin, W. Winckler, K. Shah, T. LaFramboise, W.F. Moriarty, M. Hanna, L. Tolosi, J. Rahnenführer, R. Verhaak, D. Chiang, G. Getz, M. Hellmich, J. Wolf, L. Girard, M. Peyton, B.A. Weir, T.-H.H. Chen, H. Greulich, J. Barretina, G.I. Shapiro, L.A. Garraway, A.F. Gazdar, J.D. Minna, M. Meyerson, K.-K.K. Wong, R.K. Thomas, Predicting drug susceptibility of non-small cell lung cancers based on genetic lesions, J. Clin. Investig. 119 (6) (2009) 1727–1740.

[16] J.C. Costello, et al., A community effort to assess and improve drug sensitivity prediction algorithms, Nat. Biotechnol. (2014), http://dx.doi.org/10.1038/nbt.2877.

[17] Q. Wan, R. Pal, An ensemble based top performing approach for NCI-DREAM drug sensitivity prediction challenge, PLoS ONE 9 (6) (2014) e101183.

[18] M.W. Karaman, S. Herrgard, D.K. Treiber, P. Gallant, C.E. Atteridge, B.T. Campbell, K.W. Chan, P. Ciceri, M.I. Davis, P.T. Edeen, R. Faraoni, M. Floyd, J.P. Hunt, D.J. Lockhart, Z.V. Milanov, M.J. Morrison, G. Pallares, H.K. Patel, S. Pritchard, L.M. Wodicka, P.P. Zarrinkar, A quantitative analysis of kinase inhibitor selectivity, Nat. Biotechnol. 26 (1) (2008) 127–132.

[19] B. Hasinoff, D. Patel, The lack of target specificity of small molecule anticancer kinase inhibitors is correlated with their ability to damage myocytes in vitro, Toxicol. Appl. Pharmacol. 249 (2010) 132–139.

[20] R. Kurzrock, M. Markman, Targeted Cancer Therapy (Current Clinical Oncology), Humana Press, Totowa, NJ, USA, 2008.

[21] M. Kuhn, M. Campillos, I. Letunic, L.J. Jensen, P. Bork, A side effect resource to capture phenotypic effects of drugs, Mol. Syst. Biol. 6 (2010) 343.

[22] A. Gaulton, L.J. Bellis, A.P. Bento, J. Chambers, M. Davies, A. Hersey, Y. Light, S. McGlinchey, D. Michalovich, B. Al-Lazikani, J.P. Overington, ChEMBL: a large-scale bioactivity database for drug discovery, Nucleic Acids Res. 40 (D1) (2011) gkr777–D1107.

[23] Y.Z. Chen, C.W. Yap, H. Li, Current QSAR Techniques for Toxicology, John Wiley & Sons, Inc., Hoboken, NJ, USA, 2006, pp. 217–238.

[24] R. Franke, A. Gruska, General Introduction to QSAR, CRC Press, 2003, Boca Raton, FL, USA, pp. 1–40.

[25] W. Muster, A. Breidenbach, H. Fischer, S. Kirchner, L. Mller, A. Phler, Computational toxicology in drug development, Drug Discov. Today 13 (7–8) (2008) 303–310.

[26] I. Oprisiu, E. Varlamova, E. Muratov, A. Artemenko, G. Marcou, P. Polishchuk, V. Kuz'min, A. Varnek, QSPR approach to predict nonadditive properties of mixtures. Application to bubble point temperatures of binary mixtures of liquids, Mol. Inform. 31 (6–7) (2012) 491–502.

[27] I. Oprisiu, S. Novotarskyi, I.V. Tetko, Modeling of non-additive mixture properties using the Online CHEmical database and Modeling environment (OCHEM), J. Cheminform. 5 (2013) 4.

[28] R. Altenburger, T. Backhaus, W. Boedeker, M. Faust, M. Scholze, L.H. Grimme, Predictability of the toxicity of multiple chemical mixtures to *Vibrio fischeri*: mixtures composed of similarly acting chemicals, Environ. Toxicol. Chem. 19 (9) (2000) 2341–2347.

[29] J. Lehar, et al., Synergistic drug combinations tend to improve therapeutically relevant selectivity, Nat. Biotechnol. 27 (2009) 659–666.

[30] Y. Liu, Q. Wei, G. Yu, W. Gai, Y. Li, X. Chen, DCDB 2.0: a major update of the drug combination database, Database (Oxford) 2014 (2014) bau124.

[31] H. Xu, K. Larson, Improving the efficiency of crowdsourcing contests, in: Proceedings of the 2014 International Conference on Autonomous Agents and Multi-agent Systems (AAMAS'14), International Foundation for Autonomous Agents and Multiagent Systems, Richland, SC, 2014, pp. 461–468.

[32] Y. Xie, Sociology of science. "Undemocracy": inequalities in science, Science 344 (6186) (2014) 809–810.

[33] J.A. Robertson, Extending preimplantation genetic diagnosis: the ethical debate. Ethical issues in new uses of preimplantation genetic diagnosis, Hum. Reprod. 18 (3) (2003) 465–471.

[34] E. Callaway, How elephants avoid cancer, Nature 1038 (2015) 18534.

[35] P. Liang, Y. Xu, X. Zhang, C. Ding, R. Huang, Z. Zhang, J. Lv, X. Xie, Y. Chen, Y. Li, Y. Sun, Y. Bai, Z. Songyang, W. Ma, C. Zhou, J. Huang, CRISPR/Cas9-mediated gene editing in human tripronuclear zygotes, Protein Cell 6 (5) (2015) 363–372.

[36] G. Kolata, Chinese scientists edit genes of human embryos, raising concerns, New York Times, 2015.

[37] J.E. McEwen, J.T. Boyer, K.Y. Sun, K.H. Rothenberg, N.C. Lockhart, M.S. Guyer, The Ethical, Legal, and Social Implications Program of the National Human Genome Research Institute: reflections on an ongoing experiment, Annu. Rev. Genomics Hum. Genet. 15 (2014) 481–505.

Index

Note: Page numbers followed by *b* indicates boxes, *f* indicate figures and *t* indicate tables.

Printed in the United States
By Bookmasters